Mother Russia

Mother Russia

BY MAURICE HINDUS

MCMXLIII

Doubleday, Doran and Company, Inc.

Garden City, New York

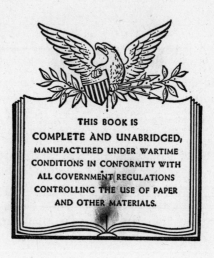

THIS BOOK IS
COMPLETE AND UNABRIDGED,
MANUFACTURED UNDER WARTIME
CONDITIONS IN CONFORMITY WITH
ALL GOVERNMENT REGULATIONS
CONTROLLING THE USE OF PAPER
AND OTHER MATERIALS.

COPYRIGHT, 1942, 1943
BY MAURICE HINDUS
ALL RIGHTS RESERVED

PRINTED IN THE UNITED STATES
AT
THE COUNTRY LIFE PRESS, GARDEN CITY, N. Y.

FIRST EDITION

PREFACE

It is twenty years since I first went to Russia on a special assignment from the late Glenn Frank, one-time editor of the now-defunct *Century Magazine*. In *Broken Earth,* the book which resulted from that trip, in stating the method and the purpose of my writing, I said:

"I went to Russia with only one purpose in view—to hear the people talk. I use the word 'people' not in the English but the Russian sense— meaning the dark masses; that is, the muzhik, the peasant. In the last analysis he is Russia; not the Russia that writes notes, issues ultimatums, signs treaties, entertains ambassadors, grants interviews to foreign correspondents, but the Russia that toils, produces, fights, and dies. . . ."

These words were written in 1926. Much has happened since then to change not only the aspect of the Russian land but the face of the "dark masses." The machine age, which has swept with fabulous speed the Russian countryside, has scraped the beards off their faces and ground the darkness out of their souls. It is out of these "dark masses" that for the first time in Russian history have come many great generals, among the greatest in the present-day Russian armies. Alexey Rodimtsev, Andrey Yeremenko, Vasily Chuikoff, Semyon Timoshenko are only a few of their names. As the defenders of Stalingrad and the victors over picked German troops there, their names will rank as high as, if not higher than, those of Kutuzov and Suvorov, both noblemen, and of Ivan the Fourth (the Terrible) and Peter the Great, both czars. The fact that two of these generals, Yeremenko and Timoshenko, are Ukrainians is of special significance in the light of the Russo-Polish controversy over the Western Ukraine.

The truth is that the peasant of old is no more. The muzhik of Tolstoy, Chekhov, Bunin, and of other leading Russian writers who have for genera-

v

tions wept over him, scolded him, loved him for his sorrows and his sim-
plicities, has vanished from the Russian scene. With his disappearance has
gone the former meaning of the word *narod*—people. The term no longer
has its historic implication and has become as all-inclusive as its English
equivalent. Today narod embraces the entire nation and all the peoples
within the vast territories of Russia.

Yet throughout the years in my writing about Russia I have deviated but
little from my original method and purpose. I have never sought the counsel
and the judgment of the elite and the eminent. I have never striven for the
front-page headlines. Exciting and enlightening as would be the adventure
of putting questions to Molotov, Stalin, and other leaders and hearing their
answers in their own words, I have never been lured by it. The official
nature of such answers would unfold little of the drama and the intimacy
of the experience which Russia as a nation, a many-peopled nation, has
known so overwhelmingly since the advent of the Soviets. Not even in these
war days have I ever gone to the Kremlin gateway with a letter to Stalin
soliciting an interview. I have always felt that a single day in a village or
a single meal with young people in a factory would tell me more of the
heart and the mind of Russia than an interview with any leader, however
exalted.

Despite the limitations of such a method it has given me guidance to
appraise calmly some of the stormiest campaigns of the Soviets and particularly the two most hotly debated events in recent Russian history—the
pact with Germany in August 1939 and the military course of the Russo-
German war.

On my return from Sweden on October 14, 1939, a reporter of the New
York *Herald Tribune* asked for an opinion of the Russo-German Pact.
In reply I said:

"I am convinced that Russia and Germany will go to war . . . the more
they co-operate now the more cause they will have to clash later on."

When the war broke out and the air was crackling with prophecies of
Russia's imminent collapse within six weeks, or at most six months, I set
to work on a book the title of which is *Hitler Cannot Conquer Russia*.
The book was published in September 1941, when the German armies were
plunging madly toward Moscow.

These opinions were no accident. They were no "hunch." Most manifestly they were no prophecies. They were logical and inevitable deductions from the mass of information I had gathered in my travels, especially
in villages, and from hearing "the people talk."

There are writers, editors, diplomats, military observers, crusaders of
social and other causes, to whom any account of Russia in the prewar days

which failed to depict the country as a land of atrocities and not much more carried little or no validity. Nor has the attitude of many of them substantially changed since the outbreak of the Russo-German conflict. Hostility has insulated them against the recognition of forces, which, even in the might of turbulence and want, have been transforming the Russian land and the Russian people.

Of course there has been terror in Russia as grim as any nation has ever witnessed. A revolution means civil war, which is the most brutal form of conflict known to man. The larger the country, the more complicated its problems, the more varied its racial or nationalist make-up, the more ambitious the revolutionary program, the more violent the opposition it stirs inside and outside its borders, the more cruel the terror is likely to be—a terror which often enough sweeps to doom the innocent no less than the guilty.

But to speak of Russia solely or chiefly in terms of terror and atrocities is like speaking of the Cromwellian revolution solely or chiefly in terms of the massacres in Ireland, or of the American Civil War solely or chiefly in terms of the depredations of the Northern armies, particularly of General Sherman's tactics, or of the French Revolution solely or chiefly in terms of the guillotine.

Here is Edmund Burke, the eminent British Liberal, than whom no European statesman was more horrified by the outrages of the French Revolution. In his memorable *Reflections on the French Revolution,* he fulminates brilliantly and achingly against these outrages.

He writes:

"France has bought undisguised calamities at a higher price than any nation has purchased the most unequivocal blessing! France has bought poverty by crime! France has not sacrificed her virtue to her interest but she has abandoned her interest, that she might prostitute her virtue. . . . France, when she let loose the reins of regal authority, doubled the license of a ferocious dissoluteness in manners and of an insolent irreligion in opinions and practices. . . . France, by perfidy of her leaders, has utterly disgraced the tone of lenient council in the cabinets of princes and disarmed it of its most potent topics. She has dark, suspicious maxims of tyrannous distrust. . . . They have found their punishment in their success. . . . Their cruelty has not even been their base result of fear. It has been the effect of their perfect sense of safety, in authorizing treasons, robberies, rapes, assassinations, slaughters, and burnings throughout their harassed lands. . . ."

How familiar these words sound and how natural, too, in the mouths of those who because of religious sentiments, economic self-interest, political prerogative, love of social order, or of any of the other virtues and by-products of a stabilized and cultivated society in a democratic or autocratic world, abhor anything so untamed and so tempestuous as a revolution, even if it springs out of antecedents which they never had known or out of historical processes which their own lands might have long ago resolved—not always in a gentle way.

Yet subsequently Edmund Burke was impelled to admit the error of his thought not because he had become reconciled to terror but because of other forces in the Revolution which had eluded his comprehension. In his *Thoughts on French Affairs* he wrote:

"The evil is stated, in my opinion, as it exists. The remedy must be where power, wisdom, and information I hope are more united with good intentions than they can be with me. I have done with this subject, I believe forever. It has given me many anxious moments for the last two years. If a great change is to be made in human affairs, the minds of men will be fitted to it; the general opinions and feelings will draw that way. Every fear, every hope will forward it; and then they who persist in opposing this mighty current in human affairs will appear rather to resist the decrees of Providence itself than the mere designs of men. They will not be resolute and firm, but perverse and obstinate."

In commenting on Burke's extraordinary reconsideration of his earlier views, Matthew Arnold remarks:

"The return of Burke upon himself has always seemed to me one of the finest things in English literature, or indeed in any literature. That is what I call living by ideas. . . . I know nothing more striking, and I must add that I know nothing more un-English."

But living by ideas in moments of the stress and the terror they may eventuate and seeing, however nebulously, beyond the havoc they stir is no simple matter. Hence Matthew Arnold's eloquent tribute to Burke's intellectual courage and integrity.

As for Russia the incontrovertible fact is that however costly the process of revolutionary change, it has not been without nation-saving reward. With an industry built largely since 1928, she has for nearly two years been effectively battling all alone on the European continent the most highly mechanized and the most powerful army the world has known, supported by the industry of Germany, France, Belgium, Holland, Czechoslovakia—of all western Europe—with a steel capacity from three to four times as high as her own.

To ignore or underestimate the meaning of this positive and astounding fact is to reckon without one of the most formidable forces of our time.

The stanchest enemies of the Soviets can no longer mock or disregard the immense creativeness which they have unleashed.

In this book, as in others I have written, my chief concern has been the people. After a quarter of a century of Sovietism and fighting the most crucial war of their lives and of their history, they are much changed. They are not like the people they were in 1917, in 1923, in 1928, hardly even like the people they were in 1936, the year of the adoption of the new Constitution. On this trip I found a Russia which in many ways was new to me— a Russia that had rediscovered her past, reappraised her present and her future, and whose people were ready to die—old and young, men and women—rather than to forswear their newly fashioned identity and their immediate and remote destiny.

More than ever did I feel certain of my initial appraisal of Sovietism not as an international but as a national, distinctively Russian revolution, with international implications that essentially must remain bounded by purely Russian strictures.

A neutral diplomat in Moscow once said to me: "Either England, America, and Russia develop a common language of action in the war and afterwards or God help us all." In these words the author heartily concurs. It is in the hope of bringing to the reader a fresh understanding of the Russian people, so that he can more clearly appreciate the full meaning of these words, that I have written this book.

MAURICE HINDUS

CONTENTS

PREFACE v

PART ONE **RUSSIA'S FLAMING YOUTH**

1. THE GREAT FACT 1
2. SHURA 9
3. LIZA 18
4. ZOYA 32

PART TWO **RUSSIA'S COMING OF AGE**

5. DESTRUCTION AND CREATION 55
6. THE BLACK CITY 64
7. LAND OF GLAMOUR 72
8. MOTHER VOLGA 83
9. REDISCOVERY OF THE PAST 93
10. RUSSIA GOES RUSSIAN 102
11. OLD PEOPLE 107
12. THE GOSPEL OF HATE 113

PART THREE **RUSSIAN CITIES**

13. TULA 123
14. MOSCOW 131
15. STALINGRAD 143

PART FOUR RUSSIA'S NEW SOCIETY

16. FACTORY OWNERSHIP 159
17. FACTORY MANAGEMENT 174
18. FACTORY LIFE 185
19. INCENTIVES 197
20. THE KOLKHOZ 208
21. RELIGION 225
22. MORALITY 235
23. ROMANCE 245
24. LOVE LETTERS 250
25. FAMILY 260
26. YOUTH AND CULTURE 274

PART FIVE RUSSIAN WOMEN

27. THE NEW ROLE 285
28. A MOTHER–IN–LAW 297
29. THE GIRL WITH THE COUGH 301
30. CAPTAIN VERA KRYLOVA 305

PART SIX RUSSIAN CHILDREN

31. LITTLE PATRIOTS 317
32. VANYA ANDRYONOV 325
33. ALEXEY ANDREYITCH 326
34. SONG OF THE NIGHTINGALE 329

PART SEVEN ON THE TRAIL OF THE ENEMY

35. TOLSTOY'S OLD HOME 333
36. "THE NEW ORDER" 350

PART EIGHT RUSSIAN QUESTS

37. "WILL WE HAVE TO FIGHT RUSSIA?" 371
38. AFTER THE WAR—WHAT? 374
39. AFTER TWENTY YEARS 388

ILLUSTRATIONS

FACING PAGE

Shura Chekalin 10

Liza Chaikina 18

Zoya Kosmodemyanskaya 34

Vera Krylova 306

"The Poor Man's Tree" by the Restored Tolstoy Home 334

Tolstoy's Bedroom, Left in Flames by the Retreating Germans . . . 335

The Domed Room in Which Tolstoy Wrote *War and Peace* . . . 335

Mother Russia

PART ONE: RUSSIA'S FLAMING YOUTH

CHAPTER 1: THE GREAT FACT

I PASSED through a cobbled courtyard, climbed two flights of dimly lighted stairs, and knocked on the door.

"Who is it?" came a familiar voice from inside.

"An old friend," I replied.

The door swung open, and Natalya Grigoryevna stood before me. She could hardly believe that it was I, though she had known me in Moscow ever since she was a little girl with pigtails. In the city of Kuibyshev in the summer of 1942 America seemed too far away for anyone to come from there. She welcomed me inside, into her one-room apartment in which she was living with her two children, a boy of five and a girl of three, and her old mother-in-law. She had been evacuated to Kuibyshev, she said, in the autumn of 1941, when the Germans were storming at the gateway of the capital.

Short, blue-eyed, with a pale, broad face framed in a mass of wavy reddish hair, she appeared only slightly changed since I had last seen her in Moscow in 1936. She had grown heavier, more disciplined, more subdued, more careworn, yet no less energetic and no less articulate. She introduced me to her mother-in-law, whom I had never met before. A bent, gray-haired woman with a remarkably smooth skin, an unpleasant dour expression, she scarcely mumbled a word in reply to my greeting. She was feeding boiled rice to an unruly, giggling little girl.

Natalya Grigoryevna, or Natasha, as we called her, apologized for the condition of her one-room apartment. It resembled less the living quarters of a family than a museum of old furniture, old pictures, children's playthings, household furnishings, kitchen utensils. She had had to flee from Moscow, she explained, as had hundreds of thousands of others, and she was lucky to have taken along as much of her personal and household effects as she had and to have found this well-lighted though crowded room. Nothing was normal in Russia, anyway. Life was hard, harder than she had ever known it, especially for a mother like herself with little children. But no matter—she had become accustomed to the disorder of her new home and the privations the war had imposed on her.

I asked about her husband, Yuri.

"Dead," she said in a low voice. "Killed on the Leningrad front."

I heard a sob. It was the old woman, Yuri's mother. I glanced at her. The shiny spoon with which she was feeding her granddaughter seemed as if frozen in her hand, which remained suspended above the plate of rice. Natasha too glanced at her. There was reproach in her eyes, but she said nothing. The spoon dropped to the floor with a clatter, and the sound seemed to waken the old woman. She stooped down, picked up the spoon, washed it, returned to her seat, and resumed feeding the little girl. Only now her hand trembled as if she had the palsy.

Disregarding this incident, Natasha proceeded to ply me with questions about America, my flight across the Atlantic, my first impressions of war-time Russia. The more she talked, the more she seemed like her old self—lively, vivacious, astir with curiosity about the outside world. Then we talked of the war. Hardly anyone she knew but had lost someone at the front. Russia was a nation of widows and orphans, millions of them. Her two brothers were fighting. One was in the Navy, and she had not heard from him in five months. She didn't think he was alive. Her other brother was a colonel of artillery. Yuri's three brothers were also fighting. They were all well, though one had been seriously wounded. But—there seemed no end to the war, and she was prepared to receive more sad news. Every woman, every family in Russia was.

I heard another sob, louder than before. Turning, I saw the old woman wipe her eyes with a faded blue apron. Her body trembled; her head swayed like fruit on a wind-blown tree. Again Natasha glanced at her but said nothing. The old woman rose, set her unruly and resisting little granddaughter on the floor, took her by the hand, and without saying a word started for the door.

"That's right, little mother," said Natasha. "Take Nina for a walk in the park. It'll do you both good to be out in the sun. But don't stay too long."

Without a word of reply, and with eyes averted as if too proud to be seen in tears, the old woman walked out of the room.

"Poor woman," said Natasha, "she cannot stand it. Yuri's death has broken her completely. She's aged ten years in the last four months since it happened. There is an old-fashioned Russian mother for you."

"And you?" I asked.

She struggled; then with a swift toss of the head, as though in triumph over inner turmoil, she said,

"I am different. I did my share of weeping—more than my share. The first weeks were unbearable. I saw Yuri everywhere. I'm not superstitious, but he was always before me. At night in the dark it was worse than in the daytime—he was as if standing before me—and waiting for me to speak." She paused, adjusted her shimmering red hair, and resumed quietly. "It was terrible, terrible. But I finally got hold of myself. I started to work, to go out, to see friends, to read, to attend meetings, and now I've become hardened. Only it hurts to see Yuri's mother so broken up. After all, we Russians, despite all our sorrows, have got a lot to be proud of, to be comforted by. As for myself, there are my children—Sasha is the image of Yuri. Sorry he's away—gone on a picnic with other children. You must come again and see him—a wonderful child, really. He says he wants to be a flier and to go to war and kill Germans. It's amazing how much our children understand—perhaps it's bad that they do, don't you think?"

As if remembering something, she rose, went over to a crude built-in shelf on the wall on which lay a mass of folded newspapers. Lifting several copies, she returned to her seat and offered me one; she asked me to read an item which she had marked heavily with red pencil.

It was a story about a Russian girl named Petrova who lived at a railroad station up north beyond the Arctic Circle. The Germans had dropped incendiary bombs near the station. A fire started, the flames leaping closer and closer to several oil tanks. Petrova saw it. To prevent the fire from reaching the tanks she ran out, flung herself on it, rolled around and around until she put it out.

"A brave girl," I remarked.

"Luckily," said Natasha, "she wasn't hurt. She might have caught fire and burned to death. But she didn't think of that. She was only thinking of saving the oil tanks."

She unfolded another paper, and pointing to another column underlined with red pencil, she asked me to read it. This was a story about five marines on the Sevastopol front. They had fought until two of them were dead and the others were out of ammunition except for anti-tank grenades. Shoving the grenades into their belts, the remaining three marines had hurled themselves under the advancing German tanks. The tanks blew up, and with them the Russian marines.

"With their own lives," said Natasha, "the dear ones broke up a tank attack—with their own lives." Slowly, devoutly, as if reciting a prayer, she read the names of the five marines: "Nikolai Filshenko, Vasily Tsibulko, Yuri Pashin, Ivan Krasnoselsky, Daniel Odin." She paused, her eyes on the paper, on the names she had just recited. Then she added, "You see why I'm hopeful? We have millions like them—millions, I tell you. In the end, no matter what the Germans do to us now, we shall win. We must win. We won't be conquered."

It was good to hear this young mother, even in the midst of great personal tragedy, speak in this high spirit. In the supreme heroism of individual soldiers and civilians she beheld a sign and a guarantee of ultimate triumph.

Nor was she alone!

Russia is the Great Fact, perhaps the greatest of our times. Politics, prejudice, wrath aside, what hope of emancipation could the conquered and devastated peoples of Europe cherish had there not been a many-millioned Russian Army to fight the Reichswehr? It matters not that Russia is fighting because she was attacked, because she has to, because not to fight meant to invite degradation and doom. It matters not that in their first onslaughts the Germans occupied vast Russian territories and inflicted powerful blows on Russian industry, Russian agriculture, Russian pride. Russia fights on and on. She is destroying German troops, German equipment. She is making it impossible for Hitler to proclaim himself conqueror of Europe and to make final disposition of the peoples and the lands he has vanquished.

History repeats itself here. In the first World War Russia, according to figures compiled by the United States War Department, mobilized an army of 12 million men. The Allies, including the United States, mustered an army of over 42 million men. Of the 1,773,700 German soldiers and officers killed in the first World War, over a million, according to Russian sources, met their death on the Russian front; these figures do not include the casualties of Germany's allies, particularly Austria-Hungary and Turkey. The Revolution broke out and, before the war ended, Russia, torn with internal strife and weary of fighting, concluded a separate peace. Yet it was because of her vast losses in Russia that Germany could not overcome the united might of the Allies; even though she was subsequently fought on only one front, in the west, she was crushed and had to sue for peace.

In this war the gigantic scale of the operations on the Russian front, the daily wreckage of German equipment, the heavy toll of German life are weakening Germany, draining more and more of her lifeblood. Whatever nation or group of nations administers the final fatal blow, Russia has played a crucial part in battering not only the tremendous might but the extravagant pride and explosive self-confidence of Hitler and the Reichswehr. This part will prove far more momentous than in the first World War, if only because, at this writing, twenty months since the outbreak of the Russo-German war, Russia is fighting the continental German Army all alone on the longest front the world has ever known.

Soviet Russia and the Anglo-Saxon world have many grievances against each other. For nearly a quarter of a century they have wrangled often and bitterly. There was no friendship between them: there was only hostility. The faith and the ways of the one were and may still be the agony and devastation of the other. There is no use underestimating, even in this moment of common ordeal and common battle unto death against a mutual enemy, the deep social and ideological gulf between the English-speaking countries and Russia. In the future—perhaps more than in the immediate Soviet and distant czarist past—old and. new grievances may flare into fresh antagonisms. In history, as in one's personal life, one can only hope for the best, but one must not be too sanguine about the fulfill-ment of this hope. Yet it is only a truism to say that no greater tragedy can come to Russia and the English-speaking world than the inability of their statesmen to compose the differences that divide them and to evolve a common. formula and a common policy, not of internal accommodation and administration—that is now unthinkable—but of external or inter-national partnership.

By uninterruptedly and violently battling the German Army, Russia is enabling America and England, as well as herself, to uphold their respec-tive faiths and respective ways of life, however divergent these may be from her own. Now, as the Russians so often say, it is not life but death that counts—the death of the enemy, of his faith and dreams, his plans and methods, above all of his human and mechanical powers, for only by his death can others, including the conquered peoples in Europe, hope to insure their own. life and their own destiny.

Alien, and aloof, far away and impenetrable, Russia, like China,. has been thought of by the more advanced and industrially developed nations in terms of masses—coming into the world in masses, enduring plague, famine, flood in masses, dying in masses, fighting and getting killed in masses.

We have heard of Confucius and Sun Yat-sen, of Tolstoy and Tchaikov-sky. We have honored and revered these names. But they are the mountain peaks that overshadow the misty plains below. Many of us have thought of the people in Russia and China as de-individualized or un-individualized agglomerations—scarcely more than living automatons with little will of their own and still less power to exercise it, and therefore with no particu-lar claim to distinction and no special capacity for personal achievement and personalized existence.

The Soviet Revolution, with its emphasis of socialized aims, crowd action, and rigid political dictatorship, has only heightened this conception of the Russian people.

Yet here is Natalya Grigoryevna, not only thrilled by the heroism of the *individual* Russian soldier, but convinced that it is a testament to Russian invincibility and a guarantee of Russian victory. In this conviction she may have been more romantic than realistic. The defeat of so powerful and highly mechanized an army as the German Reichswehr, with its wanton disregard of established amenities of warfare, requires more than the heroism of the individual soldier, however supreme.

Yet if Russia is the Great Fact of our time, the unexpected overpowering Fact, then the heroism of the individual soldier is a component and conspicuous part of this Fact. It is like the sap in the tree. Without the sap the tree dies. Without the valor of the individual Russian soldier Russia would have been trampled to extinction. The Germany of today is not only the most ruthless but the most formidable enemy Russia has known. Arrayed against the civilian population no less murderously than against the Army, she is striving for the degradation and the annihilation of both.

In times of defeat and disaster such as the Russian armies have again and again been facing, it requires untold audacity and faith to fight an army like the Reichswehr.

Every child in Russia has heard of Captain Gastello. Once he was a factory worker in Moscow. During the war he became an aviator and rose to the rank of captain. On July 3, 1941, he and his squadron were in an air battle. The war in the sky was paralleled by the war on land. A shell hit Gastello's fuel tank. His plane caught fire. He might have parachuted to safety; he might have tried to make a landing. But he didn't. He wanted to aid the Russian armies on land. It was a question of minutes before the flames would reach the cockpit and engulf him—minutes of life and desperation. The plane was falling lower and lower but was not yet out of control. Gastello couldn't rise with it, but he could hold off the moment of its fall. Observing a column of German oil trucks moving toward the German lines, he veered his plane to their direction and crashed into them. There was a thunderous explosion. Truck after truck blew up, caught fire.

Gastello perished. Since then many a Russian aviator has followed his example, and the name Gastello has become a synonym of soldierly audacity and self-sacrifice.

I do not mean to say that all Russian soldiers have attained heroic stature. Of course not. There have been men in the Army whom the roar of tanks, the blast of German noises, have terrified into flight. Violent and acrimonious have been the editorials in denunciation of these men. There have been generals who in the face of overpowering German aviation have cowered in retreat. The Russian withdrawal from Rostov in the summer of 1942 was, as one Russian has privately expressed himself, "a scandal and a disgrace." *Red Star,* official organ of the Red Army, did not attempt to spare the feelings of the generals responsible for that costly and ruinous blunder. The Russians have been as ruthless in exposing their military shortcomings and denouncing their strategic errors as they ever have been in anything they have undertaken since the establishment of the Soviets.

Consider the importance of a play called *Front,* by Korneichuk, which was more than a literary sequel to the retreat from Rostov. Scathing is its indictment of the so-called civil-war generals. Prestige, earnestness, political devotion, immeasurable personal sacrifice had elevated these men to positions of the highest eminence in the Army. But few of them had mastered the fighting technique of the advanced machine age. German pincers baffled them and German fortifications stumped them. Gruesome, says the author, has been the price Russia has paid for their incompetence. When I saw the play in the incomparable Moscow Art Theater the audience was so stirred with disdain for the civil-war generals, and with fervor for the younger men who challenged them, that in violation of the historic tradition of the theater they again and again gave vent to their violent emotions in the middle of an act.

In an army of millions, it is as natural to find men of faint heart as it is to find weeds in a field, no matter how well tilled.

But so frequent and so breathless have been the heroic exploits of individual soldiers that, like stars on a darkened night, they have again and again illumined the dismal skies that the war has cast over Russia. They have stirred the fervor and the faith of the Natalya Grigoryevnas of the country and have challenged the admiration of the seeing and feeling foreign observer.

A crew of four men and a tank find themselves in the enemy's rear. They refuse to surrender. Steel and fire pour over them. They keep on firing their guns. They know the fight is hopeless, but at the moment damage to the enemy looms more important than the preservation of their own lives. So they hold out. Their tank is afire. They still have time to surrender. They are offered the chance. They spurn the offer. The end is near, very near, they scribble a note, which later is recovered:

"These are our last minutes—kerosene is pouring over us—soon we shall be dead—good-by."

They burn to death.

Here is the village of Muravyovki, in the Orel province, within earshot of guns on the battlefield. The harvest is superb; ears are heavy with grain, the grain fat with substance. From dawn to dark the peasants are in the fields gathering the precious harvest. Neither guns nor bombs halt them;

even the children go on working. When German planes appear in the
sky, they run for cover; then the hum of motors overhead is heard no
more; they come out again, ply their sickles, bind the sheaves.

One sunny day a German plane drops leaflets over the fields—an order
from the German High Command forbidding the peasants to gather the
harvest. The peasants ignore the order. The next day a Focke-Wulf—"the
crutch," as peasants call it—flies over the fields and opens machine-gun
fire. Soon there's a roar of artillery and trench mortars; shells and mines
scream over shocks and stacks of grain. The peasants run for cover. A Russian fighter—a *"yastrebok,"* little hawk, as the peasants affectionately
speak of it—soars into the sky, chases away "the crutch." Presently a
squadron of Russian planes silences the German artillery and trench-
mortar batteries.

Tragedy comes to Muravyovki. A mine splinter strikes a sixty-year-old
woman. Blood streams down her face, reddens the swath of grain stalks in
her lifeless hands. There is a public funeral; there are tears and wails;
there are vows of defiance. A wreath is laid over the old woman's tomb—
not of the ever-present, ever-glowing cornflowers and daisies, but of the
blood-drenched grain stalks. . . .

When night comes, the people go back to their harvesting. They work
by the light of the stars. Old men, old women, children—all are in the field.
Soldiers from a near-by army post help. Sickle in hand, guns in their belts,
they keep a swath ahead of the women reapers. If parachutists come, they
are the first to battle them.

Shears in hand, children follow the reapers. They cut the stalks which the
sickle has missed. A sergeant named Salvushkin from the faraway Siberian
village of Bolshaya Shanga works with the children. His right arm is gone
—amputated. But his left arm is good. With a pair of shears in his left
hand, he also cuts standing stalks of rye. Not one stalk is missed. All
work fast, and so happy are they because no German planes hover over
them that they sing as they work—not loudly, but with more than wonted
fervor.

In the morning "the crutch" reappears. But the grain is already har-
vested.

When the Natalya Grigoryevnas hear or read of villages like Muravyovki
their blood mounts, their faith soars, they glow with pride, and in their
hearts they cry out: "We will win! We will win!" Is it an empty cry? Not
to the Natalya Grigoryevnas.

All the more rapturously do they cry these words because of the incredible heroism of Russia's children—high-school boys and girls. Those
children are writing the most moving chapter in this war. There has never

been anything like it in Russian history. No one had expected it. No one had counted on it. They were only boys and girls following prescribed courses of study in school. Except in their own homes, among their own friends, in their classes, no one had ever heard of them. Their names are new—as new as their deeds.

Yet now it is they, and not the generals or soldiers, however valiant and self-sacrificing, who evoke the highest tributes of the country. They are the pride of their families and their schools, the fame of their towns and their villages, the glory of all Russia. In the press, on the platform, they are lauded as models for other youths to emulate. They have made millions weep with grief and hope. They have stirred the imagination of the Russians, and of others who have heard of them, as no other figures in recent times. They are the war heroes of the country.

For this reason I am telling here the stories of three of the best-known of these youths—one boy and two girls. Think what we may of Sovietism, its economics, its sociology, its politics, the spirit and personalities which emerge from these stories explain, more than volumes of figures and facts, the source and the nature of Russia's unsuspected patriotism and formidable fighting power.

CHAPTER 2: SHURA

A BROAD OPEN FACE; prominent ears; a round chin; a wide forehead peering from under a tall cap; hair falling over the ears; large, shrewd, good-natured eyes overhung with straight heavy brows; a firm mouth; curves of the open throat losing themselves in a muscular neck; an earnest, happy expression, as of someone given to reflection and enjoying it; an air of self-confidence; a poise, a calm, as of someone at peace with himself and the world. Such is the portrait of Alexander or Shura Chekalin, the sixteen-year-old high-school boy whom the Germans hanged and who is one of Russia's great war heroes.

Looking at his picture, it is difficult to imagine that so young a boy managed to pack so much audacity and adventure, so much action and heroism into so brief a life. Had there been no war he might never have attained renown nationally or locally, not at any time in his life. In Pesko-vodskoye, his native village in the province of Tula, he might have remained what he was before the war—Shura Chekalin, a high-school boy popular with his classmates, loved by his father, adored by his mother, worshiped by his younger brother, Vitya—no worse, perhaps no better, than hundreds of high-spirited boys of his age in the Soviet Union. Today his name is remembered and reverenced everywhere.

I have seen his picture in city after city—in parks, schoolhouses, museums. In speech after speech I have heard Shura Chekalin lauded as a matchless and sanctified example of valor, of devotion, of self-sacrifice. Again and again the lively, audacious, militant *Komsomolskaya Pravda,* daily journal of the Soviet youth, writes of him with an eloquence and fervor which in other lands are reserved for people of mature age, of life-long attainment.

Let Shura's story speak for itself.

He was born on March 17, 1925. All around his native village were dense forests rich in all manner of wild game. They were a paradise for sportsmen like his father; for him, too, they were a source of endless delight and adventure. He loved the forest, its wild life, and frequently accompanied his father on hunting trips.

At an early age he learned to shoot. Without saying a word to his mother, he often went away all alone with a gun on his shoulder and came home loaded with rabbits and wild birds.

Fishing was another sport that absorbed Shura Chekalin. He fished with a net, a line, a gun. In the spring he parked himself somewhere on the bank of a stream, under a bridge, gazed intently at the flowing water, and on sight of the dark and elusive pike he fired his gun. He caught endless numbers of pike with bullets.

Hunting and fishing didn't exhaust the boy's diversions and adventures. He loved horses and knew how to manage them. He was one of the most skilled riders in the village. Disdaining a saddle, he would jump on the bare back of a sprightly colt and gallop away at top speed. He did it often, yet never had any accidents.

His father had an apiary, and Shura became interested in bees. He studied them, learned to take care of them. Wild bees in the forest held a special fascination for him. He learned their ways and methods of tracking them to hives inside the hollows of trees. Cutting down such trees, scooping out the honey, bringing it home to his mother was one of his supreme pleasures.

Shura was also fond of mechanics. Like American boys he liked to tinker with tools and mechanisms. Whenever anything went wrong with the electric lights in the house, Shura repaired them. If an implement on the farm broke, he knew how to mend it. He built his own radio set. When a camera was given him as a gift, he took it apart and put it together. He was fond of amateur photography and with his own hands built an enlarging apparatus. Snapshots he especially liked were enlarged in his own darkroom.

SHURA CHEKALIN

A boy of such varied interests, with such a passion for outdoor life, grows up without knowing fear. He feels as much at home in the forest as on a frequently traveled highway. His ears and eyes are keyed to the sounds and sights of nature. He develops an acute sense of direction. He is an intent and penetrating observer. He can go anywhere without feeling or getting lost. He knows how to look after himself in the midst of danger. He is quick of thought, in an emergency he is no less quick on the trigger.

With these habits and traits, acquired since his earliest years, Shura had an ideal preparation for the kind of life and type of fighting guerrillas must pursue.

Shura was also a brilliant student. He loved books. Tolstoy and Gorki were his favorite authors. He read much Russian history. Like other Russian school children in very recent years, he learned to revere the names and deeds of the men who had won military victories for Russia and had conquered foreign enemies. He and his younger brother Vitya shared a room, played and went hunting together. Shura often teased his brother on the commonplace quality of his name.

"Vitya, Vityusha," he would say. "What a silly name! But mine—Alexander! Think of it—Alexander Nevsky, Alexander Suvorov—there's a name for you!"

An energetic boy, Shura was always active. He could put his hands to anything around the house and acquit himself with credit. When his father and mother went on a trip they didn't need to worry about Shura. He took care of himself and of his younger brother. He cooked his own meals, washed his own dishes. He did the chores; watered, fed, and milked the cow, and tended the other livestock.

Shura was growing fast. He was healthy and strong, with a broad back, muscular chest, rugged neck. His eyes were dark, his hair was dark. He had a remarkable memory and could absorb a lesson after a single reading of the text. He was cheerful and sociable. Popular with friends, his home was crowded evenings with young people, and the noise and the mirth were so loud that neighbors told his mother her house was a playground rather than a home. Chairwoman of the village Soviet, the mother was fond of children and more than welcomed the gatherings of Shura's friends.

Shura's closest comrade was a boy named Andrey Izotov from the neighboring village of Green Meadow. Andrey often came to Shura's home and stayed for days. Together the two went berrying, fishing, hunting. They helped Shura's father with the bees, with the work on the land. They went off to the haymow, read books, engaged in endless discussions. They slept in the hay. Invariably Vitya went along, and the three took with them a balalaika and an accordion and entertained themselves late into the night with song and play.

"He was the joy of my life," says his mother, Nadezhda Chekalina.

She encouraged Shura in all his hobbies and adventures. She was convinced he would grow up to be a man of importance—perhaps a great engineer, perhaps a great scientist.

The war came.

"Mother," said Shura, "this will be a terrible war. I'll go with Father and fight."

The mother felt sad. She had been proud of Shura, was happy to watch him grow into a sturdy, playful, studious boy. Now he wanted to go to war. She was mayor of the village; this was her sixth year in office. It was her duty to mobilize and direct enthusiasm for the war. Men who weren't summoned to the colors she urged to join of their own accord. Women she exhorted to double and treble their efforts for the Army and the fatherland. Yet she was a mother. Shura was only sixteen—a clean-cut joyful lad. It hurt her to hear him speak of going to war. He was so earnest about it, so calm, so self-confident, as though it were nothing more serious than a hunting trip in the forest. She knew she could not talk him out of it, and she didn't try.

The Army would not have him—too young. Later, as the Germans were pushing deeper and deeper eastward, nearer and nearer to home territory, a company of volunteers was organized in the village and Shura joined them. Deeply impressed with his strength and alertness, his skill with gun and horse, his devotion and fearlessness, the company commander made him a member of a local mopping-up squad.

Shura's father served in the same squad. Their duty was to penetrate the depth of the primeval forest, search out parachutists, spies, wreckers, and wipe them out. They went off together for several days, leaving the mother and Vitya alone at home. Never did the mother know where they went. Never did they tell her. Nor did she know when they would return or how long they would be gone. Son and father maintained the closest secrecy about their movements. Returning to the village, they sang hunting, marching, and patriotic songs. Then the mother knew they were alive and well. After a brief rest son and father would be off again.

The Germans were pushing ahead with unexpected speed and fury. The Chekalins were now living in Likhvin, and the Germans were dangerously close. People were becoming more and more concerned, but Shura was growing more and more determined. He was ready for real fighting, however desperate and dangerous. One day he said to his mother:

"Please, Mamma, pack up my warm clothes. Most likely I'll be gone all winter."

The mother's heart sank. She knew what it meant—Shura was joining a guerrilla detachment. She looked at him a long time, but asked no questions, said nothing. After all, she was mayor of the village; it was her duty to stir the fighting spirit of everybody, including her own children. She felt like crying and only with great effort held back the tears.

Quietly she gathered Shura's winter clothes—padded waistcoat, felt boots, heavy underwear. She tucked into the bundle three loaves of bread and wanted to lay in a supply of meat, but Shura told her his father had killed a pig.

"We are taking a whole carcass with us, Mamma," he said, "also two large casks of honey."

Now the mother knew that not only her son but her husband was going to the forest. She remained silent. Shura bade her a cheerful farewell and went off with his father.

Five days later everybody in the village was ordered to evacuate to the rear. But Nadezhda Chekalina refused to leave until after she had again seen Shura. She had to see him, she insisted, if only for one hour. Misgiving gnawed at her heart; the worst might happen, and she simply had to see her son before departing for the rear.

Her message was transmitted to Shura. He came out of the forest displeased, sullen.

"Why have you sent for me?" he inquired.

"My dear boy," said the mother, "don't you want to say farewell to me?"

She told him she was being evacuated from the village because the Germans were threatening to occupy it.

"Of course I do, very much, Mamma. But I don't want to see you in tears. You are a brave, clever woman. You've got to be cheerful."

This time the mother broke down. She couldn't help weeping, sturdy and patriotic though she was; she knew that being a guerrilla in territory that might soon be in German hands was dangerous. Yet to Shura she mentioned no word of her premonition. She would be brave. It was her duty to be brave. She stopped weeping and talked of cheerful things. Buoyed by her fortitude, Shura recounted some of his more exciting experiences in the forest. With his bright talk and hearty manner he made his mother happy. Never had she loved him more. Never had he seemed more of a boy with all the joviality, all the excitement, all the self-confidence and adventurousness of a boy. They talked a long time, and when he was ready to leave she said:

"Take good care of yourself, son. Remember you are no trained soldier."

Laughing, Shura replied:

"Don't worry, Mother. I'm a better marksman than older people."

That was true, and the thought of it helped the mother control her grief.

Mother and son embraced, kissed and parted. Shura returned to the forest. The mother moved with her thirteen-year-old Vitya to a village eastward but not too far away from the forest.

Now began the most crucial, most exciting period of Shura's life. He was ready for any task, any adventure. The youngest member in the guerrilla squad, no one, not even his father, was so richly talented, so superbly suited for guerrilla work. He proved a magnificent scout. He could go anywhere on foot, on horseback. He could penetrate any swamp, any part of the forest and never feel lost. Again and again he went off alone to German-held territory and returned with priceless information about enemy forces, their geographic disposition, their equipment, and their behavior toward the civilian population.

Now and then the guerrillas were hungry for a taste of something sweet, and Shura went off hunting for wild honey. Seldom did he fail to find it. On his return to the dugout, the guerrillas gathered and held a celebration. They all loved Shura, and his presence in the forest and in the dugout mitigated the harshness of their everyday life.

He was the only amateur technician in the squad, and out of materials he found in the dugout and in the woods he constructed a radio receiving set. Now they could listen and be heartened by the broadcasts from Moscow, from the Russian front. They could check on the rumors that Germans were energetically spreading, particularly their loudly proclaimed boast: "Moscow *kaput* [Moscow is finished]." Shura could also cook, and he helped prepare many a meal over an open fire.

Interludes for rest and diversion in the forest were few, especially for Shura. He was so skilled, so daring a scout that the commander of the guerrilla detachment was always sending him off on expeditions. His disguises were many and effective..When he needed to penetrate deep into the enemy territory he often dressed as a German soldier. With Germans all around outside the forest, it was no difficult task for a marksman like Shura to shoot a German soldier and provide himself with a German uniform, German weapons, and make his way into the village swarming with Germans. Of course capture meant death. The sixteen-year-old Shura knew this only too well. The Germans hated and feared nothing and nobody so much as the guerrillas. That's why they usually stayed away from the forests and from the roads that wound and skirted through them. But Shura knew no fear. Not once did he complain of feeling uneasy in a village occupied by the enemy. He saw and heard what he wanted and slipped back to the dugout in the wilderness.

He was always ready for encounters with Germans, sentinels or patrol guards. Quicker than they, he never failed to shoot his way out of difficulties. If his rifle failed him, hand grenades came to his rescue, and he was as skilled with the one as with the other. Nothing gave him greater pleasure than to surprise his fellow guerrillas with an armful of captured German weapons. They all marveled how so young a boy could carry so many at a time. But he only laughed and said that next time he would bring more. He did, too, over and over again.

He had many breathless escapes. Once, while out with a scouting party, he found himself cut off from his companions. Several Germans had almost surrounded him. Life and death hung in the balance. One false miscalculation or misstep and all would be at an end. A rifle was useless. Only hand grenades could save him. Had the Germans thought as fast as he, they might have blown him to pieces. They were determined to capture him so they could have a live guerrilla from whom to obtain information about his companions in the forest. They let slip the few most priceless and decisive moments, and Shura made precious use of them. He flung one hand grenade after another and fled to safety. When he spoke of this incident to his fellow guerrillas, he showed not the least sign of disturbance. He was gay and happy, as though he had just returned from a particularly successful hunting trip.

Sensational, as his mother relates, was a subsequent encounter of Shura Chekalin with the Germans. Disguised as German soldiers, he and a group of fellow guerrillas descended on a village in which his relatives lived. He went to the home of these relatives; old folk, they were peasants who hated the Germans and were eager to help the guerrillas, particularly their kinsman Shura and his companions. The guerrillas decided to spend the night in the village, and Shura's old relatives were willing to take the chance of their not being discovered. After turning out the lights, all went to sleep.

Late in the night the Germans came. They searched the house and discovered men asleep on the floor and on top of the brick oven. They asked the old folk who they were.

"Your people," answered Shura's relatives.

Shura and the others heard the Germans but never moved, never opened their eyes, pretended to sleep soundly. The Germans flashed lights over the guerrillas and, deciding they were real Germans, motioned to the old folk that everything was in order and stretched themselves out in the available space on the floor, the benches, the table, and quickly fell asleep.

Later, when Shura described this incident to his companions in the dugout, he said his first impulse was to "treat the newcomers to a taste" of his hand grenades. Only the thought of what would happen to his relatives kept him from obeying the impulse. They were too old to flee and join the guerrillas. They would have to remain at home, and the Germans would be certain to execute them.

Thus Shura lived the life of an active, daring guerrilla. He knew little rest, no comfort. Nor did he care for either. He was young, healthy, eager for any exploit, any adventure. But constant exposure to cold and moisture and constant overwork weakened him, and he became ill. He lay on twigs in the dugout, hoping his fever and pain would abate. There was no doctor in the forest, and no drugstore. One of the guerrilla girls knew how to bind up wounds and stop the flow of blood—but nothing about internal ailments. Shura was getting worse. His temperature rose, his pain increased. He needed immediate attention, above all a dry, warm place in which to rest. He had to get back to civilization.

But where could he go? All around were Germans in every village, every town, on every road. The Russian lines were some distance away. To reach them would require time and the kind of maneuvering which in his weakened state Shura was unable to negotiate. Time was precious; every hour of delay and neglect only aggravated his condition. So the guerrillas decided to send him to his native village. It was quite near and, though Germans were holding it, his relatives there would care for him and would shield him from discovery.

In disguise, Shura reached his old home. It was quite changed now. Germans were everywhere: in the schoolhouse, in the town hall, in the streets, in the homes. Quietly Shura made his way to the home of his relatives. They gasped when they saw him. They knew what would happen when the Germans learned of his presence in the village. But they were brave people and helped Shura to the warmest place in the house—the top of the brick oven. They fed him, comforted him, assured him he was safe and needed to fear no one in the village. Shura fell asleep; but, always the guerrilla accustomed to sudden danger and possible attack, he kept his hand grenades beside him.

Try as hard as they might, his relatives could not prevent German spies from discovering Shura's presence in the village and in their home. His reputation as a formidable guerrilla had preceded him, and late at night twelve Germans armed to the teeth swooped down on the house. Shura awoke. His mind was clear. He saw himself trapped. But he had been trapped before in the forest, on the open road, on the village street, and had managed to fight or trick his way to safety. He would try again; he could not help trying any more than an animal in the forest when cornered can help trying to save itself. If he had to die, it would not be without inflicting death on those who had come to take his life. A hand grenade was beside him, ready for just the emergency he was facing. Lifting it, he flung it at his adversaries. But there was no explosion. It was a dud! It lay on the floor, dead and useless.

The Germans seized Shura and led him away.

He looked so young and sick they thought he would tell them the things they most wished to know—where his guerrilla detachment was, how many were in it, who they were. But while Shura's body was feverish, his mind was clear and alert. He refused to answer any but irrelevant questions. The German officer who examined him was so enraged that he started to denounce and swear at him, at all youths like him, at all guerrillas. Losing his temper, Shura seized an inkwell on the table before which he was sitting and flung it at his inquisitor, spilling the ink over the officer's eyes.

Shura was seized and beaten. Russian eyewitnesses told of it in detail afterwards, when the Germans were driven from the village. With bayonets the German soldiers pierced his felt boots, lacerated the calves of his legs. The boots were soaked with blood and Shura was in excruciating pain, but he remained stubborn and incommunicative—true to the code and oath of Russian guerrillas. He would not answer questions and could not be tempted to answer them by promises of leniency and forgiveness. Realizing the futility of their effort to force or bribe information out of him, the Germans sentenced him to be hanged.

Shura heard the sentence without show of alarm. Whatever his inner torment and his thoughts at this moment of darkness, whatever he felt about his mother, his father, his little brother whom he adored, Shura kept to himself. He did not weep, did not plead for mercy, did not utter a single word.

The gallows was erected in the public square where Shura often played and ran races with other boys. Peasants in the village whom he had known and who had known him since his childhood were ordered to attend the hanging. They saw Shura on his way to the public square. Though his felt boots were bloody and he was obviously in pain, he walked erect, with unbowed head. One of the Germans gave him a small laminated board and ordered him to write the words: "This is what happens to every guerrilla." Contemptuously Shura ignored the order. Turning to the Germans, he shouted defiantly: "You cannot hang us all. There are too many of us." That was all he said.

The executioner threw the noose over his head, and even then, in the last moments of his life, Shura did not lose himself. He started to sing the Russian national hymn. The words and the tune died on his lips.

The Germans tied to his breast the thin laminated board on which, in large letters and in Russian, they had printed the words: "The end of a guerrilla."

I saw this board in the History Museum in Moscow. "It will remain here," said the attendant, "as long as there is a Russia on this earth."

The Germans did not permit the peasants in the village to take down the body and bury it. They wanted it to hang on the tree in the public square as a symbol of their ruthlessness against those who resisted their invasion. "It made our blood curdle," testified a peasant from the village, "but what could we do?"

The winter blizzard, the fiercest the country had known in years, finally tore the body off the tree and buried it deep in the snow. It lay there until the Germans were driven from the village. Then it was recovered.

"We washed it," relates the mother, "dressed it in Shura's Sunday suit, and held a public funeral."

Shura was buried in the very place where he was hanged.

Now this place bears the name of Alexander Chekalin Square. The name of the village has likewise been changed from Peskovodskoye to Shura Chekalin. The Soviet Government has bestowed on him posthumously the title of "Hero of the Soviet Union," and his picture now adorns a new issue of Russian postage stamps.

The mother came to Moscow and addressed a mass meeting. A college girl who attended it said to me:

"The mother broke my heart with the story she told. But she also strengthened it with the love I felt for her wonderful boy. This is the way we all feel about that sixteen-year-old high-school youth."

CHAPTER 3: LIZA

A BLIZZARD RAGED over the village of Runa in the Kalinin province. Snow swept the streets and shrouded the homes of the people. Even dogs dared not be out in the open.

Against the blizzard's violence, the dense forests of birch, fir, and pine surrounding the villages offered no protection. As the evening wore on, the storm, instead of abating, gathered fresh fury. Bedtime came, and Axinya Prokofyevna, a middle-aged, illiterate woman, became alarmed.

"Where is Liza?" she asked her husband.

He did not know. In spite of the storm, she rushed outside, called on neighbors. No one had seen her daughter Liza or knew where she was.

Returning home, the woman broke into tears. One of a family of nine, of whom only four survived, Liza was her favorite child. Lively, imaginative, with an irrepressible hunger for books and learning, she showed promise of rising to a higher position than any of the Chaikins had ever attained—and now she had disappeared. Peasant fashion, the mother, amid tears, murmured to herself:

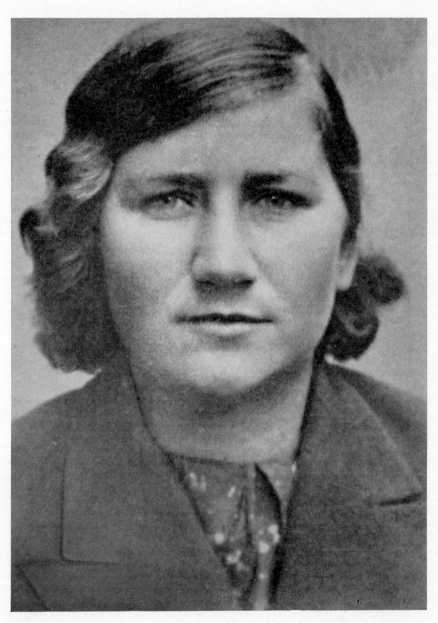

LIZA CHAIKINA

"Where are you, my dearest? Where are you, my bright little sunshine?"

She imagined the worst. Blinded by snow, Liza might have tumbled into a ravine or canyon. Swept by the gale, she might have blundered into the forest, where fugitive dogs and wolves, frenzied by hunger, might have attacked her. The mother wailed and wailed.

The father also was troubled. But, being a practical man, he walked outside, opened one of the shutters and, returning to the house, placed a red light in the unshuttered window.

"If Liza is lost in the storm," he said, "she'll see this light—in darkness, red can be seen a long distance away."

Late at night Liza returned.

Entering the vestibule on skis, she hastily took several books out of the bosom of her coat and showed them to her parents. She had gone to Zalesskaya, a neighboring village which had a library, to borrow the books. While there, she learned of a meeting of young people and stayed to attend it. That was why she was late in coming home. She spoke with such enthusiasm of the books and the meeting that the father and mother didn't bother to scold her for neglecting to tell them where she had gone.

Seven years later all Russia was to hear of Liza, talk of her, revere her. Her picture was to grace leading museums, schools, offices of generals, barracks of privates, dugouts at the front line, millions and millions of homes. She was to become one of the great heroines of the war, one of the great women in Russian history.

Because her mother was illiterate and her birthplace one of the "deafest" villages in Russia, Liza's life story, even more than that of Shura Chekalin, offers an insight into Russian home life, the mentality of Russian youth, the nature of Russian education, Russian personality, Russian patriotism, Russian morale, of which in the days before the Russo-German war little was known in the outside world and even less believed.

From her earliest years, her mother tells us, Liza was a precocious, inquisitive girl. She loved to hear her father tell stories and sing songs of old Russia, of peasant life as it was under serfdom and afterward. Many of these songs and stories depicted so vividly the peasant's "thievish destiny," as the mother so matchlessly expresses herself, that Liza would sob and mutter, "Such an unjust song!" To lift her out of such depression, her father would follow with some lively ditty describing the more comical aspect of peasant character and peasant experience.

Liza graduated from a four-year primary school, the only one in Runa. But her hunger for education was so enormous that in the midst of a raging blizzard and without saying a word to her father and mother, she dashed off on skis to a neighboring village just to borrow books.

That was Liza Chaikina at fifteen. At that time, outside of Runa and a few near-by villages, no one in Russia or in the province of Kalinin had heard of her. She was an ambitious, energetic peasant girl who, like so many other girls and boys, tried to take advantage of the new life and the new adventure which the Soviets had opened to village youth.

When her older sister Manya started going to school, Liza picked up her books, turned the pages, and asked her mother what was in them. Axinya Prokofyevna could explain nothing; she had never learned to read. That only enhanced Liza's curiosity about the mysteries of the printed word. Constantly she kept asking her mother to send her to school. At eight, the legal school age in Russia, Liza had her wish fulfilled. From the very first she was a brilliant student, attained the highest marks; her mother only wished that her son Shurik, who was twelve, had manifested as keen a desire for books and learning.

On finishing the primary school Liza started to work on the land. She sowed and planted, she pulled weeds and husked flax, she milked cows and looked after livestock. She helped her mother in the house—washing, cooking, doing other chores. Her free time she devoted to books. She never could read enough of them. There was no library in Runa; but Liza often went to Zalesskaya to read magazines, newspapers, and to borrow books.

The more books she read, the more social-minded she became and the more of her leisure she devoted to social or missionary work. She read newspapers and leaflets to her parents, to other people in Runa. She talked to them about the *kolkhoz* (collective farm) which was new in the village and which, because of its newness and its violent departure from the established mode of landholding and land tillage, had stirred doubt and discouragement in many of its members. From day to day Liza was bringing more and more "culture" to Runa. Of the older people her mother was among the first to respond to her exhortations and to help enlighten other villagers on the meaning and purposes of the kolkhoz, of the bright future it held for them if only they would slough off their apathy and disbelief and work hard to achieve its bountiful promise.

Englishmen and Americans who have often wondered how such a stupendous revolution as the kolkhoz, embracing a population of over one hundred million people—a revolution not only in agriculture but in political thought, in social manners, in everyday life—could be achieved within so brief a space of time, would do well to ponder on the importance of the work of young people like Liza. Of course she was exceptionally precocious and energetic, with a natural gift for people, for leadership, for organization. Most young people of her age possessed neither her understanding and enthusiasm nor her sense of social adventure and social application. But other young people, older than she, picked and sifted from the multitudes at the disposal of the Soviets, carried as much responsibility as she did, though not always did they acquit themselves of it with as much telling success. While not the initiator of new ideas and new practices, the youth in Russia, especially the Komsomol,[1] always are the interpreter and the disseminator of both.

Liza organized the children of Runa into Pioneers—an organization which bears enough resemblance to the Boy Scouts to invite comparison, and enough difference to challenge the resemblance. The Pioneers admit girls as the equals of boys. They also espouse a political program which is alien to the Boy Scouts or to the Camp Fire Girls.

Liza did more: she organized a farm club in Runa and propagandized new ideas of agriculture. She flung into the vocabulary of the village, so barren of scientific terms, a flock of foreign words such as "cultivator," "tractor," "combine."

Runa is situated in the heart of the flax-growing region, and Liza secured special books on advanced methods of cultivating this highly profitable crop. She read the books and discussed them with her mother and other women—with women more than with men, because since ancient times flax raising had been more of a woman's than a man's enterprise in the Russian village.

While still only fifteen, Liza, in addition to all her other duties, assumed charge of the library reading room in the village of Zalesskaya, only a short distance from Runa. She infused so much fresh energy into the institution that it became *the* gathering place of young people and of the intelligentsia in the countryside. Agronomists, livestock experts, schoolteachers, physicians, all came there evenings and holidays to read magazines and newspapers and to discuss the events of the times in the countryside and in the outside world.

Liza saw to it that there was always something new and culturally exciting to attract visitors. She arranged lectures and discussions. She started a dramatic club, a musical club, a political club. Earnest and indefatigable, she found time for these manifold activities without ceasing her work on the kolkhoz. Nor did she neglect to further her own education through self-instruction. She read omnivorously—Lermontov, Pushkin, Tolstoy, Gorki, several of the more prominent Soviet authors. She played less than other young people, though she did not always remain aloof from dances, picnics, and other purely social diversions.

In 1939, when she was twenty, Liza was chosen secretary of the district Komsomol. She traveled from kolkhoz to kolkhoz, lecturing, organizing, lifting the morale and faith of young and old. She formed Komsomol units in village after village and through them organized all manner of study

[1]Though called Union of Communist Youth, the Komsomol is a non-party organization, and anyone may apply for membership.

clubs. Always well informed on events of the day and on Soviet policy, she delivered lectures, speaking in the homely, concrete language which she had learned so well in her own home, especially from her brilliantly loquacious mother. There was not a task or institution in the district that escaped her attention. Schools, shops, the tractor station, collective farms, she visited them all, studied their work, listened to complaints, offered suggestions, strove in every way to improve their condition, increase their usefulness—all at the age of twenty! In the two years that she was district secretary of the Komsomol she doubled its membership and lifted the youth of the countryside to a level of enlightenment, social responsibility, and personal application which it had never known.

Once, shortly before the war, her mother said to her:

"Lizenka, you are twenty-two already. It's time you were married. Isn't there someone you really care for?"

To a peasant mother a daughter of twenty-two and still single is a source of concern—a warning and a premonition of calamity. In spite of the new ideas she had absorbed from her daughter, Axinya Prokofyevna had not changed in her motherly solicitude for the marital happiness of her feverishly active Liza. In reply Liza laughed.

"When I find the right man, Mamma, I'll bring him to the house. Now I have no time."

Instead of marriage, Liza was planning to attend college. In the summer of 1941 she took leave from her tutorial duties and went to Kalinin to attend summer school.

She had no more than settled down to intensive study when the war broke out. Immediately she left Kalinin and went back to the district in which her native village was located. Again she plunged into work and again she centered her energies on organizing young people. The harvest had to be gathered, and this time every potato, every strand of flax, every ear of rye was more precious than ever. There was to be no delay in the work in the fields and no waste of time—of men, animals, machinery. Local transport needed to be improved, and Liza devoted much attention to the problem. Above all, young people, boys and girls, had to be trained for war.

Liza found time for everything, including military training. She was mastering the rifle, the hand grenade, the machine gun. To a childhood friend, Nura Barsukova, she wrote:

"I shall become a soldier, a real soldier—that's my ambition now."

The Komsomol office became military headquarters. From all the villages around, boys and girls came for advice, for training, for inspiration. Liza had a personal word for each one of them, and they all looked to her, even more than in prewar times, as their leader and teacher.

"We love our country," she said in speech after speech, "and we shall defend it. We shall uphold our government and our fatherland. The enemy shall learn what it is to attack our Soviet people."

While Liza was assiduously converting the district of Peno into a powerful arm of national defense, the Red Army west of the territory was continually retreating, drawing closer and closer to Peno, to Zalesskaya, to Runa, to all the villages around.

One evening in August 1941 Liza heard loud noises in the street. She looked outside. Behind a cloud of dust she beheld an ominous procession— a herd of livestock driven by children, behind it a caravan of wagons loaded with more children and with household goods. She asked a barefoot boy who, whip in hand, was following a herd of livestock:

"Where are you from?"

Laconic and significant was his reply: "We are retreating from the Germans."

Liza was overcome with grief. The war was coming home to all these humble, hard-working· folk she had striven so hard to educate in a new way of life and work, in the use of the cultivator, the tractor, the gangplow, and many another implement of which they had never heard before. The mastery of the new machines in the village and of the collective method of exploiting land had no more than begun. Now it might all have to end. It would end when the Germans swept over the fields and forests which she knew so well, loved so deeply. No, she was not doing enough for the Army and the fatherland—no, she was not.

She went home to Runa, arriving late at night. Her mother was happy to see her. She fetched milk and bread. But Liza, always so talkative, was now solemn and reticent. Her mind was feverishly busy with plans of fighting the enemy with every weapon. All she could talk about to her mother was of the calamity falling over village after village, with the people running as if from an advancing flood and seeking refuge eastward for themselves, their cows, their pigs, their sheep.

The next day Liza went to Peno and called a meeting of the young people. Gone now were the songs, the jokes, the laughter, the liveliness of yesterday. All were serious, silent, expectant. Liza spoke:

"We will not be Hitler's slaves. He shall never break our will to fight and to be ourselves. We will stand up and battle like one man. For us young people it is an honor to be on the battlefield. All of you whose hearts are beating, rise up and fight the enemy! . . ."

She started to organize a guerrilla brigade. Within a short time she had mustered sixty-eight volunteers. Among them was the family of Leonid Grigoryev, a schoolteacher from Leningrad, who was spending the summer with his family in a near-by village. His older son Nikolai, a student in a mining college of Leningrad, his daughter Nina, and his grammar-school son Vladimir, all joined Liza's detachment.

"We lived as one family," said the schoolteacher, "now we shall fight as one family."

Before her departure for the forest, Liza went to bid her mother farewell. She arrived home at midnight. Knowing that her decision to become a guerrilla would upset her mother, she resolved to say nothing until the moment when she was actually leaving. She asked her mother to heat the bathhouse—the little wooden steam cottage which the humblest peasant in that region had and which was easy to heat. Axinya Prokofyevna built a fire in the cottage, and when the stones in the hearth immediately above the fire were heated she poured pails of water over them. The bathhouse filled with clouds of hot steam, and mother and daughter washed together as they had so often done. While in the bathhouse Liza said:

"Mother, I'll have to leave very early in the morning so nobody will see me."

The mother did not ask any questions. She had been accustomed to Liza's sudden arrivals and departures.

"Very well," said the mother, "you can leave on the first boat."

In the morning the mother rose, built a fire, and cooked breakfast for her departing daughter. As she was busying herself over the oven, the chairman of the kolkhoz called and asked Axinya to go and help thresh rye. No one appreciated more than Liza the importance of the threshing. For days she had been seeking to fire people, young and old, with a will and an energy to leave not a single stalk of rye unharvested and unthreshed. But now she was home—perhaps for the last time—and she couldn't help requesting the chairman to allow her mother to escort her to the boat. The chairman asked no questions and agreed. He knew Liza would make no such request unless there was a special reason.

After breakfast Axinya and Liza walked through the woods to the river. Raspberries were in season, and the bushes were clustered with them. Liza picked a handful and threw them into her mouth.

"Let's pick a lot more, little daughter," said the mother.

But there was no time. "No, Mother," Liza replied, "we must hurry."

The mother observed Liza's growing solemnity. Her face had never been so serious. Now it flushed, now it grew pale, as though in manifestation of or in response to the changing surge of thoughts and moods. They walked on. Liza would not hold her mother in suspense any longer.

"Mother," she said, "I am leaving for the forest. Life may be hard there— I may go hungry and cold, but I've got to go. I'm joining the guerrillas."

The very word "guerrilla" was new to the mother, and she asked Liza to explain what, as a guerrilla, she was supposed to do.

Liza answered: "Mother, you are my friend, but I won't tell you a word about guerrillas."

Secrecy is the first law of the guerrilla, and not even a mother can be taken into confidence. Yet Liza felt she had to prepare her mother for the worst and arm her with some defense in case she faced the enemy and was held culpable for her daughter's acts as a guerrilla.

"Don't be alarmed, Mother," Liza continued. "Only remember: if anything happens to me, if they capture me or I am killed and you see my body, don't say anything—pretend you never knew me, don't admit I am your daughter, or they'll burn our village."

The mother trembled with terror. She had not expected such a gruesome warning from her daughter. Of course the Germans were coming closer and closer. Masses of people were fleeing before the enemy's advance. Many of them were being hit by long-distance shells, by mines dropped from the air. But Runa was still untouched by the war. The forest looked so alive, so healthy, was so close to her, was all around her. And the Volga, the mother Volga, "cradle of Russia's liberty," was only a few steps away, as broad and shiny and as reassuring as a loving mother. Yet Liza had spoken of death—her death. How horrible!

"It would have been better, far better, daughter mine,'" said the mother, "if you hadn't said anything to me."

They walked on in silence until they came to the river. The boat was not leaving for another hour. So, with a sack of personal belongings on her shoulder, Liza escorted her mother back over the path they had just been pursuing.

It was still early morning. The dew shimmered on grass and trees. Mother and daughter walked together, then stopped. Axinya Prokofyevna sobbed, and tears came to Liza's eyes—tears which she had hardly known since the days when she was a little girl and had wept over the songs and stories her father had sung or recited. A religious woman, the mother offered Liza her blessings in the old-fashioned Orthodox way by making the sign of the cross over her daughter's body. They embraced and kissed over and over. Then, wiping her eyes with her bare hand, the mother went on her way to the village to thresh rye, and Liza returned to the pier for the journey to the forest and to the life and adventures of a guerrilla.

The military leader of the squad was a man named Filimonov.

"Being a guerrilla," he told his fellow combatants, "is not the same as angling fish on the bank of a stream. You don't have a cozy home to which to return, you have only a mighty forest in which to fight. We shall encounter multitudes of difficulties. If any of you feel it's going to be too much for you, speak up. We don't intend to hold anyone by compulsion. Guerrillas are never drafted. They are always volunteers."

No one asked to be relieved.

New and amateurish, the squad was short of weapons, of food, and was without fighting experience. Yet neither Filimonov nor Liza, who was the social and spiritual leader, was dismayed. Soldiers of fortune and misfortune whose life was as irregular as their fighting, they hoped for the best and prepared for the worst.

"The most important thing," said Filimonov, "is to remain steadfast, to look ahead, never to lose sight of the final objective." Liza applauded these stalwart fighting words.

The squad set to work. They dug caves and set up camp. Several boys and girls had studied radio mechanics, and out of the scanty materials at their disposal they constructed a radio box. . . .

The squad was ready for action.

The youngest member was a boy named Vasya, only fifteen years old. He was a brilliant scout, as fearless as Liza. One bright starry night he and Liza left on a scouting expedition. No sooner had they come out of the forest than they beheld flames in the open skies. Liza recognized the villages that were burning—Golovkino, Zamyatino, Toropetz. She knew them all.

She and Vasya continued their journey, and all along the route they saw broken tanks, burned automobiles, ruined homes, fields pockmarked with craters. They walked and scouted all night, and in the morning they heard the hum of engines. They knew German tanks were coming, and they swiftly darted into a gulch, lay there awhile, then rushed back into the woods and hid under a tree.

"Run, Vasya," whispered Liza, "and report to the squad." Vasya sprang to his feet and galloped away. Liza unfastened her holster and remained where she was, observing the road and the clearing ahead. She heard an explosive noise—it was a German motorcyclist. He sped by and was swallowed by a cloud of dust. Following him came tanks—first one, then another, then a third. Presently the motorcyclist returned. Liza saw him clearly. He was talking to the men in one of the tanks. Immediately afterward the tanks turned off the road and so did the cyclist, and they all stopped in an unmowed meadow and settled for a rest. Sheltered by trees and foliage, Liza lay without moving and watched.

Out of the tanks the Germans took bottles, blankets, pillows, napkins. They spread everything on the grass, and before a man with the stripes of a non-commissioned officer they laid out a napkin. Keenly Liza watched their movements and scrutinized their uniforms to ascertain their respective ranks.

The sun was rising, auguring a hot day, and Liza saw an officer, a tall, spindly man, take off his uniform and shirt, tie a towel round his neck, and make his way to the sandy shallows of a near-by stream. Presently Vasya returned. Lying down beside Liza, he whispered,

"They're coming."

Leaving Vasya to watch the Germans, Liza crawled quietly away, deeper and deeper into the forest. Then she looked back, and when she saw nothing but dense rows of trees which no German eyes could penetrate, she rose to her feet. Time counted, and she would not lose a single second. Hastily she took off her shoes so she could run faster. She stuffed the shoes into her sack and started to run, swinging her arms as though she were running a race.

A few minutes later she was standing behind a gigantic willow tree that rose high above the bank of the stream to which the German officer had gone. He was on the opposite bank, and Liza saw him clearly. He was in trunks, smoking a cigarette and enjoying the sun, the peace, the seclusion. She clutched at her gun and waited. It was not yet time to act.

Soon she heard loud noises—the crackle of guns, the hum of engines, the explosive bark of a motorcycle. She knew the guerrillas had come and were in action. Now was her chance. Her eyes were fixed on the long-legged German officer. Taking aim, she fired—one, two! The officer sprang to his feet as though the earth under him were afire. He threw away his cigarette and ran a few steps. Then he fell with his face to the earth, never again to rise.

In the History Museum on the Red Square in Moscow, I saw the long, heavy, dark-gray revolver with which Liza had shot the officer, and the Iron Cross he had been wearing.

The fame of the guerrilla squad spread to near-by and remote villages. Peasants knew who it was that was wrecking German military trains, blowing up bridges, making raids on villages in which Germans had settled. Readily they communicated information to the guerrillas, supplied them with flour, milk, cereals, butter, dried biscuits.

Once a group of young people from the forest went to a village for meat. Germans were in the village, and the first man the young guerrillas met was so suspicious he assumed the protective pose of "darkness" which in years and ages past had served the peasant so well in moments of trouble.

"How do I know where the Germans are," he said, "or where the guerrillas are? We here are dark people, and all we know is work on the land. As for the rest—you know more than we do—you read the newspapers."

But the young guerrillas had no difficulty in convincing the old man who they were. His manner instantly changed. He crawled into his deep cellar and soon returned, his arms loaded with heavy chunks of meat—which they carried into the forest.

Meanwhile, by her example of personal courage and her words of un-dimmed optimism, Liza kept high the morale of the detachment. So great was the respect for her that even though its members were not subject to the social or military discipline of regulars, they invariably stood up before Liza as before a superior officer. She, in turn, never missed an oppor-tunity to show personal interest in all members of the squad, cheering them to ever greater feats of courage and valor.

Once, at the end of October 1941, two girls and a youth went off on a scouting expedition. Three days passed, and they did not return. Liza was alarmed. She wondered if the Germans had captured them. The peasant population was loyal to the guerrillas. But always there were individuals, former kulaks and others, who nurtured a spirit of revenge against those who made their entrance into the kolkhoz inescapable. There were also those who were weak in character and yielded to the blandishments of Germans, especially to the promises of rich reward for acting as their agents and spies. Liza wondered if the three young guerrillas had been betrayed by a peasant renegade.

Soon the two girls returned, but the youth, whose name was Fokin, was missing. The girls reported that they had all been seized by Germans but had succeeded in escaping. They could not tell what had happened to Fokin.

Liza was attached to Fokin. She had been more than a sister to him; he had been more than a brother to her. Perhaps they were in love with each other. No one knows. Liza wondered. If the worst came—— But she couldn't imagine Fokin no longer alive. She hoped for a miracle that would bring him back to the forest—and to her.

After the squad had been in existence a month, Filimonov and Liza called a meeting. Filimonov was bristling with hope. At the start the squad had had few weapons. Now, thanks to successful raids on Germans, they were in possession of many automatic rifles, machine guns, trench mortars, two field guns. They had fought well and with success. They had blown up or burned one hundred enemy trucks loaded with military supplies. They had held up caravans of carts hauling food for the German Army. They had no reason to feel displeased with themselves and their work, though they must look forward to more serious battles and sterner trials.

After the meeting, Liza sat down on the ground by the feeble bonfire. It had to be feeble so German aviators could not detect it and discover their camp. The other guerrillas crowded around Liza, and she observed how changed the young people had become within only one month in the forest. They looked more mature, more coarse, more determined, more untamed—real warriors. She was heartened by the thought that none of them had ever voiced a word of complaint of the hard life they were living, of the severe hazards they were daily encountering.

Warm, comradely, in repose they started to sing "Yermak," an ancient ballad about a Cossack outlaw who, with a band of Cossack guerrillas, had penetrated the depths of Siberia and conquered much of it for the Russian nation. It is a long ballad, and it was all the more heartening to Liza and the others because it depicted the kind of life and the type of fighting that resembled their own. They sang in low voices, with their hands over their mouths so as to prevent the sounds from reaching the ears of German scouts. They sang slowly, too, as though to prolong the pleasure of recounting the heroic deeds of the Cossack chief and his brave guerrillas.

In the middle of November, Filimonov called Liza aside and gave her a small pamphlet. Leaning against a tree, Liza read it. The pamphlet was a welcome messenger from "the big earth"—the guerrillas' term for the lands in the rear of the Russian front lines. It was Stalin's speech delivered on the anniversary of the Revolution. The slogan of it was: "Death to the German invaders!"

Russia was in a dark mood, especially the population in the occupied territories. Nor did the Germans fail to take advantage of this mood. By word of mouth, by leaflet, by special bulletins, by motion pictures they kept spreading rumor after rumor, alarm after alarm. "Moscow *kaput* [Moscow is finished]"—was the burden of all their messages, all their propaganda. With all the means of communication, proclamation, and terrorization at their command, they sought to persuade the Russians in the occupied territory that their cause was hopeless and that it was futile of them to offer resistance to the Germans. Particularly did they seek to discourage the peasantry from aiding the guerrillas and from maintaining opposition to the "new order" which they were bringing to Russia.

It was imperative to counteract German propaganda and to sustain the fighting will and fighting power of the civilian population in the occupied territory, especially in the villages. Liza was the logical choice for this task. She was an experienced organizer and able speaker. She had friends in all the villages around. She knew the young people, the older folk. They had trusted and admired her. She had enemies, too, but now they were of no account. Above all, she understood how to address peasants in their own vocabulary and in terms of their own immediate and future welfare. It was a hazardous mission. The Germans had heard of her, had spread out a net to catch her. They might have enlisted the aid of her enemies who never had become reconciled to the kolkhoz. But Liza knew the momentousness of the mission on which she was to embark, especially at a time when the Germans were continually pushing closer and closer to

Moscow, when their rumors, however exaggerated and wild, might find some response among the suffering multitudes of the villages.

Liza bade Filimonov and her friends farewell and left on her journey all alone. In mud and rain, in wind and darkness she trudged from village to village. She gathered small groups in houses of friends and preached to them the one message she wished to burn into the soul of every Russian— "Death to the German invaders!" Though invariably she had to speak in a low voice, her words rang with fury and hate.

"Kill the Germans," she admonished, "burn them, give them no food, help the guerrillas, fight and fight the enemy without sparing yourselves. No Fritzes shall be ordering us around in our lands. . . ."

Through underground channels the mission of Liza was made known to neighboring villages, and when she arrived there people were prepared to receive her, to hear her, to shield her from discovery or betrayal.

In one village she read a bulletin which the Germans had posted on the public square. They were offering a reward of five thousand marks, a house, a garden, a cow to anyone who would disclose the whereabouts of a certain guerrilla leader. This leader was none other than Fokin—the Komsomol from her squad whom she had so deeply admired and who was supposed to have been captured by the Germans. Obviously he had escaped—more, he had organized a guerrilla squad of his own. The fact that the Germans were offering a generous reward for information about him was eloquent proof of their fear of him and of his success in fighting them.

Liza was overjoyed. Fokin was not only alive but fighting! She made inquiries about him and hastened to his hideout. The reunion was one of the happiest and most exciting events in Liza's exciting life. She was devoted to him and listened with feverish eagerness to the account of his adventures since they had last seen each other. The Germans had beaten and tortured him. But he had never given them a single shred of the information they were seeking to worm and whip out of him. Finally he had managed to escape. Within a short time he had built up a new guerrilla squad and resumed the underground war against the enemy. Liza met Fokin's associates and was happier than she had been in a long time.

Fortified with fresh fervor, she again set out on her mission to enlighten villages on the fighting spirit of Moscow and the vast Russian rear. She urged peasants not to trust a single word of German propaganda, with its emphasis on Moscow's defeatism and the certain collapse of the capital. She visited fifteen more villages, and everywhere she was accorded a hearty welcome.

One evening she went to the village of Krasnoye Pokatishtshe to visit an old friend, a girl named Marusia Kuporova. As she was entering the house she met Timofey Kolosov—a man who had bitterly resisted the kolkhoz in his village. She did not know much about him, but he, without a moment's hesitation, rushed to German headquarters and reported her arrival. Soon afterward German soldiers, led by an officer, swooped down on the house and seized Liza. They removed her shoes and most of her clothes and walked her to Peno, the seat of the district in which she had spent all her life and where she knew not only every person but every house and every tree.

The Germans started to question her. Where was her guerrilla squad? Who was in it? How many were in it? Liza made no answer. The questioning continued. Once she grew so exasperated she flung out: "Death to all of you cursed creatures!" and spat into the officer's face. They punished her severely. But she remained as incommunicative as did Shura Chekalin under similar questioning, and, like him, she neither wept nor pleaded for mercy and strove hard not to betray the physical pain she was suffering.

"Tell us where the guerrilla squad is and you'll remain alive!" said the German officer.

Liza made no answer. Then, turning to residents of Peno, where even the children knew her, he asked if they could identify her. No one spoke. He repeated the question. One woman, Arishka Kruglova, said:

"Yes, she is a guerrilla and the leading Komsomolka here."

The officer gave orders to the firing squad, and Liza said:

"I am ready. Go ahead, executioners, shoot!"

A volley of bullets whistled over her head. But not once was she hit. This was a trick to break her will and to frighten her into confession. Again the officer started to question her, but with no more success than formerly. The trick failed in its purpose. It only stiffened Liza's determination to bare none of her secrets to the enemy.

"Death to the German invaders!" she cried.

"Fire!" commanded the officer.

This time Liza was hit. Motioning with her hand, she again cried out: "I am dying for our victory——"

A third round of shots followed. Liza fell on the fresh snow. Feebly, according to eyewitnesses, she continued, "—for the fatherland—for our people!"

The detachment in the forest learned of her death the next day. In their dugout they pronounced a death sentence on the man who had betrayed her to the Germans and on the woman who had identified her in Peno. That night they started out on a punitive expedition. By morning eleven villages in which Germans were staying were set afire.

Two young members of the squad made their way to the home of Timofey Kolosov. They seized him and brought him to the forest. They questioned him, then shot him.

Later several other guerrillas found their way to the village of Peno. They entered the courtyard of Arishka Kruglova. Though there were Germans in the village, they eluded detection, and one of them knocked quietly, like a friend, at the shutters which were already closed.

"Who's there?"

"Friends."

Kruglova admitted the guerrillas to the house, and when she saw who they were she grew pale, knelt down, and, peasant fashion, started to beg for mercy.

"Precious little doves," she supplicated, "don't touch me—I didn't mean a word I said—it was only some foolish prattle on my part—that's all— your hands and feet I shall kiss if you'll only let me live——"

"Come on out."

She attempted to cry. They seized her, stuffed her mouth with a cloth, and led her outside.

In the morning the people of the village of Peno found her dead body on the village athletic field.

For seventeen days Liza's squad had sought in vain to rescue Liza's body from the Germans. Alarmed by the executions of Kolosov and Kruglova, by the fires the guerrillas had started, the Germans heavily increased their armed forces in Peno. Yet on the eighteenth day the guerrillas succeeded in recovering the body. They took it to the forest, buried it there with military honors.

Since then the grave of Liza Ivanovna Chaikina, like the tomb of Shura Chekalin, has become a national shrine.

CHAPTER 4: ZOYA

On the road between Moscow and Mozhaisk lies the village of Petrishtshevo. It is a small village rising on the steep bank of a stream and sheltered by a wood. Like nearly all villages in the vicinity of Moscow or any northern Russian city, it is noted for the trade of its people. In Petrishtshevo the trade is tailoring. During the winter months, especially, when there is little work on the land, the peasants sit in their homes or shops and ply their needles.

Until December 1941 Petrishtshevo was only one more village in the

Moscow region, like many others in all Russia, with no claim to distinction because of its geography, history, or the achievements of its citizens.

Now it is one of the most celebrated places in the country. Hardly a child of school age but has heard of it and reveres it. Were it not for the war, Russians from all over their boundless land would be trekking there on pilgrimages. When the war ends, and for as long a time as there are Russians, they will be making pilgrimages to this humble, remote, wood-sheltered village, for with it is associated the life and death of Zoya, the eighteen-year-old high-school girl who has become one of the most revered and heroic figures of the war. ·

The Germans have darkened the public squares of Russian cities and villages with forests of gallows. On these gallows they have publicly hanged thousands of Russians, especially young people. None of these executions has fanned such a blaze of wrath, such a storm of national emotion, as the hanging of the tall, dark-haired, pretty Zoya. To the Russian people she has become a symbol of their martyrdom, of the heroism of their women, of the invincibility of their spirit. To the youth of the land she is the incarnation of all that is precious and noble in the Russian character—of virtues and attainments which it must emulate and exalt.

Countless volumes will be written about this girl—biographies, novels, plays, poems. Painters are already making portraits of her for the public museums. Konstantin Simonov, one of Russia's leading playwrights (author of *The Russian People,* produced in New York this year), is writing a play about her. The composer Kovalevsky is writing an opera. The sculptors Zelinsky and Levedeva are making statues. At Alma-Ata, Russia's Hollywood, a leading director is producing a motion picture about Zoya. All this is only the beginning of the tributes which writers and artists will for ages and ages bestow upon her.

Her mother told me that her books, her school compositions, her letters, her diaries have been snatched up by the museums of the country. She herself has few specimens of her daughter's writing left. Cities, villages, schools, factories, museums will be named after the girl all over the land. In moments of trial or triumph Russian girls will be asking themselves, as they are already doing: "What would Zoya do in my place? What must I do to be like Zoya?" She has stirred the imagination of the Russian people as no other woman of our time or in all Russian history.

This eighteen-year-old high-school girl is by way of being haloed into national sainthood. In the guerrilla detachment of which she was a member she was known as Tanya, and for several weeks after her execution, though she had already become a national heroine, the public did not know her real name. Her mother was on a streetcar when she first heard the story of her martyrdom, but she did not associate it with her own daughter.

During hours of questioning Tanya, German inquisitors made violent efforts to learn her identity. In vain. She wouldn't tell them. She carried the secret to her grave, and only later was it discovered.

A teacher of literature in a village told me she was convinced that Tanya had borrowed her assumed name from Pushkin's heroine in *Eugene Onegin*. This seemed so much like the sentimentalism of Russia's ancient and highly sheltered school girls that on first thought I dismissed it as a bit of rash fancy. I said so to the teacher, and she replied:

"Come to my classes sometime and hear the students discuss Pushkin's Tatyana. You'll learn something about the sentimentalism of today's Russian youth."

Later, when I met Tanya's mother, I asked her if she concurred in the opinion of the schoolteacher.

"I never asked my daughter," the mother said, "because I never saw her after she went to war. Most likely it is true. She adored Pushkin's heroine. She could recite all of *Eugene Onegin* by heart."

Her full name was Zoya Kosmodemyanskaya, derived, like many Russian surnames, from the Christian names of saints—Kosma and Demyan (Cosmos and Damian). She was of peasant origin and was born September 13, 1923, in the village of Osinnovy Gayi, in the province of Tambov.

This province is noted for its black earth, its forests, its deep and sticky mud, its magnificent horses, its rapacious wolves. Osinnovy Gayi was a large village of some five thousand inhabitants. A trading center for the surrounding countryside, it boasted scarcely a hint of the modern machine age. On the outskirts were deep ravines which wind and rain were continually making deeper. All around were fields of rye, wheat, oats, potatoes, buckwheat, millet, barley. In winter the snows banked high, the frost bit through woolen garments, wolves sought shelter in the snowbound ravines. Desperate with hunger, they sometimes ventured into the village in search of a lamb or some equally tame and palatable prey. In summer, nature was rich and verdant with beauty. Fields and meadows gleamed with flowers.

At birth Zoya was very thin, very light, weighing only six Russian pounds, which are about five and a half American pounds. As a child her face was white, her eyes azure, her hair dark and wavy. She matured early, and at the age of a year had not only cut her teeth but talked and walked. She was a healthy, playful, obedient child and from her earliest years took pride in helping her mother with the housework and with the care of her younger brother Shura.

When she was six years old the family moved to Siberia. For the first time in her life Zoya traveled on a train, and the seven days that the journey lasted was an exhilarating adventure.

The town of Kansk, deep in the heart of Siberia, was even more of an

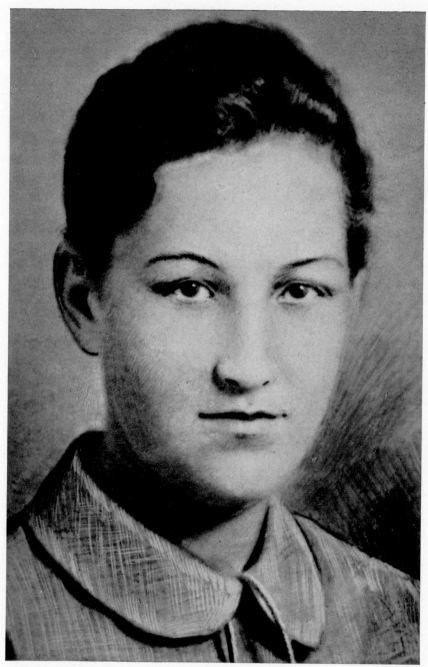

ZOYA KOSMODEMYANSKAYA

adventure than the journey. Zoya had never seen a town so profligate in the use of wood—wooden houses, wooden roofs, wooden sidewalks. In the village of her birth, even in summer, she had never known the natural grandeur in which Kansk abounded or the comfort that it boasted. No black bread—only white. In winter people wore the furs of wild animals which they themselves hunted. The river Kan, broader and swifter than any body of water she had ever seen before, was a source of endless mystery and joy. Zoya often went there with her brother to fetch the water for the house or to run along the bank and see if she could outrace the waters. The river was one of her favorite playgrounds.

A primeval forest of age-old cedars, pine, and other trees shadowed the town. Elk, wildcats, wolves, bears were among its permanent inhabitants. But Zoya had no fear of animals, domestic or wild. The forest lured her, and she often coaxed her brother and other children to accompany her on trips there. Mushrooms and raspberries were abundant in the forest, and people went with wooden pails to pick them. Bears also hunted raspberries, but Zoya never gave a thought to possible trouble with bears. She went berrying at every opportunity. Bird cherries were equally plentiful. Siberians gathered them in sacks, dried them, ground them in the mill, used the powdered product in their tea, their pastry. Zoya likewise gathered bird cherries. She reveled in the diversions and adventure that the faraway Siberian town of Kansk so abundantly provided.

Once she and her parents went on a trip to another town. On their return to the hotel Zoya was not there. The mother went to find her. She walked all over the town—but no Zoya. Returning full of apprehension, she wept. The father went to look for Zoya. Soon he came back, leading her by the hand. She had gone exploring the town and the forest, and on her way home got lost; she couldn't find her way to the hotel and was taken to the militia. Her father found her there, sitting over a glass of tea like a guest of the house and talking with animation of her native village in the province of Tambov. It was not in Zoya's nature to feel lost or become panicky. From her earliest childhood she was a stranger to fear.

After a year in Siberia, the family returned to the Tambov village. The father and mother left the children with their grandparents, and themselves went to establish a house in Moscow. Zoya was lonesome for her mother and father, and soon the family was reunited in the capital.

At eight, the school age in Russia, Zoya started her formal education. She joined the Pioneers and took to heart the teachings and precepts of this children's organization in Russia, which in some of its teachings resembles a Protestant Sunday school. Pioneers abjure smoking, telling lies, and disrespect for parents, elders, and teachers. They exhort self-control, kindliness to animals and friends, usefulness to society. To Zoya these precepts were no mere abstractions. They were an everyday guide, the moving force in her daily behavior.

"I want justice," she used to say at meetings in her classes and also at home.

"To her," says her mother, "justice was the acme of fulfillment."

From her earliest years she learned to be direct and straightforward; people who fawned or falsified roused her indignation.

She was an excellent student, never failed to do her homework; beginning with her fourth year in school, she passed her examinations with an "excellent," the highest attainable mark in Russian schools.

When Zoya was ten her father died suddenly and she became her mother's closest friend. Supporting two children in a large city was not an easy task for the mother, a schoolteacher named Lubov Timofeyevna. She loved her husband, widowhood caused her endless grief. Often she wept with misery. Invariably Zoya would put her arms around her and say: "Don't weep—we'll get along. I'll grow up and take care of you, see if I don't—you have nothing to worry about."

Such demonstrations of affection drew the mother and Zoya closer together.

At fifteen, Zoya moved from the Pioneers to the Komsomol, the Soviet youth organization. She enjoyed being attached to an organization and took her Komsomol duties as earnestly to heart as those of the Pioneers. Now she started doing social work. Their own housemaid had a child and couldn't attend literacy classes. Zoya taught her to read and write. She got other girls to promise they would teach at least ten illiterate women to read and write. She liked to feel grown up, with a sense of responsibility to herself, her mother, the outside world. The older she grew, the greater this sense of responsibility became, and the more eager she was to express it in social work.

The mother taught in a factory school.

"At one time," she told me, "I held four positions in four different factories. I'd be gone all day, and on my return home I'd find everything done—dinner ready, the floor washed, the fire going, the food and other things bought."

Zoya would have everything prepared.

When still very young, Zoya decided that when she grew up she would have a career. She couldn't imagine life for a woman without a career. She was poor in mathematics and felt she'd never make a good engineer. But she loved books and dreamed of becoming a writer, preferably a literary critic.

Once she wrote in her diary: *"Respect yourself; don't overestimate yourself; don't lock yourself in your own shell; don't be one-sided; don't shout that people neither respect nor appreciate you; work hard to develop your-self and you'll acquire greater confidence in yourself."*

These are the thoughts of an earnest, ambitious, self-willed, sensible girl to whom setbacks would bring neither dismay nor disillusionment.

She read omnivorously. She never could get enough books from one library, so she borrowed from several, or rather had her mother borrow them from the libraries of the factories in which she was teaching. Zoya loved the Russian classics, fiction and non-fiction, particularly the literary critics of the nineteenth century. Chernyshevsky, Dobrolubov, Pisarev, Belinsky—she read them all from cover to cover. She not only read what these men (whose literary criticisms were also flaming social sermons) had written, but what had been written about them, and the stories of their lives.

Russian history and Russian folklore also absorbed her attention. With increasing emphasis in Russian schools on purely national culture and national subjects, Zoya grew increasingly patriotic. She delved deep and with ever-heightening fervor into the past of her own people. Ilya Murometz, the wise, simple, generous, all-conquering hero of Russian folklore became her hero. When a junior in high school she wrote a composition picturing him as an embodiment of the gifts and virtues of the Russian people. To this day the composition remains the most brilliant piece of writing of any student in that school. In style and in thought her compositions were so outstanding that the teacher always read them to the class.

Heroes in Russian history became her heroes. Alexander Nevsky, Mikhail Kutuzov, Alexander Suvorov, Dmitry Donskoy were names she cherished and revered. She read their biographies, gloried in their achievements. Patriotism, love of Russia, her past as well as her present, became a stirring part in Zoya's make-up. *War and Peace* was her favorite novel. Often she gave talks in school on Kutuzov, who is one of the heroes of this monumental epic. The descriptions of the battle of Borodino and of the fighting of the guerrillas during the Napoleonic invasion so excited her that the better to fix them in her mind she copied them in longhand not once but several times.

Nor did she confine her reading to Russian authors and Russian stories. Byron, who since his first appearance in Russian print has been a hero of Russian student youth, was one of her idols. So was Dickens. Her mother showed me a neatly written list of books that Zoya had asked her to borrow in factory libraries. The list includes five poems by Byron, five plays by Molière, Dickens' *David Copperfield, Nicholas Nickleby,* and *A Tale of Two Cities.* She had also read Mérimée, De Maupassant, Flaubert, Mark Twain, Longfellow, Walter Scott, Rabelais, Victor Hugo, Jack London, Alphonse Daudet, Cervantes.

"She had an extraordinary memory," said the mother. "If she liked a poem she could recite it by heart after only one careful reading. She knew half of the folk poet Nekrasov by heart."

Zoya loved music and the theater as much as books.

"Mother," she often said, "don't ever pass up a chance to get tickets to the theater."

The mother didn't. The two always went to the theater together. *Eugene Onegin* was her favorite opera, Tchaikovsky her favorite composer.

"She adored Beethoven too," said her mother, "and never missed a chance to hear him. Goethe's war poem (from *Egmont*) which Beethoven set to music she knew by heart and often recited it." I saw it copied in one of her schoolbooks in a finely rounded, opulent handwriting. *Anna Karenina* she had never seen on the stage.

"We never got tickets, so great was the demand for them in Moscow," said the mother. "It was one of the plays Zoya had dreamed of seeing."

In her diary Zoya often wrote down thoughts and sentences from books which especially appealed to her. Here are a few quotations that further illumine the character of the girl:

Chekhov: *"Everything about a human being has to be beautiful—his face, his clothes, his thoughts, his soul."*

Mayakovsky: *"To be a Communist means to dare, to think, to will, to act."*

Chernyshevsky: *"Rather die than kiss without love."*

General Kutuzov: *"I wouldn't exchange one Russian for ten Frenchmen."*

Of Shakespeare's *Othello,* Zoya wrote that its theme was *"man's struggle for high ideals, for moral cleanliness . . . for victory of man's true and high emotions."* This comment only emphasizes the girl's romantic nature and her staunch puritanism.

I asked her mother if Zoya, romantic as she was, loving music, poetry, the theater, and worshiping Byron, Pushkin, Lermontov, Nekrasov, and Tolstoy, was given much to close friendships and infatuations with boys or men.

"Not much," was the reply.

Zoya had many admirers, but whether there was any one boy or man to whom she was particularly attracted, the mother never learned.

In school she never followed the practice of other girls of sending notes to boys in the same class. She opposed such frivolity. Once the mother asked Zoya's brother Shura, who was in the same school, whether he had ever known her to give special attention to any one boy. Curtly he replied that this was completely "Zoya's affair."

"At one time," said the mother, "I was worried about her. I wondered if there was something wrong with her emotional make-up. I couldn't talk to her about it, because she mightn't answer my questions or might dismiss them as irrelevant. But when she was sixteen she started looking at herself often in the mirror, taking a fresh interest in clothes. She insisted on my buying her a pair of fashionable shoes. Still I had misgivings, until one day I went to her school and saw her with other students, boys and girls. For the first time I saw her eyes afire, and I breathed in relief. I knew that she had the blood not of a fish but of a woman."

Yet there is nowhere any record of Zoya's attachment to any one boy, though older men constantly sought her company, paid her compliments, sent her flowers, invited her to go walking and skating with them. Zoya loved sports, especially light athletics and skating. Like all high-school students she took military training. She became a skilled marksman and also acquired creditable competence in the wielding of a bayonet.

A healthy girl, Zoya grew fast, and in her eighteenth year was as tall as her mother. She cut her hair short—dark, thick, wavy hair. Alert and graceful, she loved fun and laughter, but she seldom wept. A fanatic on clean living—by which she meant truth-telling, comradeship, abstention from tobacco and alcohol—she eschewed not only vodka but wine, and wouldn't touch a drop of either. Not only wouldn't she smoke, but she tried to persuade her mother to give up cigarettes.

"I only smoked when I felt low," said the mother. "Life wasn't easy for a widow like myself with two children to raise, and often I became depressed and sought forgetfulness in one cigarette after another. If Zoya saw me she would come over, put her arm around me, and say, 'Feeling badly, Mother darling? Is it because of anything Shura or I have done?' Then gently she would pull the cigarette out of my lips and throw it away."

A senior in high school, a home-loving girl, a devoted daughter, an excellent student with a will and a mind of her own, thinking earnestly of a place for herself as a woman and as a citizen in the vast, tumultuous Russian world in which she was living—such was Zoya Kosmodemyanskaya on the eve of Germany's attack on her native land.

On the morning of June 22 Zoya's mother was on an errand and did not hear Molotov's speech announcing Germany's sudden war on Russia. When she returned home Zoya and her brother, talking at once, interrupting one another, told her of the broadcast and spoke of what it might mean to Russia and to them personally. Though tears came hard to Zoya they now coursed freely down her cheeks.

"Things are going to change," she said. "Life is going to be different."

Little did she and her mother realize how tragically prophetic these words were—for both of them. Then with a show of indignation and to the annoyance of her brother, who, though younger, cherished a boy's natural feeling of responsibility for his sister, Zoya said she was sorry girls were not taken into the Army like men and sent to the battlefield to fight like men.

"I'd go, Mother," she said. "I am a crack shot."

With a sinking heart the mother glanced at Zoya and sensed something new in the girl, something that startled and frightened her.

"I had the feeling that my daughter might become a soldier," said the mother.

Then came the bombing of Moscow. Zoya neither sighed nor swore. She was cool and self-possessed. She joined a fire brigade and always remained on duty during an air raid. Not once did she go into shelter.

From day to day her mother observed definite changes in her. She was becoming increasingly reflective, mature, resolute. Often as she sat alone thinking, her lips tightened, her eyes lit up with a flame that had been alien to her—a flame of determination and battle, a flame which puzzled and disturbed the mother.

Smolensk fell. Kiev fell. Daily the Germans were drawing closer to the capital. September arrived, the chilly, drizzly September of central Russia. School closed, and Zoya, together with her brother, went to work in a factory as a wood turner. But she was not happy. The work was too tame. She yearned for something more difficult, more important. She quit the factory and joined the labor front and went to a state farm to dig potatoes. Snow came, frost seeped into the ground. It grew cold and muddy, the cold and mud of the lashing Moscow autumn. Zoya worked hard and resented laxity in Komsomols who were working beside her. Once, after a meeting in which she had upbraided "slackers," she sat down in the corner of the dormitory and by the light of a kerosene lamp wrote a letter to her mother. The letter was written on October 3, 1941.

"My darling Mamma! Forgive me for not writing you so long. There is never any time for letters. Mamma darling, you surely know that we are digging potatoes, helping the people on the state farm gather the crop. The day's output per person is 100 kilograms . . . on October 2 I gathered only 80 kilograms. Darling little Mother! How are you? I think of you all the time and feel uneasy about you. I am terribly lonely, but I shall soon return, as soon as we have finished digging the potatoes. We eat well—lots of potatoes and three glasses of milk a day. Of late we've been getting meat for dinner. Regards to Shura and all our relatives. I am keeping a diary. . . ."

Then she added a postscript:

"Mamma, dear, forgive me—the work here is very dirty and none too easy—I've torn my rubbers. But don't worry, I have plenty of footwear.

I shall return to Moscow well and cheerful." Then with true Russian self-criticism she added: "Dear Mother, I am so unworthy of you, but I remain your devoted daughter Zoya."

On her return to Moscow, Zoya looked thin but healthy, with good color in her cheeks. She had liked the hard work in the potato fields and had grown attached to the womenfolk on the land. She brought home a sack of potatoes, also a peasant recipe for making bread which she instantly taught her mother to use. To this day Lubov Timofeyevna bakes bread according to this recipe.

In the three weeks of her absence Moscow had changed, had grown more vigilant, more stern, more aware of itself. Factories, shops, office buildings were shuttered and camouflaged, changing the color, the lines, the very face of the city. The Germans were still drawing nearer and nearer. There was battle and death in the air. Soldiers in trucks were dashing to the front. Daily Zoya was growing more annoyed and disheartened with herself. . . . Once she told her mother that she felt miserable because she wasn't doing what in her heart she felt she should be doing. . . .

"But, darling," said the mother, "you are only a young girl. You've just returned from work on the labor front—that, too, was work for the country and for the Army."

"That's not enough, Mother."

That evening Zoya wrote in her diary:

"The renowned village elder Vasilisa and all peasant guerrillas (of the war of 1812) were serfs. They were treated lawlessly by their fatherland, yet of their own free will they rose in its defense. That's amazing. What greatness of spirit even in ancient times was hidden in the Russian people." Outgoing pride in Russia in the people was becoming a transcendent glory and dedication.

"Poor darling Mamma, she's had such a sorrowful life!" Zoya wrote further in her diary. "I know that I am more than a daughter to her. Since the death of Father I have been her closest friend and if anything happens to me . . ." But she did not permit the thought of catastrophe to harass her or to interfere with her resolution. Significantly she added: "But I cannot act otherwise."

Her mind was made up. To the battlefield no matter what happened!

To cheer herself or perhaps only to express joy at the cessation of internal conflict she wrote down from memory Goethe's martial poem:

> *The drum is resounding*
> *And shrill the fife plays;*
> *My love for the battle*
> *His brave troops arrays;*

He lifts his lance high
And the people he sways.
My blood is boiling,
My heart throbs pit-pat
Oh, had I a jacket
With hose and with hat!
How fondly I'd follow
And march through the gate,
Through all the wide province
I'd follow him straight.

"How I love this song," she noted. "A great composer [Beethoven] wrote the music to the words of a great poet. One hundred years have passed since then, yet the song arouses emotion."

Speaking of her newly made decision she said: "Tomorrow everything will be settled—though for me everything is already settled."

In the morning, without a word to her mother of her momentous resolution, she left the house. Her very silence aroused the mother's apprehensions. Yet Lubov Timofeyevna wouldn't permit herself to believe that her Zoya would enlist in the fighting ranks of the Army. Distraught and heavy-hearted, the mother sat down to dinner with her son Shura. Her thoughts were in a whirl. After dinner, as dusk was descending over the city, she went over to the window, looked outside. She hoped to see Zoya. But none of the girls in the street looked like Zoya, who was tall and upright like a "young poplar" and easily recognizable even in a crowd. The mother blacked out the room, turned on the light, and looked about her. Her eyes fell on Zoya's writing table. A book lay open on a page showing the portrait of a woman in an ancient Russian costume. Underneath the portrait was the inscription "Guerrilla of the People's War of 1812, Vasilisa Kozhina. Performed valiant service for Russia. Received a reward of 500 rubles." The mother turned the page—saw portraits of other guerrillas—Davidov, Fugner, Seslavin. Without shutting the book she looked at the title. Professor Tarlé's work on Napoleon. In the corner stood Zoya's rubbers. From the rack hung Zoya's worn skirt, draped in a green woolen scarf. On the dining table were the unwashed dishes. But to the mother the room looked desolate and empty.

Later in the evening Zoya returned, her cheeks flushed, her eyes aglow. She embraced her mother, looked into her eyes, then—

"Mother, I have a secret to tell you. I'm going to the front to work in the enemy's rear. It's a terribly responsible job and I'm proud I'm so trusted. Please do not say a word to anybody, not even to brother. Say I've gone to the country to visit Grandfather. Remember, darling, not a word about your daughter's going off to join the guerrillas."

Without letting her mother speak, Zoya continued:

"In two days I shall be leaving. I wish you'd get me a soldier's knapsack, several bags like those you and I sewed. Other things I'll get for myself. I shan't need much—two suits of underwear, a towel, soap, a brush, pencil and paper—that's all."

The mother was silent. She was afraid to talk lest she break down and weep. But her eyes filled with tears and she managed to mutter:

"You are taking too much on yourself—after all, you are not a boy."

"What difference does it make?"

Brought up to regard women as socially and intellectually the equal of men, Zoya was always impatient with people who pointed out differences between the sexes, especially in their obligations to the outside world.

"But must you go?" the mother said. "If you were mobilized it would be different."

"You mustn't talk like that, Mother darling," said Zoya. "You don't defend the fatherland by compulsion, do you? Guerrillas are all volunteers."

Guerrillas! The word beat like a hammer through the mother's head. She could restrain herself no longer. After all, Zoya was only a senior in high school, still her "baby"!

Grasping both her mother's hands, Zoya said: "Mamma, you yourself always told me one has to be brave and honest with oneself. How can I act differently when the Germans are advancing? You know how I am. I cannot do otherwise."

"During the following two days," says the mother, "Zoya would return home late in the evening. Where she spent her time she didn't say. And how could I ask? In those two flitting days she seemed suddenly to have grown far, far beyond her years. She said to me: 'Don't despair, Mother. If anything happens to me—you're not going to be deserted. The Soviet will take care of you.'"

She checked her kit and the things she had packed away. She wanted to take along the diary which she had begun keeping while she was digging potatoes on the state farm, but her mother advised against it. Zoya agreed.

Zoya's final night at home! The mother couldn't sleep. She kept asking herself over and over:

"Will I ever see her again? She is at home now sleeping in her bed—is it for the last time?"

Cautiously the mother rose and stepped over to her. Zoya awoke.

"What's the matter, darling?" she said. "Why aren't you asleep?"

"I got up to see the time—so I won't oversleep; and you, daughter dear, sleep some more, sleep. . . ."

The mother lay down but no sleep came to her. She wanted once more to get up, go over to Zoya, talk to her—maybe she had changed her mind? Maybe it would be best for all of them to be evacuated from Moscow—an alternative Zoya had been offered.

"But at dawn," said the mother, "as I glanced at my sleeping Zoya, at her calm face and her closely pressed stubborn lips, I said to myself, 'No, she hasn't changed her mind.' "

Early in the morning Zoya's brother Shura, ignorant of what his sister had decided to do, not suspecting he would never see her again, went off to the factory to work. The mother had saved for Zoya a piece of cheese, of which she was especially fond.

Mother and daughter drank tea together; then Zoya started to dress for her journey to the forest. The mother gave her own woolen sweater to Zoya. She refused it.

"What will you do without a warm sweater?" she protested.

But the mother finally talked her into wearing it. When Zoya was packed and dressed, mother and daughter walked together out of the house.

"Let me carry your bag," said the mother tensely. Zoya was disconcerted.

"What's the matter, Mother, you look so sad? Look at me. Ah, there are tears in your eyes. . . . Please don't see me off with tears."

The mother forced herself to smile.

"That's better, darling. Be proud you have such a daughter. Either I'll return home a heroine or I'll die a heroine."

She embraced her mother, kissed her, got on a tram and rode away.

The mother went back home. She still felt the warm presence of her daughter. But her daughter was gone—gone. Would she ever return? The question didn't leave the mother's mind, and yet she felt proud of Zoya and of herself for having such a daughter. She was hoping for the best.

Zoya reported to the military barracks—a large, gloomy room. At the table sat the leader of the squad to which she had been assigned. After looking at her long and intently he said:

"Aren't you afraid?"

"No, I'm not."

"To be all alone in the woods at night—terrible, isn't it?"

"I can stand it."

"If Germans take you prisoner and torture you?"

"I'll endure it. I shan't betray a soul."

The leader was satisfied with the answers and Zoya left Moscow, no longer as Zoya but as Tanya. Not even her mother knew of this change in her name.

Winter came. At first there was no snow, but the earth had already frozen, and it wasn't easy to find drinking water in the woods where Zoya's squad of guerrillas made their headquarters. One evening Zoya went with a kettle to fetch water out of a spring in a fir grove. In the dark she stumbled into an opening and felt as if she had touched a stair. She wondered what it could be—the lair of an animal, the dugout of a guerrilla, a German trap? She dared not light a match. It was too dangerous. Prowling Germans might see it, and there would be an end to all of them. She didn't trust herself with the investigation of the opening in the ground and returned to report the incident to her squad leader.

The fire was already out. The guerrillas lay crouched together on the bare ground which had been made warm by ashes. She waked the leader and told him what she had seen. All the guerrillas rose and all went to investigate. It proved to be the dugout not of Russian guerrillas but of German soldiers. All around was evidence of hasty escape: an iron kerosene stove, on top of it a partly filled kettle with soup, bottles, cups partly filled with wine, a pack of cards, leather gloves, a little saw, a loaded revolver.

The guerrillas rejoiced. Here was a ready-made dugout, and they could spend the night under cover. They fetched wood and started a fire. Searching further, they found a barrel of water, several cans of preserved meat, a small sack of flour.

"Wait," said Zoya, "I'll soon treat you all to noodle soup."

The guerrillas forgot that they were tired and had already been asleep. They watched Zoya as she worked. She boiled a kettle of water, she scrubbed the table, mixed the dough, rolled it out with a bottle and cut it up into thin strips. Laughing and teasing Zoya, the guerrillas waited expectantly. For ten days they had wandered in the woods without a mouthful of hot food. Now Zoya was cooking a hot supper for them—noodles with chunks of meat—a dish of dishes! The guerrillas showered Zoya with praise; this made her happy. They were all happy at the unexpected interlude in an always stern and hazardous life.

Busy as she was in the forest, Zoya found time to write to her mother. Her letter was taken across the line by guerrilla messengers. It consisted of only two lines.

"Darling Mother. I am alive and well. I am feeling splendid. How are you? Your Zoya."

Soon a second letter went off, then a third written on November 17.

"Darling Mother! How is everything at home? How are you feeling? Are you well? Mamma, if you find it possible please write me a few lines. When I return from my present assignment I'll come home for a visit. Your Zoya."

The mother was enraptured. Daily she expected her daughter. Whenever possible she laid away in the cupboard a piece of cheese, Zoya's favorite food. Days passed, long days. No Zoya. The mother grew apprehensive. She made inquiries at military headquarters. They had no information. Zoya had written of a special assignment. What could it have been and where? If only she knew! But then—guerrillas worked in the utmost secrecy. Communication with them, scattered and hidden as they were in the forests, was difficult and uncertain. Zoya would come. She must come! the mother kept telling herself over and over.

Those were days of gravest danger for Moscow. On November 16 the Germans launched a new and formidable offensive. In the rear guerrillas sought to inflict on them all the damage they could—to tire them, bleed them, blow up communications, cut their telephone wires. But the Germans felt safe in the village of Petrishtshevo. They made this village a resting place for an increasing number of their troops. One of the staff offices settled there. A battalion of the 332d Regiment, 197th Division, remained there. A large cavalry unit also stopped there. Every house was crowded with Germans.

In the tomes of documentary evidence which has been painstakingly gathered about Zoya's life as a guerrilla, it is recorded that when she learned of the Germans making of Petrishtshevo a resting place for themselves she said:

"We shall see what sort of rest they'll have."

Together with a group of young guerrillas Zoya went off deep into the German rear. Nights they scouted for guerrillas and for the Red Army. They cut telephone wires, blew up bridges, harassed German transportation. By day they remained in the forest, built a fire, warmed themselves, and slept with their backs against tree trunks. Always they had to be on the alert lest they be ambushed. They felt hungry but didn't complain; the work was too important, too exciting, and they didn't bother about personal well-being.

The time came to return to headquarters. But Zoya remembered Petrishtshevo and decided to go there and "disturb" the rest of the German vacationists.

"I may perish there," she told her companions, "but I'll take the lives of ten Germans."

Together with ten other guerrillas Zoya started for Petrishtshevo. In the dead of night they drew close to the village. Zoya's companions remained behind as guards and scouts, and she proceeded alone to her destination. Soon afterward flames rose from several buildings, the ones her commander had ordered her to set afire. Quickly she got away and joined her companions. As they were retreating from the village they could see the burning buildings. Zoya had done more—she had cut the telephone lines so that the Germans in Petrishtshevo could not at once communicate with other German units.

In the evening of the following day guerrilla scouts brought back the report that the fires which Zoya had started had caused little damage: only a few houses had burned down before the flames were put out. Whether any of the Germans who lived in these houses had perished the scouts were unable to establish. Zoya regarded her mission a failure. Disgusted with herself, she said:

"I shall go again."

"Yes," said the guerrilla commander, "but wait until they calm down—they have guards at every house now."

"I shall wait a day," said Zoya; "then I'll go."

She wouldn't let anyone talk her into waiting any longer. She had failed in her purpose and she had no right to fail. That was the way she reasoned, the same self-willed, resolute girl she had always been, impatient to perform well and promptly the assignment at hand, whether a problem in geometry, teaching a housemaid to read and to write, or setting fire to homes in which German soldiers were living.

In her diary she wrote down the line from *Hamlet:*

"Adieu, adieu, adieu! remember me!"

The next evening she made herself ready for another trip to Petrishtshevo. She dressed like a man in cotton quilted trousers, knee-high felt boots, a fur jacket, a fur cap. Over her back she slung her case in which she carried several bottles of benzine, matches, ammunition, a few personal effects. Around her waist she wore her regulation leather belt and a revolver in a leather holster. Before leaving the dugout she said to a friend named Klava:

"If anything happens to me, you must promise to write my mother."

Klava seized her hand and said:

"How can I write your mother when I don't even know your real name?"

"That's not important. Write to the Komsomol committee in the Timiryazev district in Moscow. They'll deliver the letter."

Not even in this crucial moment would she disclose her identity to a fellow guerrilla who had become her closest friend but whom she had known only a short time. Together the two girls went out of the dugout. They bade each other farewell and Zoya left. Soon she was swallowed by the woods and the darkness.

She followed the road she had followed two days earlier on her first trip to Petrishtshevo. Finally she came in sight of the village, the cottages looming out of the snow like dark spots on the horizon. She walked nearer and nearer. No sounds anywhere. No sight of guards. No lights. The village seemed fast asleep. Zoya's objective was the stable which, according to the information scouts had imparted to the commander, was sheltering two hundred horses. Whipping the gun out of her holster and holding it in her hand, she proceeded to the designated objective. Still no sound of anything, no sight of anybody. Slipping the gun inside her bosom, she knelt down. Without wasting a second she drew a bottle of benzine out of her little case, poured it over a bunch of kindling wood she had gathered, and struck a match. The match broke. She struck a second, and suddenly someone seized her from behind.

Pushing away her assailant, she drew her gun out of her bosom. But she had no time to pull the trigger. The gun was knocked out of her hand. Her shoulders were firmly held, and her hands were tied behind her back by a rope. The German guard sounded the alarm. Other soldiers swiftly came along and led Zoya to a house in which lived a peasant named Sedov. On the top of the oven lay Sedov's wife and his young daughter. Awakened by the sudden noise, they looked down and saw the Germans snatch the fur cap off "the boy" they had brought in, then the felt boots. Then they took out of the case two bottles of benzine and a box of matches. The Germans were slow in stripping the fur jacket and the sweater off their captive. They had to untie the prisoner's hands first and presently they discovered not a boy but a girl!

Barefooted, her warm clothes removed, her hands tied behind her back, Zoya was led at the point of a gun to another house, that of a peasant named Voronin. There the Germans had their local headquarters. Zoya had studied German in school. She not only read but spoke the language and therefore understood everything that was said by her captors. But she never intimated to them that she knew German.

A German officer pointed to a long bench and she sat down. Facing her was a table with a telephone, a typewriter, a radio, sets of papers. More and more German officers were arriving, among them Lieutenant Colonel Ruderer, commander of the 332d Regiment. He questioned Zoya.

"Who are you?"

"I won't tell you."

"Was it you who set fire to the stable night before last?"

"Yes."

"What was your purpose in doing this?"

"I wanted to destroy you."

"When did you cross the border?"

"Friday."

"You were quick in getting here."

"Why did you suppose I wouldn't be?"

He tried to find out who had sent her, who were her associates, where they were hiding. To all these questions the eighteen-year-old girl replied: "No, I don't know, I won't tell," or she remained silent.

Exasperated, the colonel exclaimed:

"You don't know? Soon you'll find out."

He ordered her flogged. After ten strokes he stopped the flogging and started to question her again.

"Will you tell now where the guerrillas are?"

"I will not."

"Ten more strokes!"

The woman of the house saw it, counted the strokes, then later, when the Germans were driven from Petrishtshevo, she told the story to the investigators. She wept with pity and, being a religious woman, made the sign of the cross over her body. But Zoya remained obdurate. Again and again she was asked the question that was uppermost in the mind of the German officer:

"Will you tell now where the guerrillas are?"

"I will not," Zoya said.

So she was flogged and flogged. Her clothes were soaked in blood. But she neither wept nor complained. She only bit her lips, bit them so hard they bled and swelled. Not a word of information could the Germans extract from her except on unimportant subjects, and then she did not always tell the truth. To the question where she was from she once said, "From Saratov," which is on the middle Volga and hundreds of miles away from Petrishtshevo. She would not tell what her name was or anything about her family.

After two hours of this inquisition she was led barefooted and thinly clad to still another house—to the home of a peasant named Vasily Kulik. He and his wife Praskovya were asleep on the oven. They wakened instantly. By the light of the lamp, Vasily took a good look at her. Her lips were swollen and bleeding. Her forehead was cut and full of black lumps. Her arms and legs were also swollen. She breathed hard and her hair was disheveled. Her hands were tied behind her back. Her clothes, the few she was permitted to wear, were red with blood. The guard ordered her to sit down on a bench and stationed himself at the door.

Vasily took a dipper of water to Zoya, but the guard stopped him. Seizing the lamp, he held it out to Zoya, shouting that kerosene was the drink for people like herself. Finally the guard relented, and Zoya drank two dippers of cool water.

The German soldier taunted her, shook his fists at her. Vasily begged him to leave the girl alone, if only for the sake of his little children, but the guard went on abusing the helpless girl. Zoya endured it all in silence.

The guard remained with her from ten in the evening until two in the morning. Every hour, at the point of his gun, he ordered Zoya into the street. Though barefooted and in scarcely more than her underwear, she obeyed without a murmur of protest. Following her with a gun at her back, the guard made her walk the street from fifteen to twenty minutes.

Another guard came. Finding an opportune moment the peasant woman Praskovya went over and talked to Zoya, gave her water which she drank. "She ought to lie down," said Praskovya, glancing at the German. He shrugged, and the woman said to the girl, "Lie down." Zoya's feet were numb and her hands were still tied.

"Untie my hands," she said to the soldier in German.

The soldier surveyed her. She looked helpless enough, and he granted her request. Zoya lay down, and Praskovya covered her with a blanket.

"Who are you?" asked Praskovya.

"Why do you want to know?" Vigilant and distrustful, Zoya was on her guard even against Russians she didn't know.

"Have you a mother?" the peasant woman asked.

No answer.

"Was it you who were here two days ago? Speak. Don't be afraid. He"— pointing at the German with her eyes—"doesn't understand our language."

Zoya answered: "It was I."

"You set the houses on fire?"

"Yes."

"Why?"

"Those were my instructions. To burn Germans, destroy their military supplies. How many houses have I burned?"

"Three."

Zoya sighed: "So few. What else burned?"

"Twenty of their horses and—what do you call it?—the telephone cable."

"Have any Germans been burned?"

"Only one."

"Only one. What a pity."

Through the few remaining hours of the night Zoya slept.

In the morning the German lieutenant colonel and other officers came again and started to question her. She remained as mute and stubborn as the night before. Not one word of information could they wring from her. She complained that the soldiers had stripped her almost naked, and the colonel said, "Give her back her clothes." A soldier brought only a part of her things—the waistcoat, the trousers, the stockings, her bag in which she had kept matches, sugar, salt. The fur cap, the fur jacket, the felt boots, the knitted sweater were missing, presumably appropriated by a soldier or officer. Her gloves were given to the redheaded cook in the officers' messroom.

Zoya was permitted to dress, but her spine wouldn't bend and her fingers refused to obey her will. Praskovya helped her. The Germans made one more effort to question her. It was useless. They gave up.

The gallows in the public square was ready. From the crosspole on the top dangled a rope. Beneath were two boxes laid on top of each other. Zoya was led to the place of execution. Over her breast the Germans had hung one of her bottles of benzine and a little board with the inscription "Incendiary of homes."

Several hundred German soldiers gathered on the square. Ten German cavalrymen with drawn swords were stationed round the gallows as if to fight off a sudden raid of guerrillas in an attempt to rescue their Tanya. The peasants in the village were ordered to attend the hanging. Not many came and some of these quietly slipped away. Zoya was lifted by German soldiers to the top box. The noose was thrown over her neck.

One officer brought out his camera to make a photographic record of the execution. It took him some time before he obtained all the views he wanted. Zoya was waiting quietly, and the peasants who remained on the square turned their eyes away and sobbed. . . . Then, taking advantage of the waiting, Zoya turned to her people and said:

"Here, comrades! Why do you look so gloomy? Be brave, fight on, kill Germans, burn them, poison them!"

The executioner pulled the rope. The knot started to choke Zoya and with a superhuman effort she loosened it with her hands and shouted:

"Farewell, people! Fight on, fight on! Stalin is with us!"

These were her last words.

With his heavy boot the executioner kicked the box from under Zoya's feet, and she remained suspended in the air.

That was on the fifth day of December. For three weeks, or until December 25, frozen and blown about by wind and snow, the body remained swinging on the gallows. With a view to terrorizing Russians into submissiveness the Germans wouldn't permit the peasantry of the village to take it down. Then they relented. The peasants carried the body not to the village cemetery but to the rear of the schoolhouse. There, under the shadow of outspread willows, they chopped out a grave in the frozen earth and without speeches or songs, without any mark of solemnity or public demonstration, with the noose frozen to her neck, they laid Zoya to rest.

On my arrival in Moscow I looked up Zoya's mother. She is a tall woman of about forty with slightly stooping shoulders, a pale, handsome face, singularly unwrinkled, and small, finely rounded brown eyes. Her broad chin, somewhat lofty forehead, long hands, give one an impression of energy and will power. Her eyes blinked as she talked; she smoked incessantly. She talked freely, fluently, though now and then she paused, lowered her head, swallowed. My mind teemed with questions, but many of them I never asked.

The mother showed me photographs of Zoya at various stages in her life —when she was small, when she entered school, with various classes in school, with the family when the father was still alive, and the very last one taken for her passport before she left for the forest. This photograph is the most recent and shows Zoya with a broad, open face, a bright expression, narrow, penetrating eyes, capricious lips, and dark wavy hair rising crestlike over a lofty forehead. It is this photograph that one sees on display all over the country and that artists are using as a model from which to paint her portrait.

"Do you know," said the mother, "her body was disinterred three times." I asked why.

"Another woman claimed Tanya was her daughter, and the authorities wanted to establish Zoya's identity beyond any doubt. Her teachers journeyed to Petrishtshevo, and I went down there. Of course it was Zoya. The body was frozen, but the face was well preserved. It was white and beautiful, very beautiful."

We talked of Zoya's brother Shura.

"As soon as he learned of his sister's death," said the mother, "he enlisted. He was only sixteen, but he is very tall and athletic and he couldn't remain a civilian any more—he wanted to fight. . . ."

"What branch of service is he in?"

"He is in a tank school," the mother answered.

At the time of this conversation, tank battles were raging all over Russia's south. The newspapers carried long and vivid accounts of the ferocity of the battles and long, impassioned appeals to tankmen to fight to the last breath, with their flesh when necessary, so as to deprive Germans of the advantage of superiority in numbers. . . . Yet the mother spoke with no show of trepidation or alarm. "In his last letter," she went on, "he wrote he will soon graduate. He'll get a captain's commission and will command a tank brigade."

On August 23, at the Iraqui restaurant, Carman, a well-known Russian writer and film photographer, gave a dinner in honor of Eisenstein and Pudovkin, Russia's noted motion-picture directors, who happened to be in Moscow for a few days. I sat next to a young Russian journalist and scenario writer who had spent two and a half months with guerrillas in the

Leningrad regions. He told exciting tales of his experiences. Once he and a leader of a guerrilla detachment were driving in a bobsled over a country road. The Germans were patrolling this road from the air, and soon far in the distance and flying low the leader beheld a German plane. Swiftly he grasped the journalist by the neck and pushed him out into the snow. Then he jumped out, struck the horse with his whip, and dropped into the snow beside the stunned journalist, who at first didn't understand the rough action of his companion. The horse raced on and on. The guerrilla leader deliberately left the fur lap robe in the sled so the German pilot would mistake it for passengers. The horse galloped toward the plane; the plane was flying toward the horse. Soon the two men heard the sound of shots. Later, after making sure no German plane was patrolling the road, they rose and walked on. They came to the bobsled. The horse was dead, the robe was riddled with bullets. The two men had a good laugh at the expense of the German pilot. From the tales the journalist narrated it was evident that the guerrillas battle with their wits as much as with their guns.

"If my companion hadn't used his wits quickly we might both have been dead. In this war one or two minutes make the difference between defeat and victory, between being killed and remaining alive."

Discussion of the guerrillas brought up Zoya. I told him I thought there was something Biblical and Shakespearean in the heroic life and death of this girl.

"I attended her funeral," said the Russian, "here in Moscow. Her body was brought from Petrishtshevo, cremated, and buried in the Devitche monastery."

I asked him to tell me about the funeral.

"It was almost a private affair," he said. "Only invited guests attended. The times were such we didn't want to make a public demonstration of it. A lot of school children were present: Zoya's school friends, Komsomols, her history teacher, a little fellow, rather pathetic beside the mighty stature of his famous pupil, the mother, and some others. It was a rainy day, and as we walked to the burial ground soaked to the skin I couldn't help thinking of Zoya and of the lesson of her martyrdom. Think of it, the girl wouldn't even tell the Germans what her real name was. She couldn't possibly have done anyone harm by telling it. Tanya, Zoya, Zoya, Tanya—what difference would it make to the Germans or the Red Army if her inquisitors knew which it was? None. Yet she refused to let them know. She wouldn't tell them anything even when the answer might be as harmless as the revelation of her name. That's why her action was so momentous and holds so much meaning for our people and the world, now and for ages to come. She seems to say, 'Never allow yourself to be subjected.

Don't surrender, no matter how helpless you are. Let the enemy hang you, quarter you, don't answer his questions. Make him realize he can never subjugate you, no matter how horrible the torment or the death he holds out to you.'

"A girl like Zoya is a challenge to Singapore, to Paris, to all those who lose faith in themselves and give up while they can still fight, even when they have power to win. That's the way Sevastopol fought. That's the way Leningrad held out during the hunger blockade of the winter of 1941. That's the way we've got to go on fighting; all of us—you in America as well as we in Russia."

On December 1, 1942, all Russia was cheered by the following announcement which appeared in *Izvestia:*

"In the course of the first few days [of the Rzhev offensive] the soldiers of General Povetkin smashed several units of German troops. Among them, and now completely devastated, was the regiment whose soldiers had executed Zoya Kosmodemyanskaya."

PART TWO: RUSSIA'S COMING OF AGE

CHAPTER 5: DESTRUCTION AND CREATION

NOTHING IS EASIER, even for people who have never been in Russia and do not know the Russian language or Russian history, than to draw up an indictment of the 1917 revolution. For nearly a quarter of a century the outside world has done just that. The literature on Soviet Russia in the English language, as well as in other foreign languages, has been preponderantly a denunciation of the Soviet creed, Soviet policies, Soviet manners, Soviet good faith, Soviet achievement. Books by foreign observers —particularly by those who have gone to Russia as socialists or communists and have come out disillusioned—have painted the country as a land of chaos and cruelty, horror and doom—and nothing more. Russians and others who have escaped from Soviet penal colonies and labor camps have given the world horrifying tales of the conditions of these places. Former Soviet diplomats and secret agents have added immeasurably to the picture of Russia as a land of terror and hopelessness.

Since the day the Soviets leaped to power, foreign visitors and resident foreign journalists in Russia have never been at a loss for facts with which to support whatever charge they wished to make against Soviet methods and Soviet beliefs. Every stage of Soviet advance, every phase of Soviet development, every campaign Russia launched was marked by methods which have been banned by stabilized nations, especially the democracies, and which have been deemed inhuman by contented or even discontented individuals in a stabilized society.

Once I discussed the violence of the Soviet campaign of collectivization with a highly educated Russian official. He said:

"Remember, in a revolution the fight is not always between right and

55

wrong but between two rights, an old right and a new right, your right and my right. We who fight for the new right, for our right, cannot disturb ourselves about the old right which we want to destroy, any more than a soldier on the battlefield can bother about the life of the foe he is facing."

A frank enough admission, having its own logic, its own moral code, its own law of growth and devastation, of life and death. In every stage of Soviet history we see the operation of this code and this law, and people who have been living in a stabilized society with a stabilized code of morality and a stabilized body of law with personal wants and ambitions more or less gratified, find it easier to apprehend the devastation and the death than the new growth and the new life. After all, we are creatures of environment. We cannot easily or at all detach ourselves from our own habits or traditions, our own geography and history and the opinions and orientation which they engender.

Consider the chief stages in Soviet history! The civil war, the Nep (New Economic Policy), the First Five-Year Plan, the Constitution, the Purge, the War. In the first five periods the disregard of "old rights" was so conspicuously and dramatically apparent that it roused indignation and wrath in those who had neither the desire nor the patience to search out the nature and the meaning of the "new rights" as the Russians understood them and applied them, in their own way, in everyday life.

The civil war, like all such conflicts, including those of France and the United States, was accompanied by a ferocity which is not always present in international wars, except in the one which Germany is now fighting. Brother fighting brother is usually more vindictive than soldier fighting soldier. Military communism, which enabled the Soviets to marshal all the resources of the country for the fight against the Whites and the foreign invaders, only intensified the cruelty of the internecine conflict.

Foreign eyewitnesses of that event were so appalled by it that with few exceptions they pictured the Revolution of the early days as uninterrupted and unmitigated villainy. The goal didn't concern them, especially the far-reaching aims of the new government. Its immediate methods, its immediate action, its immediate repression and mutilation were all that they beheld in the Soviet dictatorship.

Nep brought a breathing spell. The civil war was at an end, or seemingly so. The White armies, the foreign armies were gone from the Russian lands. Impoverishment of the people moved Lenin to bring back private enterprise, though only on a limited scale and confined chiefly to individual tillage of land and to small-scale trading and manufacture. The ideological disdain of private enterprise, with all the social contumely that had been leveled on those engaged in it, was not abated. The word "nepman" became as much a term of opprobrium as the word "bourgeois" during the civil war.

The Stalin-Trotsky battle—which, like the lightning and thunder in a Wagnerian opera, accompanied the social vicissitudes and the human drama of Nep—provided ample ammunition for intellectuals in Russia and outside to denounce the Revolution as a Stalinist reaction, a betrayal of the original Soviet aims and a brutalization of the Russian masses. Outwitted, outmarshaled and ousted from positions of power and influence, Trotsky and his followers inside and outside of Russia raised a cry of tyranny against Stalin. This is not the place to discuss and examine the issues in the historic controversy between the two men and the theory of "permanent revolution" of the one and of "socialism in one country" of the other. Let time and history pronounce their own judgments, their own condemnations, and their own vindications.

I, for one, fail to find in Trotsky's record during the time he enjoyed position and power in Russia the slightest indication that in his dealings with the "enemies of the Revolution," whoever they might be, he followed the golden rule. Nor can I discover that in the espousal of his political policies he took cognizance of democratic practice or democratic usage, or tolerated the immediate mitigations and relaxations that these so often make possible. After all, he was the author of the pamphlet *In Defense of Terrorism,* which is the most complete and brilliant dissertation on the subject that any revolutionary has ever penned. Stalin has dealt harshly with those he termed Trotskyites. But what reason is there to assume that had Trotsky been victor in that historical controversy, the author of *In Defense of Terrorism*—which justifies and advocates every method of combat, including the taking of hostages against the enemy—would have dealt less harshly with Stalinites or others of his enemies, whoever they might have been?

The controversy was marked by violent rhetoric and loud acrimony and only added at that time to the anger of the outside world, especially of certain groups of intellectuals, and enhanced its disappointment and its detestation of Soviet policy.

In retrospect the First Five-Year Plan looms as a resumption, perhaps only as an intensification, of the civil war, though in a new guise and with new weapons. Bent on the complete annihilation of private enterprise in industry and agriculture, determined to root out all semblance of individual landholding and individual land tillage, the Soviets ruthlessly suppressed any opposition. The exile of kulaks; "the dizziness from success," as Stalin termed the brutal excesses of the early organizers of collectives; the train of misfortunes that followed these excesses—such as the slaughter of livestock in the villages, the sabotage of work on the land, the resulting famine in 1932–33, the pitched battles here and there between party organizers and infuriated peasants, the exile of several entire Cossack settlements, the grim shortages of food in the cities, especially in some of the newly opened industrial communities—in short, the price that the Russian people were called on to pay for the destruction of the old and the enthronement of the new collectivized system of economic control stirred an avalanche of fury and hate in the outside world.

I was in Russia during those violent years. American writers—especially, I must emphasize, those who came to Moscow as socialists or communists and expected to find a ready-made socialist or communist society, or something approaching the blissful abundance and unimpeded liberations which these ideas imply and promise—became as scathing in their denunciation as their predecessors who had eyewitnessed the civil war. They perceived in the Five-Year Plan cruelty and carnage and not much more. They judged Russia in terms of the blueprints they had studied in socialist and communist literature and at least as much in terms of the political liberties and economic satisfactions they had enjoyed but never appreciated in their homeland. They beheld not distant long-range objectives but immediate short-range consequences with their heavy toll of human sacrifice. The bringing of the age of steel and the modern machine age to the remotest parts of Russia, including Asia, not through indebtedness to foreign banks or foreign governments but through Russia's own substance and blood, conveyed no meaning to them. The torrential inculcation of engine-mindedness in the many-millioned backward peasantry, so that in time of war it could acquit itself with skill in the handling of modern highly mechanized weapons of warfare, likewise held no lesson for them. The establishing of innumerable schools and colleges, especially in technological subjects, so as to prepare masses of modern-minded builders of industry and masses of modern-minded military commanders, hardly existed for them.

The Second Five-Year Plan, especially in its later period, was marked by visible relaxation of political severity and an increasing rise in the standard of living. It culminated in the adoption of the Constitution. In his pronouncement in those days—"life has become better and more cheerful"—Stalin enunciated not only a text for millions of sermons but stated a fact. Compared to that under the preceding plan, life became immeasurably better, unforgettably more cheerful.

I traveled extensively in those days, especially in the Ukraine. The condition of life and the temper of the people had changed beyond recognition. They felt more free, more hopeful. At last they were beginning to reap concrete dividends for their labors and their sacrifices. The worst of the internal fighting and hating was over. The chief of police in the Ukrainian village of Reshitilovka boasted to me that constitutional democracy was a reality in Russia. Months before the adoption of the Constitution he received instructions to make no arrests without a warrant from the judge or the prosecuting attorney. He spoke of a chief of police in a near-by district who had disregarded the instruction and was made to answer for it.

Yet even before the constitutional convention had been summoned, the purge began. Kamenev and Zinoviev and twelve others, leaders of note in their day, were tried in Moscow. They made astounding confessions of guilt and were sentenced to "the highest measure of social defense"— death by shooting.

A few days later, while in Sochi, the renowned summer resort, I read on the back page of a newspaper a brief announcement that the sentence on the Kamenev-Zinoviev group had been carried out.

From Sochi I went to the Kuban Cossack country and I was amazed at the lack of excitement over the trial and the executions. For several years Zinoviev and Kamenev had been out of the public eye, and the new generation that was coming to power knew little of them. In a revolutionary country the surge of life is as swift as a mountain stream and even more uproarious. A year is an age. The campaign of yesterday is drowned out in the tumult of the moment, and the leader of yesterday may be proclaimed the traitor of today.

The country was harvesting a superb crop. Industry was churning out increasing amounts of consumption goods. When Joe Phillips of the *Herald Tribune* and I visited a village store in the Ukraine, we were amazed at the sight of shelves loaded with goods, from towels and linens to groceries and housefurnishings. At railroad stations, on the streets of cities, government salesmen were hawking more ice cream than the Russian masses had ever known, and so people did not want to remember—not long, anyway—the political eruption which the Kamenev-Zinoviev trial and execution had signalized.

Then came the constitutional convention. What a holiday and what a triumph it was! The morning of the convention was gray and cloudy. A light snow was falling as I was on my way to the Kremlin to attend the first session. Delegates had come from all over the far-flung Soviet empire—men and women of many races, many nationalities, many climates, in all manner of national and historical garb. The buffet outside the hall was loaded with fruits, pastries, sandwiches, and it was amusing to watch delegates from the far north buying freshly received tangerines from the Caucasus by the handful and eating them with white bread or sandwiches.

Foreign correspondents sat by themselves in a box in the gallery overlooking the auditorium below and in full view of the stage that was decked out with banners and bunting. Back of them were the ascending rows of seats crowded with delegates and visitors and in the very forefront freshly arrived representatives of Loyalist Spain. We waited expectantly. The German press had released a rumor only a few days earlier that Stalin was ill with a serious and incurable heart ailment. A German heart specialist was supposed to have examined him and to have given up hope of recovery or even of the prolongation of his life. We wondered how true this rumor was. Stalin was announced to address the conventions, and we speculated among ourselves as to whether he would do so.

Soon Stalin came out, dressed as always in knee-high leather boots, khaki trousers, a khaki tunic with neither collar nor tie. From the distance he looked well. Then he got up to speak. It was one of the longest speeches I had ever heard, lasting a little over two and a half hours. At length he was explaining the basis, the nature, and the promise of the new constitution. Leaning over a rostrum, he never moved a step away from it. He spoke in a conversational tone, never once lifting it to an oratorical outburst. He punctuated the speech with only one gesture, a diagonal swing of the right hand. He drank endless mouthfuls of water as though speaking made his throat dry. Neither in voice nor in manner did he betray the least sign or symptom of the incurable heart ailment which the German press had so clamorously ascribed to him.

There was nothing stiff or formal about the occasion. As Stalin spoke I saw delegates lean over now and then and whisper to each other. Others walked out to stretch their legs or to partake of a bite of pastry and tangerine. No police stopped them. Nobody was required to stand or sit at attention. Nobody interfered with freedom of movement. Nowhere was there the least sign of the awe and tension that hovers over any place where Hitler makes a public appearance.

When Stalin finished his speech and the applause died down he appeared nervous. As the audience rose to sing he pulled out his pipe, started to smoke. But after a few puffs he realized where he was and quickly put out the light and hid away his pipe.

After the adoption of the Constitution in December 1936 people looked forward to a winter of comparative ease and comfort.

Yet scarcely had the delegates returned to their homes when trouble began: arrests, searches, exile of outstanding leaders, trials, executions. The purge started in earnest, swept the country like a tornado, and made people wonder what was happening and when it would end. The trials and confessions baffled as much as they shocked the world and stirred fresh outcries of disdain, fresh protests. That was one of the grimmest periods in the Revolution, in all Russian history; and to the outside world it seemed more than ever that despite high-sounding phrases and eloquent blueprints of a new society, Russia, in the words of an American college professor, was groping in the dark for a light that never could be found because there was nothing to generate it. Yet though impeded and delayed the creation of a new economic order and a new civilization never halted, not for a single day, not for a single hour.

Much about the trials and the confessions is still unknown. It may be years before all the facts are made available to the world. There can be no doubt now that a conspiracy was brewing or was already launched for a violent overthrow of the existing leadership. This might have resulted in a protracted and bloody civil war. No one knows, no one need claim the omniscience to know what the consequences of such a conspiracy would be. It was choked before it could show its teeth.

The period which followed the purge and which embraces the pact with Germany and the war with Finland only further disenchanted the outside world and deepened the distrust and hate of Russia. Not since the civil war was Russia so isolated and detested, so often spoken of even by former friends of the Revolution as the lost hope of the Russian people.

Thus at every stage of its young and turbulent life—during the civil war, the Nep, the First Five-Year Plan, the purge—it was the destructive power of the Soviet that impressed itself most on the mind and imagination of the world. It was the attack on the "bourgeois," the nepman, the kulak, the wrecker, the diversionist, the foreign spy, the fascist agent, or whatever the label hurled at the immediate enemy in the way or under sentence of liquidation, that the outside world chiefly beheld, lamented, and condemned with all the power of its aroused emotion. The tragedy of the Revolution clouded for many observers almost to invisibility its creative energy and its tempestuous effort to rebuild a nation and transform a people so it could save itself from the fate of forcible obliteration.

Yet when the war broke out and Russia refused to submit to the sentence of doom which leading generals and diplomats had so lavishly pronounced on her, the world wondered, and still is wondering, what there was in the people, the country, the Soviets that made them fight on. Despite frightful reverses on the battlefield they have managed to muster will and power to withstand all alone the organized might of the German Army, supported as it was or has been by the entire might of western and central European industry and by an ever-increasing number of freshly recruited legions from Rumania, Hungary, Italy, Finland, Slovakia. . . .

For twenty years I have been periodically visiting Russia. In these years I have seen more personal tragedy than any of the former communists who had journeyed to Russia in search of paradise and had come away cursing communism, which Russia never had and has not now, and proclaiming Russia a land of fire and brimstone and nothing more. Yet in the midst of the cruelest campaigns against so-called enemies, whoever they might be, I have always perceived a process of creation, not an international but a national, a purely Russian process, dedicated to purely Russian aims, guided by a purely Russian mentality, colored by purely Russian geography and history.

The primary purpose of this process was to lift Russia swiftly to the level of physical development, both human and mechanical, on which in the event of an attack by a highly advanced industrial nation she could strike back blow for blow and ward off conquest and annihilation. It was a costly process for one reason because Russia was so backward. She had to make up for lost time and do so quickly. Of all Stalin's utterances since he has been piloting the destiny of the Russian people none reveals so much of the man, of Russia, of Soviet policies and campaigns, their nature, their purpose, the intensity of violence that accompanied them, as the speech he delivered on February 4, 1931, at the conference of Managers of Soviet Industry.

Though I once before included it in a book (*Hitler Cannot Conquer Russia*) it is such an important document that I am reprinting it again.

Let it be noted that 1931 was one of the harshest moments in the First Five-Year Plan. At the behest of the Kremlin Russia was straining every atom of energy, every particle of national substance for the fulfillment of an ambitious industrial program. The people keenly felt the burden and agony of the drive. There was much talk in whispers and aloud of the need for a breathing spell which would allow people to catch up with themselves, their food, their thoughts, their sleep, their good humor. Only too well did Stalin know the mood and expectations of large sections of the people, and of some of his most trusted lieutenants. Yet this is what he said:

"It is sometimes asked whether it is not possible to slow down a bit in tempo, to retard the movement. No, comrades, this is impossible. It is impossible to reduce the tempo! On the contrary, it is necessary as far as possible to *accelerate* it. [All italics mine.] To slacken the tempo means to fall behind. And the backward are always beaten. *But we do not want to be beaten.* No, we do not want this! . . . The history of old Russia is the history of defeats due to backwardness.

"She was beaten by the Mongol Khans.

"She was beaten by the Turkish beys.

"She was beaten by the Swedish feudal barons.

"She was beaten by the Polish-Lithuanian squires.

"She was beaten by the Anglo-French capitalists.

"She was beaten by the Japanese barons.

"All beat her for her backwardness—for military backwardness, for cultural backwardness, for governmental backwardness, for industrial back-wardness, for agricultural backwardness. She was beaten because to beat her was profitable and could be done with impunity.

"Do you remember the words of the prerevolutionary poet [Nikolay Nekrasov]: 'You are both poor and abundant, you are both powerful and helpless, mother Russia'? These words of the old poet were well known to those gentlemen. They beat her, saying: 'You are abundant, so we can enrich ourselves at your expense.' They beat her saying, 'You are poor and helpless, so you can be beaten and plundered with impunity.' Such is the law of capitalism—to beat the backward and weak. The jungle law of capitalism. You are backward, you are weak, so you are wrong, hence you can be beaten and enslaved. You are mighty, so you are right, hence we must be wary of you. That is why we must no longer be backward. . . . Do you want our Socialist fatherland to be beaten and to lose its independence? If you do not want this you must put an end to this backwardness as speedily as possible and develop genuine Bolshevik speed in building up the socialist system of economy. There are no other ways. That is what Lenin said during the October Revolution: 'Either death, or we must overtake and surpass the advanced capitalist countries.' We are fifty to a hundred years behind the advanced countries. We must cover this distance in ten years. Either we do this or they will crush us."

Commenting on this speech, I wrote:

"These are cruel and poignant words, and only a man who completely disregarded the cries, the immediate tears, the immediate sacrifices of the people could utter them. . . . But just ten years later—to be precise, ten years and three and a half months—Nazi Germany, in command of the most powerful army that any Western nation has ever had, in complete control of the most advanced armament industries in the west—in her own land, in France, in Austria, in Czechoslovakia, in Poland, in Holland, in Belgium, in Norway, in Denmark, in Italy, in Yugoslavia, with free access to the high-quality iron ores in Sweden, with an army of cheap labor recruited in conquered countries to operate these industries—this Germany flung her mighty legions, equipped with the best modern implements of death, against Russia. And the Russia of fierce dictatorship, of almost complete martial law, of a constitution which was a paper document—this Russia alone of the continental powers in Europe has battled fiercely at enormous sacrifices to her armies, her civilian population, her lands, her factories, but also at mammoth cost in life and substance to the invader."

Only now in the light of Russia's stupendous resistance does the world begin to appreciate the epochal momentousness of the three Plans, which have given Russia a new industry, a new agriculture, a new mentality, a new patriotism, a new energy, a new skill and, what is equally important, a new discipline and a new sense of organization, without which she never could have stood up as sturdily as she has against the highly mechanized German Army.

In Kuibyshev I met an elderly man whom I had seen on several occasions during earlier visits to Moscow. Life, he said, was terribly hard in Russia, but the spirit of the people was magnificent and there was no thought among them of capitulation to Germany. Then he told me that great tragedy had come to him and his wife during the purge. One of their sons, a brilliant young scientist, was executed. He was convinced the son was innocent—but "the execrable Yezhov" (he was referring to the G.P.U. chief in charge of the purge, who was later himself purged) had run away with himself. To their dying days, he said, neither he nor his wife would forget the unjust death of their son. "But," he went on, "what would we have done without the Plans? We paid plenty for them, but they are saving us. With her own and all western industry Germany would have exterminated us. Nobody would have stopped her. . . . Often my wife and I talk about it. Our hair stands up, I tell you, when we think of what Germany would have done with us. The Plans are saving us, yes, the Plans, my friend, make no mistake about it. If there is a Russia in the world now and if there is going to be a Russia in the world tomorrow, it is only because of what the Plans have done for us."

Not a man or woman I have met in Russia, however deep their personal grievances and however heavy the hardships they have endured, but speaks in a similar vein.

The Russia of 1942, the Russia that fought and toiled, that marched and sang, that tightened its belt and shook its fist, that talked and laughed— yes, laughed in the theater, at the circus, at choral concerts, in private homes—this Russia revealed itself to me in a new garb, with a new image, a new discipline, a new dedication, a new power and a new invincibility.

CHAPTER 6: **THE BLACK CITY**

IN THE SIX YEARS I had been away from Russia, changes had taken place even in the outward appearance of people which were as striking as they were significant. At first sight men and women even in Baku, the rich, world-renowned oil city, appeared more poorly dressed than formerly. The cap for men, the kerchief for women were still the fashion and the symbol of the revolutionary era.

Footwear, which, out of long habit, I automatically scrutinize whenever I travel in a foreign country because it is always a guide to living standards, presented a sorry sight. No one except an occasional street urchin went barefooted. No one draped his feet in rags as Russians often did in the years immediately following the civil war. Sailors and soldiers walked around in well-made boots and shoes gleaming with fresh polish. But civilians wore sandals and misshapen shoes.

But surface appearances were somewhat misleading. People in Baku wore old clothes not because they had none other but because they wanted to save their best clothes for special occasions and for the future. The war might last a long time. In the summer of 1942 who could tell when it would end? Textile factories might be too crowded with orders for the Army to fill the needs of the civilian population. Shoe factories might be obliged to pursue a similar policy. For years it might be impossible to buy a new pair of shoes, a new dress, a new suit of clothes. Therefore they would wear out all their old clothes, even those they had planned to cast away. They would make their wardrobes last on through possible years of trial and uncertainty. They would not be caught unawares, as people were in the civil-war years when they were left helpless and ragged.

All the more readily did they wear their oldest clothes because except for invalids and mothers with large families everybody worked, was mobilized for work. The workday was long, and the longer the workday the greater the strain on wearing apparel. The eight-hour day was only a memory. Eleven hours and longer was the law and the practice of wartime Russia. So people went to work in their oldest clothes—in the suits, dresses, tunics, aprons they had laid aside for emergencies.

"We've learned to plan," said a native of the city, "for days, for years ahead. But," he added, "if you want to see the Russian public well dressed go to the theater."

Russia has become the most theater-minded country in the world. By theater I mean not only the spoken stage but the concert hall, the ballet, the opera, the vaudeville house. Everybody who can find a place or a person from whom to buy a ticket rushes to the playhouse, any playhouse. This is as true of Baku as of Moscow or any city in the country. With the decline of religion, theater-going has become an exalted occasion, almost a rite.

Baku has a population of nearly a million and boasts an opera house, a philharmonic orchestra, theaters in the Ayzerbydzhanturki, Russian, Armenian languages, a youth theater, and a vaudeville house. In addition it has a multitude of amateur little theaters in the schools, in the factories, in the oil fields. Battles in Sevastopol, in Stalingrad, in the Caucasus might rage with explosive fury, the black-out might be so dense that in the absence of flashlights people have to walk with outstretched arms so as to avoid bumping into pedestrians, buildings, or lamp posts. But the citizens flock to hear *Rigoletto, Eugene Onegin,* or the opera *Shah Ismael* by the composer Magomayev. They rush to see and to weep over Turgenev's *A Nest of Gentlefolk* and Chekhov's *Three Sisters,* to laugh at the old ballads and comic lyrics by any of the countless national choirs or to be inflamed to deadly hate against Germans and fascists by a performance of Simonov's *The Russian People.* Whatever the performance the seats are always full.

A visit to the theater in any city quickly dispells the impression of shabbiness in dress and evokes, as if by magic, a Russia bright with color, shiny with fashion, not at all the most modern in fabric and design but appropriate and becoming. Never before on any of my previous visits to the country had I known Russians to dress with such taste as now, when they went to the theater. A captain in the American Army who had last visited Russia five years earlier and whom I met in the foyer of the Kuibyshev Opera House, was as astonished as I was at the change in the dress of people, especially the women, as seen not in the street but in the theater. Fabrics might be of inferior quality, but the style, even to this writer's masculine eye, was reminiscent of New York not at its boldest but at its modest best.

If as a consequence of the war and of personal planning the Russian public on workdays made a disappointing appearance, its behavior in the street and in public places showed a surprising improvement in manners. Queues have returned not so much for bread but for pastry, cosmetics, textiles, newspapers, soda water, or some other commodity grown scarce and on sale only whenever a fresh supply arrives in the shops. But never before had I seen such orderly queues. There was hardly any of the old shouting, the old arguing, the old quarreling.

The same is true of trolleys and busses. They were overcrowded, there was seldom enough space for the passengers who somehow managed to squeeze or bounce their way inside, especially during the hours when there was a change of shift in factories or when offices closed for the day. With the best intentions people could not help stepping on each other, bumping into each other. Now and then a particularly audacious and unmannerly youth would invite the reprimand or the denunciation of an older person. Now and then one heard an exchange of harsh words. But the violent rhetoric of former times was seldom heard.

No less surprising was the sense and habit of discipline which I observed and which Russians had seldom before manifested in the years I had visited the country. The momentous speed-up campaigns, the vigor and assiduity with which they were conducted, did something to the mind, the nerves, the muscles, the very being of the Russian. They conditioned him to a pattern of behavior which was as new as it was refreshing. Perhaps the war and the realization of the national difficulties confronting all of them have helped to subdue the impatience and explosiveness of former times. Russians appeared not only more disciplined but more tolerant, more obliging in the little ways and little things of everyday life far more than in former times.

War conscious as they are and trained as they have been to perpetual vigilance, they can be curt to the point of rudeness. Together with two young Americans who had traveled with me from Teheran to Baku, I walked into a bookshop and asked the saleswoman whether she would sell us a map of the city. We should have known better than to ask for a map—in wartime. Fresh from America, which less than any other nation in the world understands the urgency of secrecy and vigilance, we thought no more of it than if we had asked for a copy of the multiplication table. But the blonde, blue-eyed saleswoman thought differently.

"We have no such maps," she snapped. And then, as if for special emphasis, she added, "We wouldn't sell them anyway."

Formerly, of course, maps and guides of Baku in Russian as well as in English, French, German could be had at the hotel for the asking. Now they have disappeared. Not only is Baku the leading port on the Caspian Sea, one of the main railway centers of the country, the seat of the richest oil fields, but it is the gateway to Asia and to Africa. Because of its oil it was known to ancient fire-worshiping tribes. It has had a late development in spite of the overwhelming amount of new building, still looks as though its destiny is only beginning. It is one of the richest possessions of the Soviets. Naturally no map of it would be at the disposal of customers, especially foreigners.

Baku was once known as the "Black City" because of the taint of oil and soot over streets, buildings, people. The comparative absence of grass and trees made the blackness all the more conspicuous. Since oil and vegetation, so the authorities once held, wouldn't mix, it was well to let oil and everything pertaining to it dominate not only the economics but the landscape of the city. The dry winds from the treeless hills and the scanty precipitation—averaging about nine and a half inches a year and falling as if for spite chiefly in winter—only accentuated the difficulty of growing vegetation. So the authorities of old remained content to let nature follow its own arid, cheerless course.

Now it is all changed. Baku still needs grass and trees. More, it needs scrubbing, painting, all manner of landscaping. The old streets have the look and breath of antiquity. Compared to other cities, it is still largely an arid place. But the ancient battle between oil and trees is at an end. Trees have scored a visible triumph. Street after street is not only flanked with them but is green with little parks. The trees are young and flourish well; the grass is thick though not too well kept; the flowers are lusty and add to the color and freshness of the city.

Everywhere these little parks swarm with children. There are so many of them, they are so gay and so alive, so neatly though simply dressed, that one cannot help pausing to watch them. They march along the well-kept lanes. They sing. They drill. They run races. They dig in the sand piles and boxes. They shout and laugh. They look happy, act happy. Always they are in crowds. A playground director watches and guides the movements of those who are of kindergarten age.

Older children of high-school age are more earnest and more adventurous. War conscious, they play war games. They march in strict military formation. They drill with guns. They learn to throw hand grenades— standing, kneeling, lying flat on their stomachs. They seek to perfect themselves in the art and science of crawling on the ground without being seen by the enemy. They do all these things with energy and zest, girls and boys together. An army officer or a factory worker or a stalwart guerrilla of the civil-war days acts as their instructor.

Watching these children, the little ones and the older ones, unwittingly brings to mind the purely racial and biological aspects of the Russo-German war. To German politicians and generals, Russia's extraordinary birth rate has always been a source of vexation. To men like Hitler and Rosenberg, the most violent Russophobes and Slavophobes in history, it has meant sheer exasperation. Through propaganda and subsidy, through song and poem, through encouragement of early marriage and the disregard of marriage conventions, the Nazis have sought to lift the German birth rate to the highest possible level. But whether for reasons of environment or of heredity, the fecundity of the German woman does not match that of the Russian woman. To a Hitler, a Rosenberg, any Nazi, the Russian birth rate is still the great menace to their far-reaching plan to make the Germans sole masters of Europe. The birth rate in Berlin in 1936 was 14.1 per thousand, but in Baku in 1938 it rose to 33.9 per thousand.

Russians are convinced that the brutality with which the German Army in Russia is treating the civilian population in occupied parts of the country and Russian prisoners of war is a deliberate effort to reduce their numbers by death. Hasn't Rosenberg repeatedly said that Russia must be driven from Europe? Hasn't Hitler again and again proclaimed that Europe must someday be populated by two hundred million Germans? In point of numbers the Slavs, and particularly the Russians, who have so rapidly been acquiring a mastery of the power and the weapons of the machine age, are the sternest obstacle to the fulfillment of this aim. Hence, say the Russians, the deliberate German policy of repression, impoverishment, degradation, starvation, and extermination of the civilian Russian population and of Russian war prisoners.

In this battle unto death between Russia and Germany, between Slav and Teuton, the birth rate not only of today but of tomorrow and of the distant future is as important as guns. And the crowds of children in the parks and the courtyards of Baku testify to Russia's overwhelming biological superiority to the "superior race."

True, the Baku population is less than one third racially Russian; about one half are Tucos. But the stamp of Russian and Soviet influence is ever present not in the speech but most decidedly in the dress, manner, and social habits of the people. The new marriage law of 1935 operates with no less force for Armenians or the Jews in Baku than it does for the Russians in Moscow or Kuibyshev. Banning abortions, frowning on birth control, restricting the conditions of divorce, the government encourages large families for all peoples in the Soviet Union. The unceasing, sentimental laudation of motherhood and the subsidies of large families have borne striking results.

As I look around this ever-changing, ever-growing unfinished city, I am more than ever impressed with the physical sturdiness of the Russian, whether he is of Armenian, Slav, or Turkish origin. There is not a fat man among them, not one—neither in the streets nor in the hotel. Of medium height, big-boned, with broad backs, deep chests, short but muscular necks, they appear possessed of extraordinary endurance like the camels that so often amble along on the outskirts of Baku. They are wiry and rugged.

Yet there is nothing slow or leisurely about them. Despite war and substantial belt-tightening, they show neither fatigue nor lassitude. There is new vigor in their movements, new briskness in their steps. I have never known them to walk so fast. Like Americans they are always in a hurry to get somewhere. The campaigns since the start of the First Five-Year Plan for speed, discipline, alertness have shaken the fat off them and have made them fleeter of foot than they had ever been.

These campaigns have done more: they have made Russians time-minded, more so than they have ever been in their history. In a modern factory time is precious—every hour, every minute, every second counts. The assembly belt brooks no delays; the engine tolerates no procrastination. Once started, the work must push on and man must be there to help the machine. Factory management and Soviet law deal harshly with those who think or act otherwise. Especially severe is the wartime enforcement of punctuality. Obloquy and degradation are the least of the penalties imposed on the idlers and the loafers. And so the men who work in factories, mines, oil fields, offices know as never before the meaning and importance of time.

Especially is this true of the youth. I have had appointments with many young people. Invariably they appeared on time. On several occasions I was late and they kidded me about my American punctuality.

The conspicuous exception is intellectuals. They seem as lax in the appreciation of time as they have always been. An hour late, two hours late, a promise broken—what of it? This seems to be their attitude. Once I reproached a writer for being two hours late to an appointment. Dismissing my protest with the ancient and good-humored *nichevo*—which means no matter or there's lots of time—he said, "I'll tell you a story. Once Bismarck came to Russia to visit Czar Alexander III. The two went on a sleigh ride and when they were deep in the country the thills of their sleigh knocked down a bearded peasant. The German chancellor and the emperor got out of the sled to see what had happened to the muzhik. But he only clambered to his feet and when asked how he felt, he shook the snow off himself and replied 'nichevo.' Bismarck was so struck by this reply that he had the word engraved on a fob which he wore on his watch and chain for the rest of his life. Now if the German chancellor was impressed by this word why are you impatient with it?"

This spirit of jolly irresponsibility is alien and repugnant to the present-day Russian youth. In this one respect the Soviet intellectual remains recklessly behind in the procession for a new culture and a new way of life.

As if to symbolize their time-mindedness, many people in Baku sport wrist watches, large ones, often covering the breadth of the wrist. On my last visit to Russia, few people had had any. Now, in Baku anyway, few seemed without them. As I observed Russians walking along the streets with cigarettes in their mouths, searching for a match or a light, I had the feeling that in Baku it was easier to obtain a wrist watch than a box of matches. No wonder people have gone back to the flint.

Russian smokers are taking to the pipe. This, too, is something new, because, except for older folk in the villages and now and then in the city, Russians have been essentially a cigarette-smoking people. The pipe, like the cigar, was always associated with foreign custom and foreign privilege. The pipe-smoker to them, like the monocle-wearer to an American, represented an Englishman. Now, owing to the shortage of cigarettes and cigarette paper, the pipe is becoming a vogue. Because so many army officers and soldiers use it, it bids fair to become an institution as it is in England.

Yet smoking is markedly, almost sensationally, on the decline. On earlier visits, especially in the twenties and early thirties, everywhere I went I saw women and girls smoking almost as much as men, sometimes more. College girls, factory girls seemed to feel it their revolutionary duty to smoke, especially in public. In theaters during intermissions the smoking lounges were as crowded with young women as with young men. The college campus was littered with cigarette butts, but not today. At a gathering in a small town at which forty-three girls of high-school and college age were present, I asked how many of them smoked. Not one, was the reply. Then I asked, how many of them would like to smoke? And the answer was, not one!

Not too frequent nowadays is the college girl who smokes. The same is true of factory girls. There is no political ban on smoking, no social taboo. But there might as well be as far as girls and young women are concerned. Boys, too, smoke less than formerly. The shortage of tobacco, matches, and cigarette paper has had something to do with it, but not nearly so much as the intense propaganda for what in English-speaking countries we call "clean living," and the striking changes in social manners and social attitudes which the three Five-Year Plans have wrought and the war has crystallized.

In Moscow and in Kuibyshev, Russian mothers told me they were worried because boys were again taking up smoking. The war tension has doubtless loosened the prewar discipline. Even so, compared to former years smoking among adolescents and women has declined markedly.

Drinking is much less prevalent. By comparison with former times drunkenness is scarcely noticeable. There is no prohibition in the country. But the grain and potatoes from which vodka is made are needed for food. Besides, much alcohol is diverted to war purposes, especially to the manufacture of synthetic rubber. Little of it can be spared for civilian consumption. A bottle of vodka is a precious possession and commands a high premium in the market place.

In former years, when liquor was scarce or was banned by the law, peasants in the villages brewed their own. In the early twenties the Russian village reeked with *samogon*—homebrew. At weddings I saw guests bring it freely as gifts for the bride. When I stayed overnight with a peasant he invariably set before me, as a mark of special favor, a bottle of samogon. Now that the kolkhoz dominates the village, indulgence in homebrew involves too many risks. Law and official public opinion are rigidly, fiercely arrayed against it. Older peasants sigh with regret and reminisce of the time when vodka could be had in government shops for the asking.

But younger people, especially girls, rarely mention the subject. Most girls have been brought up to regard strong drink as *improper*. They are not fanatical on the subject. They take no temperance vows, make no promises of abstinence to mother, father, to anyone. They will drink wine. At a celebration they will partake of vodka. But they are against orgies, against drunkenness, say so openly, definitely. There are, of course, exceptions, though only in the cities. But it is not of them I am speaking. Two factory girls in Moscow once invited a British journalist and myself to a party.

"We'll bring vodka," said the Britisher. "We can get it."

"No doubt you can get it," came the response, "but we won't drink it."

"What about wine?" I asked.

"Wine, yes. We'll drink that. But not vodka."

There can be no doubt that Russia has one of the most sober young generations it has ever known.

There is much then in Russia that even on first sight marks a return to past values and past usages. But there is much that is strikingly new and marks a striking break with the past in belief, in usage, in ambition, even in personal appearance. Consider beards, for example, once so celebrated an institution all over Russia, the pride and sanctity of fathers and grandfathers of the Russians of today, of priest and layman, of merchant and college professor, of minister and general. Peter the Great on his return from Holland shocked his boyars when he ordered them to cut their beards or when with as much mischief as zeal he clutched a handful of a visitor's beard and clipped it off with his own shears.

In the Western world and particularly in America the very word Russian suggests a man with a beard. In American vaudeville, motion pictures, stage plays, the Russian beard has become as much of a stage fixture as Charlie Chaplin's baggy trousers. But now in Russia the beard is no more, is rarely, very rarely seen. Russia is becoming one of the most clean-shaven nations in the world. Despite atrocious razor blades, the daily shave is gaining in appreciation and in vogue. The mustache is following the beard into oblivion. I have yet to see a college man in or out of uniform who wears either. This is true of the village almost as much as the city.

In their external appearance the Russians are becoming more and more Westernized, more and more attuned to the machine age which they have so indefatigably espoused and which they so openly exalt.

CHAPTER 7: **LAND OF GLAMOUR**

LATE IN THE AFTERNOON we started on the Baku–Moscow express for Stalingrad. The train was crowded with passengers, chiefly soldiers, some just out of the hospital, others fresh from training camps and on the way to the front. There were no demonstrations at the station, no music, no loud talk, no plaintive farewells—nothing but the waving of hands and handkerchiefs, last-minute messages and salutations. It might have been only an excursion train bound for the capital.

The one international sleeping car was so bright and spacious that merely seeing it, being in it, accentuated the peacetime feeling of comfort and confidence. Never in prewar days had I journeyed in a cozier and neater sleeping car. Doors and windows sparkled with cleanliness. In Teheran I had bought a number of fly swatters, an indispensable weapon of defense in many a part of Europe, including Russia, during the summer months. But so free of flies was the car that I had no use for it until long after we had left Baku. To add to the convenience of passengers there was the usual samovar in the car with the team of conductors—a man and a woman ready to serve as many glasses of tea, without sugar, as anyone chose to drink.

With the quiet hum conducive to ease and slumber which marks the movement of continental trains, we rolled out of the station. I sat down on the folding bench beside a window and looked outside, eager to observe the changes and transformations which had taken place since my last visit here. Especially impressive was the amount of new construction on the outskirts of the city. There were acres and acres, miles and miles of new buildings—small, large, single, and in chains—which in the distance loomed like military communities or new townships. There was nothing ornate about any of them. Many seemed as if hastily poured out of clumsy molds. Again and again the grounds were still dug up or unkempt. Here and there an effort was made to grow grass, flowers, and in other ways to invoke the power of nature for the obliteration of the coarse untidiness. Not always was the effort successful. Obviously not ornateness but utility was the supreme consideration, not physical comeliness but productive energy with which to dig out of the earth fresh stores of raw materials and convert them into machines for still further production or into instruments of warfare, as has always been the foremost aim of the planners.

We were moving into the North Caucasus—land of sun-flooded valleys and snow-capped mountains. To the Russian, whether a Pushkin, a Lermontov, a Tolstoy, or a Soviet girl from a textile mill in Vladimir or Ivanovo, this land has ever been a source of romance and grandeur. The mere mention of the word Caucasus rouses joyful emotions, a sense of adventure, yet of harmony with nature, with life, with one's own soul.

Of late years crowds of young people from all over the land, bareheaded, barelegged, barebacked, with packs on their shoulders, with hand-carved canes in their hands, wandered all summer over the highways and mountain passes, the sun-scorched steppes and the wooded mountain slopes in search of the rugged adventure and the romantic exhilaration which they had always associated with the country. On their return home, whether to a collective farm in the Siberian plains or a factory in the Urals, they never tired of narrating to friends the joys and glories of their excursion. Land of battle and legend, of dreams and beauty, to the people of the plains and lowlands where the vast multitudes of Russians live, have always lived, will always live, the Caucasus north or south is nature's richest gift to man.

Comprising a territory only a little larger than the state of Missouri, it is replete with wealth. Here is oil, an abundance of it and, as in Maikop and Grozny, of as high a quality as any in the world. Here are zinc and lead and a most bountiful agriculture—grapes and sunflowers, peaches and apricots, melons and honey, tobacco and tea. In the Kuban alone before the war there were 175,000 acres of orchard, one fourth in grapes. Wild fruit trees have been grafted with choicest apples and pears. The tomatoes of Maikop, thin-skinned and so red that they seem like containers of blood ready to burst open and drench the earth, have attained national fame. Here are new plantations of cotton and rice, corn and wheat lands. On the mountain slopes browse flocks of sheep tended by shepherds, some of whom are centenarians, who neither know nor care to know the valleys below and the life they contain. Herds of goats and cattle as choice as any in Russia wander over fat pastures. Here, too, are the most celebrated health resorts in the country—Pyatigorsk, Yessentuki, Kislovodsk, Zeheleznovodsk, Mineralnya Vody, and many others whose waters for drinking and bathing are renowned all over Russia and abroad.

Many are the peoples that inhabit this rugged and beautiful land. There are the new peoples who, in search of a better life, have been lured from adversity by the magnetic power of the North Caucasus. They are Russians, chiefly peasants, Ukrainians, Tatars, Jews; now and then there are Greeks, Poles, Letts, Lithuanians. There are old faces, some of them so old that they know neither their racial origin nor the place from which they stem. Melodious in Russian are their names—Ingushy, Ossetins, Balkarians, Kabardins, Chechens, Adigeytsui, Cherkessy, and numerous tribes in Daghestan, all with a physiognomy as distinct as their volcanic temper, all still clinging to native garb and native tongue and often, too often, to family and blood feuds.

"What would an American do," asked a middle-aged Chechenets whom I visited, "if on returning home from a trip he found a strange man talking to his wife?"

"It depends," I said, "who the man was and what he wanted."

"Really?" My answer puzzled my host and after a brief pause he added, "Only a few weeks ago a neighbor of mine on returning home from a trip and finding a strange man sitting there and talking to his wife, pierced him with his dagger. That's what he did without even bothering to ask the man or the wife who he was, where he was from, or what he had come for." This was in the year 1937 after the North Caucasus, like other remote territories and all Russia, had been swept by more than a decade of Soviet propaganda with its fierce assault on family feuds, tribal wars, explosive outbursts of masculine jealousy and masculine possessiveness. . . .

The train was gliding along level lands farther and farther north. To the west, out of a wall of blue mist, rose the mountains high and white. To the east, in waves of sealike green, rose the wheatlands. Here are no forests, and rainfall is usually scanty. But this year, when it was especially needed because of the war, rain was more than abundant. The wheatlands were luxuriantly rich—as rich as any I had seen in America.

The mere sight of them undulating in to the far horizons brought to mind only too vividly the stupendous agricultural revolution that has swept Russia since the coming of the Plans. Gone completely are the small individual farms with their checkerboard patches of individual holdings. Collectivization has blown them tornado-like out of existence. The only small fields are the garden plots that embroider the landscape along highway and railroad track. Because of the campaign for individual gardens they were especially numerous now. But they were mere specks on the landscape, drowned out by the surge of from one to fifty thousand and more acres in one continuous field. Here was a new kind of grandeur on the Russian scene—the grandeur of wheat. It dazzled and challenged the eye as well as the mind. No country in the world, neither America nor Canada nor the Argentine, has attained such uniformly large-scale wheat growing as Russia. As I had not seen this country since the days before collectivization, it all seemed like a new world with hardly a vestige anywhere of the world I had formerly known.

Unwittingly I thought of America's momentous contribution to this breath-taking agrarian revolution. Had it not been for the American-invented tractor, gang plow, combine, grain drill, which Russians have so painstakingly and successfully copied and mastered, the North Caucasus and the rest of Russia would still be a mass of quiltlike little farms with each family jealously and laboriously toiling over its own strip of land and its own little crop.

Only twelve years earlier peasants in a village in the Archangel province on seeing a tractor for the first time were so overawed by it that they deemed it the invention of the devil. Remote from railroads and engines, they felt certain that the reason it moved without horses was that an evil spirit inside was manipulating it. At night they swooped down on it with axes, hammers, and crowbars and smashed it.

Nor was this the only instance of violent reckoning with the tractor. So new and so overpowering was its appearance that in the really "deaf" country, farthest removed from the sight and sound of the modern machine, the peasants were often as bewildered as they were frightened by the presence of the mechanized monster.

Now in the light of its achievement it can be truly maintained that the tractor was the mightiest single weapon of the Revolution in the village. It has wrought havoc there not only with old methods of farming and land-holding but with century-old traditions and modes of thinking and living. More than any other invention of science, more than all the Bolshevik propaganda, it has blasted much though not all superstition from the village. It has forged a link of steel between the village and the city, the land and the factory. It has given the village a new conception of its own place in the world, in nature, in society. Older peasants might curse the tractor and collectivization, but they were powerless to halt the transformation both were achieving in agriculture; and despite themselves in their own life and mode of thought.

The tractor made the forest-minded muzhik engine-minded and in time, as with a surgeon's scalpel, it scraped out of him every trace of muzhik spirit. Gone is the muzhik from the Russian scene, gone almost is the word from the Russian vocabulary. Readers of Chekhov and Tolstoy, of Turgenev and Pisemsky, of Ivan Bunin and other Russian authors who have written of the peasant with tenderness and acrimony, with pity and hopelessness, will search in vain for the Ivans, Stepans, Varvaras, and Anisyas over whom they have sorrowed, perhaps wept. . . . The peasant hut has remained as humble and as dingy as it has always been, but the peasant soul has been refashioned, and it was the tractor more than any other weapon, more in this writer's judgment than all the rhetoric of lecturers and propagandists, that has wrought the refashioning. Deep and mighty have been its powers of transformation. Above all, it was the tractor that prepared peasant youths by the million for the tank and for other mechanized weapons of warfare. Without the American-invented tractor, whose formidable power of transforming agriculture and human behavior Bolshevik leaders had so clearly and poignantly envisaged, the Red armies could not have halted nor fought the superlatively trained and brilliantly mechanized Nazi Reichswehr. No wonder that in the early days of its appearance in Russia many songs and poems were dedicated to the tractor, and one enthusiastic artist even sought (though unsuccessfully) to popularize it as a design in textiles.

Buildings all along signalized the changes in the outward scene of the country only a little less momentously than the immense wheatlands. There were many more of them than I had seen on my previous visit, thousands and thousands more. The millions of virgin acres that have been turned by the plow have drawn thousands of new workers. New settlements by the score, by the hundred seem to have bobbed out of the earth—state farms, collective farms, experiment stations, agricultural schools, and tractor stations which supply machinery to the farms. Here not a cottage, not a building, whether barn, tool shed, or milkhouse, but gleams white. Nowhere, not even in the ornate Ukraine, is external decorativeness so universal as in the valleys of the North Caucasus. However new and commodious the cottage, it is not home unless it has received the proper coating of whitewash—even now in wartime. Nowhere else in Russia do villages look so trim and prosperous and livable.

The new barns are more impressive than the cottages. Long, with many windows, flanked by modern silos, they are as white and ornate as the cottages. They mark as epochal a change in livestock farming as the wheatlands do in grain-growing. The individual peasant now has but one cow for his personal use, but the collectives and the state farms, especially in the North Caucasus, vie with one another for the size of the herds they can raise. The large windows and the ventilators that jut out of the red-tiled roofs testify to a process of modernization practically unknown in the old days.

Apiaries, too, have multiplied. Here again one notes American influence. A mere glance at the type of hives and the manner in which they are set out brings to mind the late A. I. Root of Medina, Ohio, whose *ABC on Bee-keeping* has been studied and applied by many a Russian beeman.

Large orchards are another new feature of the changing landscape. Small orchards, often woefully neglected, have always been part of the peasant farmyard. They provided shade from the hot sun and yielded fruit for the family. But commercial orchards were rare. Now they are frequent, stretching for hundreds and thousands of acres. Most of those I saw were newly set out but poorly tended. It was the one aspect of large-scale agriculture that cried for improvement. Trees were bent, limbs were untrimmed, tops were nibbled off by browsing goats and cattle. Unfenced as were all fields, they proved too tempting a morsel for pasturing livestock that had obviously succeeded in eluding the shepherd's crook or the vigilance of the accompanying dog.

Seeing these wheatlands, these multitudes of whitewashed cottages, these immense barns and even-rowed apiaries, these far-stretching though neglected orchards, gave one a fresh appreciation of the riches of the country and of the monumental changes that it had undergone. Millions of new acres had been broken into cultivation. Millions more still awaited the tractor and the gang plow. Here, as in no other part of Russia except Siberia, there still lay immense wastelands.

Old as it is in history and legend, rich as it is in beauty and drama, the North Caucasus is only now emerging into the world. It makes one think of the American Middle West after the Civil War, except that the pioneering here is more intense and state-sponsored. No wonder that the Germans have always coveted this land and were soon to fight for it, Nazi and worker, peasant and student, junker and shopkeeper.

"What are you doing here so far away from home?" a German war prisoner, a mechanic in the GI Dye Trust, was asked by a Russian examiner in the North Caucasus.

"I was sent here," was the reply.

"To fight us Russian peasants and workers?"

"I had to fight."

Thereupon the Russian pulled out of his case a thick book in German on agriculture in Russia. Pointing to penciled passages, especially on pig-raising and wheat-growing, the Russian examiner inquired if it was his intention to become a landlord in Russia. The Nazi mechanic remained as silent as his book but no less revealing in his muteness. . . .

At the time I was making my journey, no sound of guns and no blinding flares disturbed the tranquillity of this wonderland. Peace and growth reigned supreme, as if sanctified and glorified by the never-ceasing song of visible and invisible skylarks.

Yet signs of war were not lacking. Few were the young men one saw in the fields. Save for old men and boys the workers everywhere were women. They operated tractors and cultivators. They drove horses and oxen. At railroad stations they sold eggs, cheese, bunches of summer radishes, cold griddle cakes, and at prices that to anyone from America sounded fantastic. They scoffed and sulked when they were reprimanded for profiteering. But they tended the lands. In their many-hued kerchiefs and dresses—with white, red, and blue predominating—they added color and ornateness to the scene. Sometimes they sang as they worked, as in prewar times, their wailing, long-drawn-out voices echoing far and wide over the steppe.

Some of them carried rifles on their backs. They guarded elevators and flour mills. Above all, they were on the lookout for German parachute troops. Parachute-conscious as no other people in the world, Russians would take no chances on being caught unaware—neither in the North Caucasus nor in other part of the land.

We passed many a station that has since leaped into the headlines of the leading newspapers of the world—Grozny, Mozdok, Prokhladnaya, Pyatigorsk, Mineralnya Vody, Tikhoretskaya, Salsk, many others. War was still far away, and these people were energetically pursuing their daily tasks. Yet everywhere I observed signs of war-consciousness. No crowds of vacationists lingered at the railroad stations. Nowhere was there any loud laughter or holiday dress or any playing of accordions. Empty were most of the dining rooms in the railroad stations. Food stalls—formerly among the very best in the country—were almost as empty. The dining rooms that were open served soldiers more than civilians. The choice of dishes was limited to bread, salted fish, hot vegetable soup, now and then cheese or sausage. Nowhere was there butter or pastry. But passengers were ready to buy anything.

The bookstands were almost as empty as the food stalls, looked even more desolate. Now and then, when newspapers were on sale, crowds gathered, brandished paper money, and shouted for a chance to buy *Pravda, Izvestia,* or some other local publication. They made one think of an unruly bazaar. Few were the books on sale, and these mostly on military subjects and on agriculture. Not a novel anywhere, not a trace of fiction.

"Don't you get any more fiction?" I asked the attendant in Grozny.

"Yes, now and then," was the reply, "but it is snatched up as soon as I lay it out on the stand."

"What about magazines?" I asked.

"I never get enough of them—not even on technical subjects."

Even while I was talking people came over and asked for books and magazines. "No, no," was the only reply their demand elicited. Great was the hunger for reading matter and small the means of satisfying it.

The ominous emptiness of Mineralnya Vody, one of the liveliest vacation resorts in peacetime Russia, was particularly striking. Empty and barren were the largest dining rooms and most of the food stalls. They seemed as if swept clean even of reminders of former abundance and joviality. Broken windowpanes remained unrepaired, peeling paint and sagging doors remained untouched. I walked inside one. Sharp was the smell of staleness as in an abandoned house. Dust lay over everything— over floor and furniture—the little that was left. The impassioned war posters on the walls looked like sleeping ghosts. It seemed as if the very spirit of the place, as I had known it on former visits, had been evacuated.

At Salsk the train was to stop for half an hour, but it stayed longer. Salsk is a famous place, an early landmark, one of the earliest, in Russia's agricultural revolution. It was here, under the supervision of American experts, that the "Giant," the largest wheat farm in the world, was launched. So immense was it that the Soviet manager traveled by plane from one part to another. At that time the world stirred with the news of this unprecedentedly huge wheat farm. Russians boasted of the achievement, the like of which no capitalist land could match. But soon the boasting ceased. The Russians learned that bigness in agriculture as well as in industry could be more of a liability than an asset. They broke up the "Giant" into several more workable farms.

A junction point of some magnitude, the railroad yards and sidings at Salsk were crowded with trains and locomotives. Passengers tumbled out of the coaches, strolled up and down the platform. Moscow papers had just arrived, and a long queue formed before the newsstand, but only men in uniform were favored. Suddenly a shrill siren rent the air. One engine seemed to vie with another in the shrillness of their sirens.

"An air-raid alarm," said one of the soldiers, and calmly started to chew a cold boiled potato. It was my first experience in an air-raid alarm and I searched the milling crowd for guidance as to what to do. But no one paid any attention to the ear-splitting warnings. People lifted their heads and searched the skies for the sight of German planes, and as none were in sight they remained calm and at ease. The soldiers who had queued up for a Moscow paper did not leave their places. The men and women who were leaning out of the coach windows did not bother to rush outside. They stayed where they were—as unperturbed and indifferent as though the sirens were only factory whistles announcing lunch or the end of the day's work.

"Isn't it amazing," I said to the conductor, "how calm people are?"

He was one of the calmest men on the platform, and stood almost serenely at the steps of the sleeping car.

"After you have been in as many air-raid alarms as we Russians, you kind of get used to them," was his reply. "In Moscow," he added, "there haven't been any alarms in three months." Then he told me that at the height of the battle of Moscow he had been conductor on an international sleeping car which was the headquarters of a commanding general. "Shells and bombs often fell all around us," he went on; "so what is a mere siren to me? You see, we Russians have powerful nerves. Look at the people here—nobody is running, nobody is excited."

"But if a bomb fell?"

"Plenty of bombs fell in Moscow at the time the Germans tried to take the city, and, would you believe it, people got so toughened to the raids that if they stood in a queue before a shop they refused to go into an air-raid shelter and the militia had to threaten arrest to get them to obey orders."

I shrugged and said nothing, wondering what it was about these Russians that made them so indifferent to the air-raid alarm. As if in answer to my bewilderment the conductor added proudly, "Our nerve is the one thing Hitler will never break, never."

The sirens were blowing frantically, but the passing crowds were as unperturbed as the conductor. Hardly anyone looked at the skies any more. Slowly passengers walked about, talking, eating black bread and salted fish, asking each other about the price of goats and sheep and cheese that the peasant women were selling and obviously unmindful of the sirens and the dangers of which they warned. There was no denying the proud boast of the conductor that Russians had powerful nerves—too powerful perhaps for their own good in an emergency!

At last the sirens were silent and our train moved on. A few stations outside of Salsk it stopped again and stayed for over half an hour. It was waiting for a clearance signal. Meanwhile passengers tumbled out again and sat or lay down on the grass. I sat down beside a group of young officers. It had just rained; the ground was still damp and smelled pleasantly of grass and fresh earth. The Russians did not mind the dampness—the grass was so inviting. Skylarks filled the air with happy melody, and the officers beside whom I was sitting were in a gay mood. They smoked hand-rolled cigarettes and teased each other. Several women with willow-plaited baskets slung over their forearms passed by, and the officers stopped them.

"How is life, Auntie?" said one, addressing himself to a middle-aged barefooted woman with a shimmering red kerchief on her head.

"We work and work and are raising good crops so your little sons and we can have bread."

"Do you belong here?" asked another officer.

"No, we are from the Orel province."

"Orel?" exclaimed still another officer, the youngest in the group. "I, too, am from Orel. How did you get here?"

"A fine question for a soldier to ask. It was not because we wanted to leave our native villages."

"I mean," said the officer apologetically, "how did you make your way here so far from home?"

"When you have to make your way you must make it—you have no choice, have you?" said the woman with the red kerchief.

She surveyed the officers with eager, questioning eyes and started to interrogate the younger one about his village, his people and what had become of them when the Germans came.

"I don't know, Auntie. I wish I did."

"Ah!"

"You don't happen to know Vasily N —— in the village of P ——?" asked the officer. "It is very near your native village."

"And why shouldn't I know Vasily? He's my brother-in-law."

The officer leaped to his knees, eyed the woman with excited, triumphant eyes. Then he pulled out a little notebook, turned the pages, and read slowly and significantly:

"Vasily N —— from the village of P ——, a man with a black mustache and a wart on his right cheek, hid me in his barn for twelve days and the Germans never found me. Twice a day he brought me food—bread, kasha

—sometimes cold potatoes and twice a handful of raw eggs. I owe my life to Vasily N —— of the village of P ——."

"There now, Auntie," said another officer, "see what a brave brother-in-law you have."

But the woman hardly heard him.

"What do you suppose has happened to him, little son?" she asked.

"I wish I knew, Auntie. Late one night he came and told me to run, and when I asked him to come along he said he couldn't leave his wife and children."

Again silence. The woman sighed, shook her head, said nothing. In her basket, unwrapped, lay several thick slices of white bread. Slowly she lifted them out and held them at the edge of the basket.

"My dear ones," she said, "I have only three slices—if I'd known there were six of you——"

"No, Auntie, we don't need it," said the soldier from Orel.

"Even if you don't need it, little son—just so you'd remember Vasily's sister-in-law."

"No, no," protested the officer.

"What do you mean no, no?" snapped another—a man with a face dusky with sunburn and with a small reddish mustache. Looking up to the woman, he said quickly, "I am from Kaluga and I know folk from Orel don't like to be insulted, so I won't insult you, Auntie, like your own countryman—here."

He reached out for the bread. She dropped it into his hand. With child-like triumph he flaunted it before his companion from Orel and went on tauntingly:

"Even if you want a piece, you won't get it, because you are only from Orel and I am from Kaluga."

There was a roar of laughter in which the woman joined.

"If the train stays here long enough, I'll bring you more white bread," she said.

"Bring it, Auntie, bring it," the dusky-faced officer said. "If our train leaves, another is sure to follow, and when soldiers come down ask them whether they are from Kaluga or from Orel, and if they are from Orel say, 'You can't have it,' but if they are from Kaluga say, 'Take it, beloved, take it, and eat it with God's blessing.'"

Another burst of laughter and I couldn't help thinking of how universal sectional pride is and of how endless a subject for good-natured, mirth-provoking banter. Who ever invented the fiction that Russians have no sense of humor, don't jest, don't laugh, don't attempt to lighten the burden of their daily struggles or brighten the ordeal of their daily life with wit and gaiety?

The whistle blew and we returned to the train again. I leaned against the window of my compartment and looked outside. We passed an irrigation pond in which two peasants were fishing with umbrella-like rod nets. A group of bareheaded, barefooted little girls with smiling faces waved handfuls of flowers at the train. A sturdy woman with a rifle on her shoulder was guarding a grain elevator. All around were fields and fields of wheat as green as the ocean and as lost in the faraway horizon.

CHAPTER 8: MOTHER VOLGA

Volga!

What memories and sentiments the word evokes in Russians! With what tenderness and sanctity old and young speak of it, sing of it. In times past they even prayed to it.

Little Mother! Beloved Mother! Darling Mother! Cradle of Liberty! Giver of food! Giver of drink!

No other name, no other place in Russia's far-flung geography, not even Moscow, rouses so much sentiment and so much love. Haunting are the songs, stirring the stories with which Russian folk have always glorified this wondrous river. Hardly a personal sorrow of hunger or love, hardly a personal triumph in life or battle, birth or death, but has found its echo in these songs and stories.

No good Russia has known, no evil she has endured but has, spiritlike, floated over the Volga's surface or come out of its depths. Ivan the Fourth (once known as the Terrible) and Stenka Razin; Tatar and German; Czarist and Revolutionary; bandit and saint; Mohammedan and Christian; White and Red—all have sought to master the river. Failure spelled irretrievable doom.

Without the Volga there is no Russia—in Europe. With the Volga, Russia is never conquered—in Europe. Hitler and the German generals knew it only too well. Hence their lust in the summer of 1942 for its conquest and their feverish push in the midst of fire and blood toward its banks. Hence also Russia's mighty, spectacular resistance.

For centuries the Volga has nourished the heart of Russia's lands in Europe and the sentiments of the Russian people—everywhere. Exiles in China, in America, in other lands dreamed of the Volga, sang of the Volga, glorified the Volga, wept for the Volga.

Mother and mistress, comrade and goddess, such is the Volga in Russian folklore. Such to this day is its appeal to Russians wherever they may be.

In Europe the Volga is the longest river, winding through plain and hill, forest and swamp, for a distance of 2,325 miles. In Russia it is fifth in length. The Ob, the Yenisei, the Lena, the Amur—each of these Siberian rivers

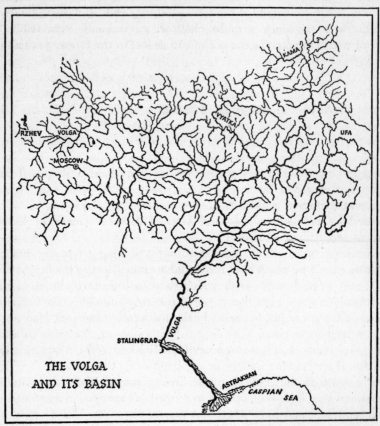

From Bolshaya Encyclopedia

On the map the Volga appears "like a gigantic tree—with bare trunk at bottom, an immense crown of limbs and foliage at the top."

outdistances the Volga by many miles. But none has played so intimate and so sturdy a part in Russian history, in Russian life, from the days of the Tatar invasion to the days of the Five-Year Plans as has the Volga.

With its clusters of tributaries, large and small, it appears on the map like a gigantic tree—with bare trunk at the bottom, an immense crown of limbs and foliage at the top. This "tree" shadows and shelters one third of the land in European Russia—a territory large enough to hold all of prewar

Germany, prewar France, and three present-day Englands combined. More than one third of the population in European Russia lives on this land.

Its wealth is as immeasurable as its glory. It abounds in forest and agriculture, in industry and fisheries, in oil and minerals. To the north are flax and potatoes, among the finest in Russia, and also dense forests. In the middle is the black-earth belt with its incomparable wheatlands, and also oil and minerals. There is fruit, there is livestock. Below Stalingrad are fisheries among the richest in the world—sturgeon and pike, lamprey and sterlet, pike perch and gudgeon—sixty-nine varieties of fish, thirty-two of which have been commercialized. The choicest caviar comes from the region where the Volga flows into the Caspian Sea. The Volga has always meant bread, timber, fish, and now means oil as well.

On the banks of the river rise some of Russia's oldest and most renowned cities, which in recent months have been made familiar to the outside world by the daily press. Rzhev, for example, and Kalinin (formerly Tver) and of course Stalingrad. Then there are Rybinsk and Yaroslavl, whose very names breathe history and romance, Kostroma and Gorki (formerly Nizhnii Novgorod), Kazan and Ulyanovsk (formerly Simbirsk, birthplace of Lenin); Kuibyshev (formerly Samara) and Syzran—seat of the oil fields; Saratov and Kamyshin and Astrakhan. A newly dug canal links Moscow with the Volga and all these cities. During the height of the Stalingrad siege it was largely the newly built factories in these cities which supplied weapons for the Russian armies. Down the Volga they floated in barges, in all manner of craft, including rowboats. Had not the Plans armed these cities with shops and chimney stacks, with lathes and forges, hundreds of them—Stalingrad, perhaps all of the Volga, might by this time have become a German possession.

At Stalingrad the Don is only forty-five miles from the Volga—the longest and most fatal miles the German Army has ever traversed in its pursuit of pillage and conquest. When the Russians complete the canal uniting these two rivers, the Sea of Azov, the Caspian, and the Black seas will be linked with them and with the Dnieper, and Russia will boast one of the most far-reaching waterways in the world.

I had been on the Volga many times. I had taken to heart the Russian saying, "If you want to know Russia, the character and the soul of the Russian people, go on a trip on the Volga." Kazan and Saratov, Gorki and Stalingrad I had known before. I had swum and rowed in the Volga, I had fished and feasted there. Yet when I reached it now, early in June 1942, it loomed more impressive and more momentous than I had ever known it. The battles of Kharkov and Sevastopol were at their height, and the question naturally arose, will the Germans reach the Volga, and if so, then what?

At Stalingrad, where I boarded the steamer, no sound of guns was yet heard.

As the boat glided along the high, shimmering waters I looked around—from one bank to the other. Save for the many oil tanks with which the right bank was studded there was nothing new anywhere. The east bank was low and flat as always—meadowland the Russians call it; the west bank was high and steep as always, and now I could not help thinking of it as a fortress towering over the river.

"If Hitler," I said to a young lieutenant beside me on deck, "ever comes here the Red Army can give him quite a trouncing."

He laughed rather incredulously and replied:

"What makes you think Hitler will ever get here? German corpses will never pollute these waters." Then he added solemnly, "This is the Volga," as if to impress on me that the Volga was not merely a river but, like Moscow, a sanctuary from which invaders are forever barred.

It was only a little later that another Russian soldier named Andreyev, a private in the ranks, wrote a memorable appeal to his comrades in arms:

"Look, you simple-hearted Russian young man! The enemy is creeping into the depth of our land. Drunk with gore, the Germans are pushing beyond the Don to the broad immensities of the steppes, seeking to break their way to the Volga. They are advancing. We have no right to retreat; we dare not allow the Germans to reach the Volga. Sacred to us is this great Russian river and it shall remain unpolluted. Without the Volga Russia is a body without a soul."

The sun was bright and hot, the river was alive with vessels—oil tanks, the longest and broadest I had ever seen with only the rims of their decks unsubmerged; freight barges drawn by wobbly, whistling little steamboats; rafts of timber floating along on the current, with tentlike hovels for the attendants and their families and with children waving their hands at us; and, of course, passenger steamers. As we passed these steamers I noted that they bore the names of revolutionary heroes, dead and living, and also of celebrated Russian authors—Lermontov, Turgenev, Uspensky.

My boat, one of the newest and largest, was named *Josef Stalin*. The cabins were small, well lit, with running water and ample bedding. They were as neat as the sleeping compartment on the train on which I had come from Baku to Stalingrad. The sight of pillows and sheets contradicted the stories I had heard in gossip-ridden Cairo that sheets and pillowcases in Russia, especially on trains and boats, had been seized for bandages.

On the lower deck, as in prewar times, two gigantic brass kettles were seething with hot water—for tea, for boiling eggs, and for the five Americans and Englishmen aboard, none of whom was accustomed to shaving with cold water. The public kitchen was crowded as always with men and women, chiefly civilians, each with a pot in which to cook soup or porridge—Russia's favorite hot dishes. Few had meat, but quite a number had fish, mostly salted, with which to season the soup. Yet as in Baku the people seemed changed—they were magnificently orderly, not at all like the hot-tempered Russians I had known on previous visits. The war, like some chemical substance, seemed to have allayed their explosiveness. The only person grumbling was the kitchen attendant—a stalwart old woman sawing wood for the kitchen stove. The saw, she wailed, was too dull, it wouldn't cut the knotty chunks, and none of the mechanics on the ship would grind it for her—the rascals!

"Grind it yourself," someone remarked.

She glared at the speaker and lashed out hotly:

"If a fish had wings it would fly, wouldn't it?"

"It wouldn't be a fish, Grandma," a boy replied, and there was a burst of laughter.

"Grind it yourself!" growled the old woman. "If only I knew how!"

The boat was crowded with passengers, not as in peacetime, with peasants on the lower decks and vacationists on the upper ones, but mainly with soldiers. There were hundreds of them, officers and men. All had seen action at the front, all were fresh from hospitals and on their way to resorts for further recuperation or on furloughs home.

They were young, sturdy, cheerful. Only a few bore visible marks of injury. One man had lost an arm, another hobbled on crutches. The others, save now and then for a bandaged hand or head, looked normal. They acted normal too. Stripped to the waist, some of them were sitting or lying down for a sun bath. Others were strolling around the deck surveying with special curiosity and friendliness the few foreigners on board. Still others were playing chess, Russia's favorite game, or dominoes. I saw nobody playing cards. Some lying prone on the deck or leaning against the wall were reading books, and the two most popular titles were Tolstoy's *War and Peace* and Steinbeck's *Grapes of Wrath*. On the lower deck at starboard officers and men were busy with all manner of chores, shaving each other, not only the face but the heads; washing linen, shining shoes, repairing clothes. Teamwork was common and was liked, and in the performance of chores differences between officers and men existed only in the insignia on their uniforms. Some of the men had guitars and strummed them; one had an accordion, and lightly—as if for his own amusement or in a wish not to disturb others—he played a lilting tune and hummed it quietly.

Many of them, wrapped snugly in their greatcoats, were fast asleep.

Many more were eating, and their fare was superior to anything I had seen below decks in the public kitchen where civilians did their own cooking. They had plenty of bread, white and black, also eggs, butter, sausage, and, above all, dry and salted fish. And how they ate! Merely watching them made one hungry. Whatever else the war may have done to the Russian soldiers it has not curbed their lusty Russian appetites. The severe climate, the heavy work, the biological sturdiness of the people demand no small amount of food. These soldiers had it—the simple, satisfying foods that have always made up the chief items in the diet of common folk in Russia.

Some of the soldiers were ambitious enough to cook their own soup and their own porridge.

There was a dining room on the boat. It was a dismal imitation of the peacetime dining rooms. No butter, no cheese, no fruits, fresh or preserved, no sweets except an occasional tasteless compote. The soup was thin and the meat, in small portions that were served in lieu of salted fish, was lean and coarse and not easy to chew. Obviously it came from the oldest cattle, for only such were being slaughtered. Young stock was being bred to replenish devastated herds and to build reserves for ruined collectives and the equally ruined peasantry.

When the news spread that Americans were among them the soldiers crowded around us. The two young American diplomats knew little Russian, but the soldiers talked to them anyway. Insatiable was their curiosity about American people, army, everyday life. Enormous was their interest in President Roosevelt, Mrs. Roosevelt, their family, the President's ideas, his character. More than one man asked if the President was planning to visit Russia. General MacArthur was another subject of endless questions. Did the American public share his appreciation of the fighting powers of the Red Army as expressed in his message on the occasion of Red Army Day in Russia? They were pleased to hear that the English-speaking peoples held the Red Army in high esteem and wished it no end of success in the fight against Nazis and Germans.

They were pleased but not flattered. This was true of Russians everywhere. I observed no self-adulation, no sign or sound of braggadocio, no chip-on-the-shoulder boastfulness. With the enemy deep in Russian territory, pressing them hard in the new offensive, they were in no mood for self-laudation. Instead they teemed with curiosity about the second front. When would it come? Why the delay? If one third or even one fourth of the German soldiers were withdrawn from Russia to fight elsewhere Hitler and his "master race" would soon be on the run—back home—to Berlin. It would be a Napoleonic retreat all over again; as fraught with hate and death for the fleeing Fritzes—for even children would be aiming rifles at them and showering them with bullets. . . . Why, then, was there no second front?

There was no acrimony, no suggestion of bitterness, in the questions and comments—not then. There was only disappointment and a frank disposition not to accept the explanation that military difficulties and no other considerations motivated the action of the Allies. But they were spirited youths and were pleased to see foreigners among them, allies, who had come to live in their country. Not one evinced the least sign of personal suspicion or personal distrust. One after the other said he regretted that the two young American diplomats knew so little Russian, for they were so likable—so tall, so friendly, so clean-cut. Were all Americans like that? Yes, they must learn Russian and they might as well begin on the boat; so a crowd gathered about Haupt, a native of Oneonta, New York, and another around McGargar, from somewhere in California. Amid fun and laughter and with the help of a Russian-English study book, they proceeded to enlighten the diplomats on the vagaries and the intimacies of the Russian language. They were as gay and hearty as American college students out on a frolic.

Yet in all these conversations I observed a striking change in the psychology of Russian youths. Not once was any of us asked what was our social origin, whether we came from families of factory workers, kulaks, intelligentsia, or the bourgeoisie. On former trips to Russia this was one of the first questions Russians, especially young people, asked foreigners. They were so replete with class-consciousness they could not keep it to themselves. This time, when I told them I was writing for the New York *Herald Tribune,* they did not even ask what was the paper's "political line"—as though the subject no longer mattered. It certainly was not in the forefront of their consciousness.

What had happened in the six years that I had been away? Clearly a new crystallization of ideas and a resulting new mentality. There were no kulaks any more in the country, no "bourzhuis" of any kind—neither bankers, nor manufacturers, nor shopkeepers. The campaigns to liquidate "the enemy classes" were over and the emotions and the vocabulary that the process of liquidation had stirred and kept afire were now as dead as the campaigns. Not that they could not be resurrected: they could, but only in the face of the rise of the former classes and all that they symbolized—which, under Sovietism, is unthinkable. It was not only significant, it was almost startling to talk to young Russian soldiers and officers, party men and Komsomols, and not hear the vocabulary of class-consciousness and class-emotion. Also they addressed us not as *gospodin* —mister—but as *tovarishtsh*—which, I must emphasize, is not an invention of Bolsheviks but an old and popular folk word.

One afternoon we talked of war prisoners. These soldiers had fought against Germans and Italians and had captured or seen others capture both.

"Crybabies, that's what the Italians are," said one army man.

"Yes, yes," came an assenting chorus.

"Last winter," said a freckle-faced lieutenant, "we captured twenty-three of them, three officers and twenty privates. The officers were sullen and silent. But the privates—I never saw anything like it. They bawled and bawled. Wrapped in pillows and kerchiefs which they had stolen from peasants, they looked like scarecrows. I asked them what they were doing in Russia, and they said they never wanted to fight. 'Why, then, have you come here?' I said. In answer they bawled and bawled. When we gave them food they bawled harder than ever—the privates, not the officers."

"Yes, Italians are pathetic," said another officer. "They cannot stand the cold and the kind of fighting they have to do here—and so they bawl like whipped children."

"But Germans are different," said a dusky-faced sergeant.

"Yes indeed," rose several voices. Now came a deluge of comments one after another. "Germans never bawl, they are too tricky and too cruel for tears."

"When they are captured they show their calloused hands to indicate they are 'workers'—like ourselves."

"They always say the word *rabochi*—worker—over and over."

"They show pictures of their wives and children and smile to show how they love their families."

"They even tell us they are communists and when we ask them why they fight us they say they have to—that's the kind of communists they are."

"Tricky devils!"

"I captured three of them near Kharkov," said a junior lieutenant so young that his face seemed as though it had never yet been touched by a razor, "and every one of them was all the time putting up his hands and pointing to his fingers to indicate the number of children he had. One of them kept lowering and lifting his hands to indicate the sizes of his children. So I said to them, 'Why don't you think of our children? Why do you Germans strip them of warm clothes and take food away from them?' And they said only officers were doing it. In the knapsack of one I found stolen kerchiefs and several baby caps, so I said, 'Are you an officer?' and the others instantly spoke up and said, 'No, he is a thief.'"

"What did you do with him?" I asked.

"What do you suppose?"

Then a dreamy, Mongol-faced sergeant spoke. He had been listening in silence with scarcely an expression on his face. He was a Kazak and spoke Russian with a harsh accent.

"We reoccupied a village in the Ukraine, and on the public square we found an old man and an old woman, and beside them a child, all shot and frozen. We shook with wrath, and as we were standing there a little girl ran up and said, 'Uncle, over there'—pointing to a house—'are three Germans!' Quickly we rounded up the Germans. Local guerrillas seized nineteen more. We brought them all together, these twenty-two Fritzes, and one after another showed his calloused hands and repeated the word rabochi. So infuriated was one of our men that he shouted, 'Vermin! That's what you are.' But they kept showing their calloused hands and repeating the word rabochi. I packed them quickly into a truck and sent them off to staff headquarters—I couldn't have stopped our soldiers from avenging the death of the old couple and the child. As they were leaving one Russian cried out, 'We should at least have shown them the corpses.' For an instant I was inclined to agree with the suggestion and almost called to the truck to come back. But I knew that I myself might want to pierce the Fritzes with my bayonet if we led them to the dead bodies. So I only motioned with my hand and, choking down my wrath, I cried out 'Attention.' The company fell into line and we marched off. But all the time, as I kept looking at the gleaming bayonets of the men, I wished I hadn't acted so properly."

For three days and three nights we sailed up the river on our way to Kuibyshev. Daytime the weather was superb and the soldiers were always on deck eager to talk, to make friends, to ask questions about America, England, ourselves. One officer, a general's adjutant, grew so attached to us that he came to our rooms, talked at length of his experiences, and was always offering to do things for us. When he saw a fly swatter on my desk he asked:

"What is this?" I picked it up and hit a fly.

"Interesting, interesting," he said; "you kill flies and I kill Fritzes." After a long pause he added, "I wish I could kill only flies."

When we were approaching Saratov a sergeant came up and asked for a cigarette, saying he would soon leave the boat. I had bought a supply in Teheran, so I gave him a box. He lit one and kept swallowing one mouthful of smoke after another. I had often talked to him and knew his life story. He had been in several battles and told many exciting tales of the fighting. He was thirty-one years old with short cropped hair, a pockmarked face, and large, dark eyes that had a piercing stare. He had lived on a collective farm with his wife and three children. He was Jewish but his wife was Ukrainian. When the Germans came to their village she had already fled with the children but with hardly enough clothes on them to look respectable. They were now living in a village in the former German Volga Republic, where a lot of Ukrainians had been sent. He had not seen them in a year and was on a furlough to pay them a visit.

"I suppose," I said, "you are glad you get off at Saratov."

"Yes," he said, but not with enthusiasm. The night was dark, and though we were leaning over the railing we could scarcely see the gleam of the water.

"It looks like a storm," I said.

He made no answer, as though he did not hear my words. But I saw him light another cigarette and puff away with zest.

"Do you know," he said, "I am a living corpse."

I laughed, but he quickly interrupted me.

"It's true. I'm not supposed to be alive. I was given up for dead not once but three times. A soldier friend wrote my wife that I was no longer alive."

"Doesn't she know you are coming home?"

"No."

"Why didn't you write to her or send her a telegram?"

"She is very ill—has a bad heart. Escaping from Germans with three children nearly killed her. That much I know from the letter she wrote me when she got to the Volga Republic."

"But," I said, "coming home so suddenly and in the night—a dark night like this—without warning her——"

"That's why I am worried. The shock may prove too much for her."

He was silent. At starboard someone was playing an accordion and soldiers were singing a lively melody, their voices rising high over the river.

"You know what I'll do?" he said.

"What?"

"I'll go to the chairman of the village Soviet and stay with him overnight and in the morning I'll send him to see my wife and prepare her and the children for my return. Don't you think that's a good way of doing it?"

"Excellent," I said.

The singing and the playing died down. Most of the passengers went to sleep—stretched out on deck in rows and wrapped in their greatcoats.

"Weird, isn't it," he said, "that your own wife and children should think you are dead when you are alive and on your way to see them?"

"Yes, quite strange," I said.

"Well, you are an American—you are lucky—you don't know what war really is."

CHAPTER 9: **REDISCOVERY OF THE PAST**

"A YEAR HAS PASSED and friends meet again. Outwardly nothing has changed in the familiar schoolhouse and in the familiar schoolyard. As a year ago, the clock on the wall is running behind, and on the familiar map, in the same places, remain the ink spots.

"Yes, all is as it was, excepting the friends themselves. They have changed. The year of war has made them more mature, more earnest. Their hands so used to pencils, to skate keys, to books, have learned to hold a sickle, to dig potatoes, to run a combine.

"The men in the family have gone to war and heavy cares have fallen on the shoulders of children. Yet, while helping grownups on the kolkhoz, while gathering medicinal herbs and plants, they often thought of the schoolhouse as of something close to their hearts and indispensable to life.

"Now its doors have opened. The pupils take their seats. They await the beginning of the first lesson. . . . The bell rings—it is so long since they have heard the heartening ring—and Elizaveta Alexandrovna Archangelskaya, the teacher of Russian language and literature, enters the classroom. The pupils arise, then sit down. As always, Elizaveta Alexandrovna has several books under her arms. Sitting down at her desk, she opens the heaviest volume.

" 'Today,' she says, 'we shall have a literary reading.' Bending her gray head over the page, she starts to read: ' "The Lord help you!" said Bagration in a resolute, sonorous voice. For an instant he turned to the front line and slightly swinging his arms and with the heavy and clumsy gait of a man accustomed to horseback, he walked ahead over the uneven ground. Prince Andrey felt as if some irresistible power was drawing him forward, and a great surge of happiness came over him.'

"The slanting rays of the setting sun are sliding over the faces of the pupils. Silence reigns in the room. When the teacher interrupts the reading and looks at the children, she beholds in their eyes a deep and stirring attention. Once more she bends over the book and a little hastily and in an agitated voice she reads the concluding lines of the chapter.

" 'Hurrah, hura-a-a-a—ah!' rang out along our line in a prolonged roar and getting ahead of Bagration and one another, not in order any more, but in an eager and joyous crowd, our men ran downhill and routed the French.'

"The pages smelling of blood and powder and victory close. Out of the past rise the figures of the noble heroes of the first people's war—the war of 1812. . . . My native land. . . . And louder beat the hearts of the children. . . ."

This story appeared in one of the Moscow dailies on October 9, 1942. To me it was a significant story, and I could not help reflecting on its meaning and implications. Here was this gray-haired teacher, obviously a woman of patriotic fervor and high literary appreciation. On the first day she met her class after a year's enforced absence she read to the pupils a stirring passage from Tolstoy's *War and Peace*.

Dramatically, nothing could be more fitting. The mood of the pupils was lifted. The teacher's own spirits, if only because of the breathless attention the reading had elicited, were raised. The emotional bond between teacher and pupil, between classroom and nation, between past and present, was enhanced, at least accentuated. But——

Some years ago, in the days before the First Five-Year Plan, the Party Secretary in the Moscow Commissariat of Education was a man named Glebov. Once, on meeting a friend who was not a member of the Party but who had been a literary editor of important publishing houses under the Czar and under the Soviets, he asked what he really thought of Tolstoy. Said the editor: "As a philosopher he doesn't impress me. As a creative writer and novelist he has not his equal in the world's literature." Glebov motioned with his hand to indicate complete disagreement with such an opinion. Judging literature and life as the overwhelming mass of Bolsheviks did in those days—though not all the leaders, certainly not Lenin—exclusively in terms of class struggle and the social origin of authors, Glebov beheld in Tolstoy no more than a *pomeshthchik,* a landlord, who in his living and in his writing had retained the interests and the soul of a landlord and therefore—so Glebov proclaimed loudly and positively—could not possibly have done any distinguished writing!

It is over a decade since the Soviets abandoned this approach to the art of the prerevolutionary days. A Glebov would now be regarded as a monstrous ignoramus, more as "an enemy of the people." Tolstoy is now one of the great immortals of Russia—a great writer, a great patriot, a great Russian. Also, the most hardheaded Bolsheviks no longer want to be reminded of the former attitude and the former condemnations.

"Don't forget," said one of them, "the Soviets are now twenty-five years old. We have come of age. So why bring up what we did when we were quite young?"

Yet for people in the outside world it is well to observe the radical departures, which the war has only reinforced, from original beliefs and original practices. It is all the more significant in this instance because the passage read is concerned with a soldier who was a prince, and who lived a long, long time ago.

Born in 1765, Prince Bagration was a descendant of a Georgian but Russified family of noblemen. He was a pupil of Alexander Suvorov, the most distinguished soldier Russia had had until the outbreak of the present war. At an early age Bagration distinguished himself as a brilliant tactician and fearless commander. He led Russian armies in Italy and Switzerland against Napoleon. He participated in one hundred and fifty skirmishes and battles.

"Napoleon," he prophesied, "will find in Russia another Egypt and—his doom."

Though an aristocrat he, like his chief Suvorov, stood close to the common soldier. In his addresses to his men he said,

"Russia is our mother. With your breast you must bar the path of the enemy."

On August 26, 1812, during the battle of Borodino, he was killed. He was only forty-seven years old at the time. He loved Russia, he fought for Russia, he died for little Mother Russia.

Tolstoy has left an immortal picture of the man—the Russian, the commander, the patriot—and so Bagration, not a peasant, not a proletarian, but a prince, with no hint of revolutionary proclivity, with nowhere a suggestion of doubt in the rightness of Czarism and serfdom, with not a trace of protest against the immeasurable privileges of his own class and the immeasurable lawlessness with which masters again and again treated peasants, is now a hero and an idol. Never before on my visits to Russia had I heard his name mentioned in conversation or seen his picture in any public place. Now I often read his name in the press, saw his pictures in public places. I heard eulogies of him from the platform and in private conversations, especially from school teachers. In *Propagandist* (issue of September 1942), an official journal of the Russian Communist party and guide to orators and lecturers all over the country, at the front and in the rear, I read a long account of his life and his career which was also an eloquent tribute to his person and his patriotism.

The astounding fact is that never before, since the coming of the Soviets, has Russia been so indefatigably and enthusiastically rediscovering herself in her past, her Russian past, reinterpreting it, reinvesting it with fresh meaning and fresh glory. In the early days of the Revolution this past was deemed scarcely more than an excrescence of the ages, to be emptied out of Russian life and wiped out of the Russian mind. How often in those days did I hear young Bolsheviks exclaim, "October[1] is the beginning of history!" All that preceded it was wicked and futile. . . . There was no

[1]October 1917, the month the Bolsheviks came into power.

good in it at all; its ideas, its institutions, its morals, its manners, its art, its literature—all of it was fit only for the scrap heap or the faggot pile. How like the exclamation of the French revolutionary Comte de Segour, who in 1789, the year of the French Revolution, exclaimed: "This is the year one in history!"

Quickly enough the French discovered the error of their repudiation, and so have the Russians. Now the past, the Russian past, is sacred in Russia. True there were villains in the past. There were loafers. There were men who echoed the words of Oblomov: "I am a gentleman and can do absolutely nothing." To which the present-day Russian would reply with the words of Oblomov's servant: "Why, then, were you born?"

True, there were men and women in the past who, like the characters in Chekhov's *Three Sisters,* felt themselves spiritually strangled because they were only idlers, and about all they did to ease or lift the strangulation was to talk of work without ever doing any of it, or doing very little.

There were tyranny and wickedness, futility and desolation, in the past. But there were also deeds and achievements, triumphs and inspirations. There were villains, but also heroes, and always there were the people. With their labors, their longings, their dreams, their valor, their battles, their sacrifices, their blood, the people had fertilized the Russian land and affirmed the Russian spirit, leaving an inheritance as precious as the breath of life. Never in all their history have Russians been so conscious of their past and of themselves as reflected in it—particularly in the heroic battles for national survival and in the dreams of a happy and magnificent tomorrow.

In September 1836, in a magazine called *The Telescope,* a young Russian nobleman named Pyotr Chadayev published, under the title of "A Philosophical Letter," a scathing indictment of Russia and Russian civilization. Among other things he said:

"We do not belong to any of the great families of humanity. We belong neither to the West nor to the East, and we have the traditions neither of the one nor of the other. . . . Alone in the world we have given it nothing, taught it nothing."

The magazine was at once suppressed. The editor, whose name was Nadezhdin, was exiled, the author was officially proclaimed insane. But the letter stirred a storm of discussion. Alexander Herzen, a leading author and populist, spoke of it as a "shot into the dark night. . . ."

This letter is included in the literature textbooks of today for pupils of high-school age. When I asked a Soviet school director in a village why this was done she said:

"It is too important a document to leave out and it isn't true, though in those times, because of serfdom and the repressions it had heaped on the people, it might have seemed the only truth about Russia. Here——"

As if in substantiation of what she meant she pointed to a magnificent war poster on the wall with a portrait of Kutuzov, who defeated Napoleon in 1812, and underneath in flaming red letters a quotation from Stalin which read: "Let the valorous example of our great ancestors inspire you in this war."

In his speech of November 7, 1941, Stalin mentions these ancestors by name. They are: Alexander Nevsky, Dmitry Donskoy, Kuzma Minin, Dmitry Pozharsky, Alexander Suvorov, and Mikhail Kutuzov. Not one is a worker. Not one is a peasant. Most of them are princes and only one, Kuzma Minin, is a tradesman.

These men led Russian armies to victory and, in critical moments of Russian history, saved Russia from foreign conquest and foreign subjugation. All lived and died before the time of Pyotr Chadayev. Therefore, to the Russians of today, whatever the wrongs and outrages that rankled in the heart of Chadayev, his violent denunciation of Russia's past as barren of achievement is untrue and unjust. Great beyond praise are the ancestors who shattered the attempts of Tatars, Germans, Swedes, Poles, and Frenchmen to push Russia off the map of Europe, to crowd her out of the pages of history.

"Alexander Nevsky," I heard a British diplomat in Kuibyshev say, "is a greater hero in Russia now than Karl Marx." Not that the Russians have repudiated Marx and Engels. They have not. Let there be no misunderstanding on the subject. But in this moment of a life-and-death struggle for national survival with the most formidable and the most brutal enemy Russia has ever known, Alexander Nevsky rouses more natural and more fervent emotions than any foreign ideologist, even though his name is Karl Marx and his teachings of economics, by the admission of the leaders, are the foundation of Russia's new society.

Alexander Yaroslavitch, as Alexander Nevsky was originally known, was a prince, a Russian, a leader and ruler of the Russian people. Seven centuries ago, when the Teutonic knights invaded Russia, he met them in battle, routed them, saved Russia from subjugation. So there is a bond of blood, old and precious, between him and the Russians of today.

In 1941 a Russian writer named Borodin published a novel under the title *Dmitry Donskoy*. *Pravda* gave it one of the longest reviews I have ever seen in that vigorously political, severely authoritative journal. This review bore the significant title "A Book on the Great Ancestor of the Russian People."

The theme of the novel is the wars of liberation which Dmitry Donskoy led against the Tatars in 1378 and again in 1380. Both times the Russians defeated the Tatars and thereby broke the "Tatar yoke," which had lain heavily on the Russian people for over two centuries. It is a patriotic novel brimming over with lyrical appreciations of the heroism of Grand Duke Dmitry and the Russian people of his time.

What gives this novel special value and importance is the high tribute it pays the Russian Church and Russian churchmen for the brilliant and unfaltering part they played in that dark moment of Russia's history. Two eminent church leaders are singled out for special eulogy—the Metropolitan Alexey and Sergey Radonezhsky, or Saint Sergey, the greatest saint of all times in the Russian Church.

The Metropolitan Alexey was Dmitry's preceptor. From the boy's earliest years he sought to inculcate in him a spirit of resoluteness and militancy so as to prepare him mentally and emotionally for the leadership that he was later to assume. "Rid the Russian soil," Alexey preached, "of the pagan yoke, Dmitry, and if thou wilt snap but one band of this yoke, thou shalt be blessed. And leave it as thine behest to thy successor to do away with the cursed yoke utterly. A free people is always strong, but an oppressed people grows weaker from day to day."

Monks everywhere were waking the people to their duty, their mission, their hour of destiny. In their cloisters they were gathering war supplies, mustering volunteers. They were energizing the people with the will to fight and win and be free of the alien invader.

Most noteworthy was the help that Sergey gave the Moscow grand duke. At *great* length, with fidelity to historical fact and with brilliance of style, the author depicts the rise and growth of the Troitsko-Sergeyevsky abbey which Sergey had founded deep in the forest and which had stored much wealth for the war. Far and wide over the land echoed Sergey's eloquent summons to unity and to arms. Prince and peasant heeded this call and for once Russians faced the enemy as a united people.

In 1380, when the hour of combat was drawing near, Sergey in a last word of prayer and inspiration said to the grand duke:

"The enemy will be desperate, because the horde, if it comes back defeated, is doomed. For them this is the decisive battle and so it is for us. The earth will flow with blood, and if the enemy wins our life is at an end. Neither towns nor monasteries will remain. Falter not before losses or bloodshed. . . . Bear up, my son, my lord Dmitry Ivanovitch, take heart."

"Yield, I shall not," answered Dmitry stanchly. "I myself, Father Sergey, see there is no place for yielding."

The veneration of the deeds of the Church and churchmen in the past is as revealing a feature of this vivid and patriotic novel as is the glorification of the hero after whom it is named.

"How old is our fatherland?" asks a Russian scholar named Vladimir

Kholodowsky. "As old," he answers, "as the Scythian mounds in the steppes. A thousand years will not measure the road it has trodden, will not encompass its age-old glory."

Eloquently and dramatically the author continues his account.

"Memory turns the pages of the centuries. Listen—and the silence of the times that have gone will bring to your ears the muffled clang of the *veche* [town meeting] bell; the whine of the Pecheneg[2] arrows; the ballads of the blind dulcimer player.

"Look—and in the mist you will behold the march of Svyatoslav's[3] legions . . . the magic city [Kiev] on the Dnieper—capital of the first princes. Truly it was said of them, 'they were lords of a land neither unknown nor unvisited—lords of the Russian land which is known and which has been heard of in all corners of the earth.'"

Nor is Kholodowsky alone in his rhapsodic appreciation of Russia's distant and turbulent past. Books without end have been written—fiction, biography, history—recalling and extolling it as a time of purpose and achievement.

How often have I read in Russian publications and heard Russian speakers tell of their historic victories over the Germans. Impressive and monumental is the record and the meaning of these victories. They begin early in Russian history, for since earliest times one German tribe after another has coveted the Russian land with its forests, its fields, its rivers, its untold wealth. In 1214 and again in 1217 Teutonic Knights from ancient Livonia stormed into Russia and were driven out.

In 1224, when Russia was prostrate under the yoke of the Tatars, the Livonian knights again sought Russian conquests. They seized the city of Yuryev but finally were ousted. In 1234 they captured Pskov. They robbed and pillaged the city. Russia was still under the Tatar yoke, but Alexander Nevsky appeared on the scene. In 1242 on the ice of Chudskoye Lake he gave battle to the "canine knights," drowned them in the lake, slaughtered them on the ice. To the survivors he said,

"Go and tell all in alien lands that Russia lives. Without fear anyone who wishes may come here as a guest. But if anyone ventures here with the sword he shall perish by the sword. This is the basis on which Russia is founded and on which she shall forever stand." This speech is one of the most widely quoted *battle cries* of the Russia of today.

But the conflict between Russia and Germany did not abate. In 1269 the Germans again embarked on Russian conquest and suffered defeat. Peace followed until 1501–02. Then came a cruel war in which, according to an ancient Russian chronicler, "the Russians fought back the vile Germans for

[2]Turkish tribe often at war with Russia in old times.
[3]One of the eminent rulers of ancient Russia (942–972).

ten versts. . . . Muscovites didn't mow them down with shining sabers, but slaughtered them like swine with picks."

A fresh conflict brews in the days of Ivan the Fourth. But the climax of Russian triumphs over the Germans, as modern Russian writers and historians interpret it, was the Seven Years War, when Frederick the Great ruled over the destinies of the German armies and the German people. Like Hitler, Frederick boasted, after early engagements with the Russians, of complete victory. Yet it was at their hands he suffered his most ignominious defeats. On August 12, 1759, Alexander Suvorov, then only twenty-nine years old, helped shape the strategy of the Russian Army at Kunersdorf, Germany. After a ferocious battle, which, according to Russian sources, lasted fifteen hours, the Russians overpowered the Germans. A little over a year later, on October 8, 1760, headed by Suvorov's garrison, the Russians occupied Berlin. After a stay of three days and the collection of an indemnity of a million and a half talers, they abandoned the German capital.

During the Napoleonic invasion the Germans participated with a corps of 20,000—of whom only a few survived the holocaust.

In the first World War the Germans defeated Russia, but the Russian armies inflicted the heaviest casualties that the Germans suffered in the fighting and enabled the Allies, even after Russia had made a separate peace with Germany (of which, incidentally, mention is not made in the present-day accounts of Russo-German wars), to bring the Germans to their knees. . . .

On the one subject which means most to the Russia and the Russians of today—the war with Germany—history and the past unfold a record of victories that sustains courage and reinforces faith in eventual triumph. In hours of greatest peril and greatest trial, when the Germans stood before Stalingrad, I heard Russians say, "We've smashed the Germans before, *we've always smashed them,* we shall smash them again."

Particularly impressive and illuminating is the new attitude toward Ivan the Fourth, also known as Ivan the Terrible, of whom in prerevolutionary Russia many a gore-drenched book was written. Here is what Russian children learn about this czar as he is described in a new textbook on Russian history:

"He was a talented and wise man. For his time he was well educated. He loved to write. . . . He had a sharp, subtle mind. In the internal and external life of Russia he skillfully and correctly formulated his problems and purposes and pursued them with consistency and perseverance. His notion of the necessity for Russia to possess a foothold on the Baltic is a tribute to his farsightedness." In the new appraisal of his place in Russian history Ivan's errors and excesses are not ignored. His flaws are not overlooked. But his virtues are rescued, emphasized, eulogized. Alexey Tolstoy, one of Russia's leading novelists, has written a play about him. On finishing it he gave an interview to the Russian press in which he said:

"Ivan the Terrible is one of the most remarkable figures in Russian history . . . he represents the Russian in all his grandiose ambitions, his fervid will, his inexhaustible capacities, his powers, and his shortcomings."

Greater tribute could hardly be bestowed on a man whom the world had come to regard as the very acme of despotism and cruelty.

To present-day Russia Ivan is a statesman of extraordinary caliber and acumen. He consolidated the Russian lands and their administration. He drove the Tatars out of Kazan and out of Astrakhan and made the Volga a Russian river. He annexed the newly conquered lands in eastern Siberia. The boyars, who opposed his plans and sought to subordinate the good of the state to their individual powers and advancement, he ruthlessly cast aside or annihilated. He wrote brilliant prose, and his state papers are considered literary and political documents of priceless value. Above all, like Peter, who came later, he symbolized the idea of a united nation advancing amid ignorance and struggle, pain and sacrifice, toward a heroic destiny.

Sergey Eisenstein, one of Russia's leading motion-picture producers, has been at work for more than a year on a film portraying the life, the times, the achievements of this formidable Russian czar. . . .

But it is not the military triumphs and the victorious military leaders alone in Russia's past that rouse the attention and the eulogy of the Russia of today. Literature, science, art, folklore, above all the growth of the Russian nation, the heroism of the people, excite their emotions. Not a personage of Russian history, whatever his social origin and whatever his official position, if his contribution to any phase of national growth is of significance, but is now lifted to a pedestal of national appreciation and national pride. At the time of the celebration of the four hundred and fiftieth anniversary of Columbus' discovery of America, a Russian journal printed a long article under the significant heading of "Russian Columbuses." It was the story of Russian explorers and what they had achieved throughout the centuries. New and enlightening were the names and the facts that the story revealed.

The Soviets have forged a new economic and social order in complete repudiation of anything Russia or any other land has known. But this order is now linked with the past, its events and personages, its dreams and deeds, its sorrows and triumphs. Now the past is no longer a nuisance or a nightmare. No one would now say that "October is the beginning of history." Now even children know that October was a sharp break with the past only for the purpose of ultimately achieving with it a more powerful bond than Russia has ever known and a more sanctified appreciation of the good it has yielded.

CHAPTER 10: **RUSSIA GOES RUSSIAN**

IN GROZNY, the renowned oil town in the North Caucasus, I bought two pamphlets at the railroad station. One was by V. Panov, of whom I had never heard, and bears the title *Peter the First*. The other was by V. Kruzhkov, of whom I likewise had never heard but of whom I was subsequently to hear much of as one of the rising young men in Soviet life and Soviet politics. Kruzhkov's pamphlet was on Pisarev, the noted literary critic of the nineteenth century.

On first thought no two persons could be so unlike and so far apart as the mighty Russian czar and the iconoclastic critic; yet as I read these pamphlets I was amazed at a similarity in the spirit of the writing. On the first two pages of the one on Peter I read the following expressions: "In the ranks of the *glorious Russian* patriots . . . symbolizing the will of the *great Russian* people . . . the *mighty Russian* people . . . *Russia* under the guidance of Peter . . . Peter created an army worthy of the *great Russian* people . . . the kernel of the newly born *Russian* Army. . . ." All the way through in this eloquent eulogy of Peter the emphasis was on the words *Russia* and *Russian*.

Kruzhkov's pamphlet concerns itself not with politics but with literature, and here, too, the emphasis was on the words "Russia" and "Russian." In the introduction I read the following significant excerpts: "Belinsky, Herzen, Chernyshevsky, Dobrolubov—outstanding *Russian* revolutionaries, publicists and literary critics. Pisarev rendered the *Russian* people an inestimable service. . . . He is dear to us because while hating *Russian* czarism and serfdom he passionately loved the *Russian* people and *Russia*—his fatherland. Whatever he wrote his thoughts were always aimed at the *Russian* people, were always replete with the wish to see *Russia* a *free, mighty,* and *cultured* people." The introduction ends with the words: "Over one hundred years have elapsed since the day of Pisarev's birth, and seventy-four since the day of his death. This is a long period, but the name of Pisarev belongs to the number of those who are never forgotten."

After my absence of six years the phrases and sentences which I have quoted and many others of a similar nature in both pamphlets carried to me much meaning and purpose. They indicated that Russia was not only rediscovering and glorifying its past but was consciously and dramatically popularizing it. It was furthering and fostering purely nationalist throughts and nationalist sentiments.

Let it be noted that in the early years of the Revolution Pisarev, like Peter the First, was not a name of any consequence. Both belonged to the past, so it was then assumed, the past that existed no more, that would never return, that would be remembered, if at all, as an agony and an execration. One of the reasons for the banning and the excoriation of the late Mikhail Pokrovsky, once accepted by the Bolsheviks as the leading Russian Marxist historian, was, in the words of the Soviet encyclopedia, because he disregarded "the progressive significance of the deeds of Peter the First." According to the same authority the works of Pisarev appeared for the first time under the Soviets in 1934–35. At first rejected, then rediscovered, finally embraced, such has been the lot of Peter and Pisarev.

This process was already marked when I was last in Russia. But the height that it has since attained, as exemplified by these pamphlets and by all the current literature, was startling. It was quite evident that whatever Stalin and the Communist party in Russia might be thinking of internationalism, of "proletarians of the world unite," of the brotherhood of workers or peoples the world over, they were indefatigably emphasizing Russia and everything Russian.

Not that they were excluding interest in foreign countries, foreign civilizations, foreign history.

Not that they were seeking to Russify any of the peoples that make up the Russian nation. During my stay in Moscow I met Samed Vurgun, the leading poet of Azerbaijan. He talked of his native land—its language, its music, its customs, its past—with the zest of a Californian depicting the glories of his state. . . . At the front I met two soldiers, a Tadzhik and a Kazak. Both expressed the hope that foreign correspondents would visit them in their homes and become acquainted with their peoples, their lands, their cultures, and make the outside world aware of them as peoples with a language, a civilization, a destiny of their own.

There are in Russia about one hundred different nationalities—more including the many little-known ethnographic groups that are scattered over the more remote territories, especially in the north of Russian Asia and Russian Europe. Many of them never had a written language. They have one now. The Soviets gave it to them. The language, the history, the purely national aspects of the life of any and all of these nationalities are not only respected but are vigorously encouraged by Moscow. The Soviets want them to preserve their identity, insist on it.

All Russia has for a number of years been surging with revivals of folklore. When I was in Teheran, capital of Iran, on my way to Moscow, I attended a concert by a visiting group of singers and musicians from Soviet

Uzbekistan. They wore native costumes, spoke in their native tongue, played on ancient native instruments, danced ancient native dances, and unfolded before the audience a picture of a purely native and national civilization.

In their economics and their politics all nationalities are bound to Moscow. Any deviation is ruthlessly suppressed. But in their purely national existence they follow their own national pattern. This policy of respecting and encouraging the languages, the folklore, the purely national or group civilizations of the multitude of peoples in Russia has been a source of immeasurable strength in the present war.

One of the remarkable documents I obtained before leaving Moscow was a list of decorations that soldiers who had distinguished themselves in the fighting had received during the first eighteen months of the war. They were catalogued according to nationality; sixty-nine nationalities were represented on the list!

But according to the census of 1939 the Russians make up 58 per cent of the population. Together with the Ukrainians and White Russians, the three leading Slav peoples, they compose 78 per cent or nearly four fifths of the population. The leading language, the leading nationality, the leading civilization, now as throughout the centuries, have been Russian. The influence of Russia—especially of Russian literature, which though passionately Russian is also passionately humanitarian—has been beyond calculation.

Russian is the first foreign language in the schools of minor nationalities and republics. That is only natural. With a knowledge of Russian the Kazak, the Yakut, the Tadzhik, the Armenian can travel all over Russia and make himself understood. He can attend the universities of Moscow and Leningrad. The mere usefulness of the Russian language, because it is spoken by a great majority of the people all over the vast land, has led to increasing emphasis by Moscow and by local governments on its study. . . .

Yet if only for the Russians alone, the rediscovery of Russia's past, the revaluation of the word Russian, is fraught with more meaning than either the Russians or the outside world is now in a position to estimate. It is one of the most amazing social phenomena of Sovietism. Yet it is as real as it is universal and is a fresh and unexpected testimony of the power of nationality in our times—power to arouse the deepest longings, the greatest self-sacrifice.

"For my beloved Russia," a young sergeant wrote to the girl he loved, "I'd be happy to offer a hundred lives if only I had them." Never before had I known young Russians to be so overwhelmed with a love of Russia.

The Russian language, Russian history, Russian achievement, the Russian people are now subjects of constant laudation and sentimentalization. Here, for example, is *Pravda,* official organ of the Communist party, printing in the issue of September 30, 1942, a rhapsodic editorial on the theme of "The Russian Rifle." After tracing the history of the three-barreled Russian rifle the editorial says:

"Over half a century it has been at the service of the Russian warrior. In the steppes, in the mountains, on the immensities of the seas; in days of burning heat and ferocious frost, it has done its work accurately and obediently. Exalted in song, it has to this day preserved its fighting power. . . . No less gloriously reverberates the praise of our Russian bayonet. . . . Our bayonet attacks and our bayonet battles have always been considered the most terrible and most unconquerable weapon of the Russian armies. In the war with the German invaders the Russian bayonet, as of old, sows death in the ranks of the enemy and makes fascists tremble with fright. . . . The Russian rifle and the Russian bayonet . . . are terrible fighting weapons . . . faithful friends."

So even the Russian rifle and the Russian bayonet are endowed with a might and a glory all their own, and by none other than *Pravda!*

The great names in Russian literature have been invoked in this reappraisal and glorification of Russia. Pushkin, Gogol, Lermontov, Nekrasov, Dostoevski, Tolstoy, Turgenev, Gorki, Chekhov. They all sorrowed and wept, loved and hoped for Russia, and Russians of today must know it, feel it, respond to it with all the power of their souls. "All Russia is a cherry orchard," says the student Trofimov in Chekhov's famous play. When I heard these words at a production in the Moscow Art Theater, they conveyed a meaning not only to Russians but to observing foreigners which in the light of the war and the deep emotions with which the word Russia has been invested, they had never before spelled. . . . Again and again I have read, for example, Turgenev's famous tribute to the Russian language. "In days of doubt, in days of heavy-hearted contemplation of the fatherland, you alone, O you great, mighty, just, and free Russian tongue, have offered me support and solace. It is impossible to believe that such a language was not given to a people that is great."

Writing on the same theme Alexey Tolstoy, whose writings throb and flame with Russian emotion, says:

"Behind grew and multiplied the graves of his [the Russian's] fathers and grandfathers; ahead grew and multiplied his people. With magic power he has spun the invisible net of the Russian tongue—bright as the rainbow after a spring shower, swift as an arrow, as touching as a cradle-song and as rich and melodious."

"Russia," exclaims Nikolay Tikhonov, "is our joy and our liberty, our past and our future. our heart and our soul. Russia was, is, and shall be.

Her life is our life and as our people is immortal, so Russia is immortal."

Never have I known Russians to be so sentimental about everything Russian. The city of Kuibyshev is one of the most backward and desolate in the country, with none of the energy, the enterprise, the vividness of Stalingrad as it was before the war, or of present day Baku, or of Ivanovo or of other cities which the Plans have re-created. But it is a Volga city.

All round are woods, hills, fields, flowers; and as I went walking with Russians along the river front or talked to them of the Volga, their language spilled over with love for this mighty and beautiful Russian river and everything within its environs. Often they resented my comments that Kuibyshev was a wretched city with battered sidewalks, mudholes in the side streets after a rain. Invariably their answer was, "But look at our landscape and our nature." Daisies in the meadows, cornflowers in the ryefields or along the roadsides, birches in the forests, the steppes, the rivers, the thatched huts in the villages, the smoke rolling out of their black chimneys—everything is beautiful, everything is sacred.

I asked the handsome and youthful Olga Mishakova, secretary of the Central Committee of the Komsomol, what kind of human being her organization was seeking to mold. Her answer was all the more important because of the high office she holds.

"First and foremost," she said, "we want our youth to be patriotic, to love their country, their people, all the good things they did in the past and all the good things they are seeking to achieve under the Soviets. Love of country is one of the most basic things in life, and one of the most blessed. *Without it a man has no place in society."*

I have italicized the last sentence to emphasize the extraordinary shift in Russian ideology since 1917 with regard to the subject of patriotism. Of course Russia is not the country it was then. It is a land in which first and foremost collectivized enterprise has shattered the foundations of private enterprise. If Russia is swept by a new nationalism it is well to point out that this nationalism means not only the Russian land, Russian history, Russian geography, the Russian tongue, the Russian people, Russian songs, Russian forests, Russian skylarks, Russian winters—everything, in short, that Russians of old understood by the term Russia and loved about Russia, but also, among other things, the new economic usage with its ban of the institution of income-yielding private property. Yet it is Russia—the source of the new inspiration and the new dedication.

That is why the word "fatherland" has such new meaning. Fatherland! A holy word it is now to any and every Russian. Pouring his heart out in his *Letters to a Comrade* Boris Gorbatov, an eminent Russian writer, says:

"Fatherland! What a powerful word! It embraces twenty-one million kilometers and two hundred million fellow countrymen. But to every man the fatherland begins in the place and house in which he was born. To me and to you it begins in the Donets Basin mines. There are our huts under the common gray weed, mine and yours. There we spent our happy youth, you and I. There the steppes are boundless like a sea, the sky is grim; and all over the earth there are none better than the Donbas boys, no sunset prettier than sunset over the pile engine, and no smell more homey than the bitter-bitter unto sweet odor of coal and smoke.

"There in the Donbas our fatherland began for us, but there are no bounds to it. Gradually the many-tongued regions of our fatherland entered our hearts: Aspheron sand tongues; the black oil derricks of Baku, which you visited with a miner's brigade; the rusty steppe of Magnitogorsk, and the Siberian snow. And though you never were on the North Pole, your heart was there with Papanin. Because there, on the floe, were our people, Russians, Soviet people, our glory!"

Perhaps no Russian has so dramatically depicted the new Russian consciousness as Ilya Erenburg. More than Alexey Tolstoy, than Sholokhov, than Tikhonov, than any other writer has he stirred the emotions of people, especially of soldiers, with his lyrical, prayerlike invocations to Russia. "Together with you," reads one of his impassioned appeals to the soldiers, "marches the frail little girl Tanya (Zoya Kosmodemyanskaya) and the stern marines of Sevastopol. Together with you march your ancestors who welded together this land of Russia—the knights of Prince Igor, the legions of Dmitry. Together with you march the soldiers of 1812 who routed the invincible Napoleon. Together with you march Budenny's troops, Chapayev's volunteers, barefooted, hungry, and all-conquering. Together with you march your children, your mother, your wife. They bless you. You shall win for your mother a tranquil old age; for your wife the joy of your return and happiness for your children. Soldier, together with you marches Russia! She is beside you. Listen to her winged steps. In the moment of battle she will cheer you with a glad word. If you waver, she will uphold you. If you conquer, she will embrace you."

CHAPTER II: OLD PEOPLE

OF THE OLD VALUES which after long submergence have again come to the surface, esteem and reverence for age is as moving as it is significant. The Russian word *starik,* old person, like the word *predki,* ancestors, and the meaning that is now attached to both, signalize as dramatic a change in Russian thinking as the constant emphasis of the words *Russia* and *Russian.*

There was a time when the word starik was a term of opprobrium, almost disgrace. It signified someone incorrigibly old-fashioned, hopelessly unregenerate. There were no special laws against old people. Age as such did not enter into the considerations and calculations of the new lawmakers. But excepting factory workers nurtured in underground revolutionary doctrine and landless peasants who saw in the Soviets a fulfillment of their lifelong yearning for land, older folk, by the very nature of the Revolution with its violent repudiation of the old world, found it hard to reconcile themselves to its demands on them.

"For some a prologue, for some an epilogue," says one of the characters in Bulgakov's *Days of the Turbines,* the most moving play that has yet come out of Soviet Russia. The Revolution in its early years spelled an epilogue for large groups of people in Russia, regardless of age, but for none so sweepingly and implacably as for old people. They saw only doom ahead, not only for themselves but for their descendants—the generations to come. They said so openly, though not always loudly except when they were peasants. They wept the most sorrowful tears that were seen or heard in those tempestuous times.

In the villages, where peasants with rare exceptions have always been irrepressibly articulate, it was the older people, men and women, who were loudest in their complaints, hottest in their rancor.

In the early days of collectivization I was making a journey through villages in the province of Ryazan. In one village I made a long stop in the co-operative store. The clerk was a young man with a shock of peppery hair and small blue eyes. In the store were his father, who was over seventy, and another old man, a boyhood friend of the father. Collectivization was in full sway, and both men were eager to unburden themselves of their thoughts.

Collectivization, they said, would bring only ruin and doom to the Russian village, to all Russia. As they talked and wailed and cursed, the young man listened without the least sign of perturbation. He had recently completed a course of political education in Ryazan, and the words of his own father sounded alien and absurd.

"Look at him," mocked the father, "he is no better than the crows who've never cawed so loudly or with such joy because of the heaps of carrion they'll soon have to glut themselves with." Even then the young man did not lose his temper or show resentment. He only said:

"*Stariki*—old men—what do you expect of them? Talk and talk is all they do and all they are fit for." Then thumping his forehead he added: "But new things, better things, better than the peasants here have ever known—these they are no more capable of understanding than are the crows of which Father has just spoken."

Grudging tolerance from their own sons was the most they received in those desperate fighting times—those benighted, lonely, heartbroken stariki. . . .

What a difference now! How cleared is the Russian air, the Russian temper, the Russian vocabulary of the least hint of slight for age. The very word starik seems invested with sanctity. It spells no weakness but strength, the strength of wisdom and experience, of resolve and dedication. No wonder that in current war fiction stariki and not youths are so often the heroes.

Here is Alexander Dovzhenko, a tall, red-cheeked, gray-haired Ukrainian. He is one of Russia's leading motion-picture directors. He is gifted not only with dramatic sense but in the use of language. He writes his own scenarios, both story and dialogue. Occasionally he condenses a scenario into a short story and prints it in *Red Star* or in some other journal. His story, "The Night before Battle," is one of the most powerful pieces of war fiction that has appeared in Russia. It is a story about the heroism of the old pilots on the river boats of the Ukraine.

"Hardly a river," writes Dovzhenko, "but was a drama, and the old pilots were like the good spirits of the river. They were daring, these grandfathers, they were stern, and they had no fear of death." And again, "I looked at Grandfather Platon and breathlessly listened to his every word. He believed in our victory. For me he was the living and awesome voice of our intrepid and invincible people."

A starik—the symbol of the intrepidity, the mouthpiece of the invincibility, and, one might say, of the goodness of the people!

War is always hard on old people. It is hard on them physically, especially in a country like Russia, where for years industrial construction has taken precedence over immediate personal satisfaction. The shortage of fats, of sugar, of white flour, of rice, of dried fruits is a special blow to old people. Their dietetic requirements often enough are poorly met by the coarse foods available for everyday consumption.

The war has been no less severe a strain on old people emotionally, far more so than on young people. The casualties have hit them harder and they are more given to tears. In their declining years, with health ebbing and capacity for physical enjoyment lessening, pride in family, attachment to kin acquire a new importance, taking the place of satisfactions no longer available or even desired. Therefore, bereavement rouses in them deep and lasting pangs.

Yet even they—if they are not invalids—keep doggedly to their daily tasks. The legal retirement age in Russia for women is fifty-five, for men sixty. Because of the low pensions and because continued work neither disbars a citizen from a pension nor lessens its amount, men and women who are capable of working remain at their jobs in offices, in factories, and on farms, even after they have reached the age of retirement. But never have they worked as arduously as now or for such long hours.

During my stay in the village of Russkiya Lipyagi one of the most interesting people I met was the old woman of the house in which I was staying. Though she was about seventy she was first to rise in the morning. She milked the cow, fed the calf, tended the chickens, started the fires; in between she looked in on the four sleeping children, drew the sheets over them, brushed away the flies, shut a door or a window if a draft blew over them.

When I arose the samovar was ready and, as if by a miracle, breakfast appeared on the table. Her daughter-in-law helped, but the old woman did most of the work, in the house and outside, because the younger woman was busy with executive work in the village.

One morning as I was strolling in the garden admiring the finely cultivated beds of potatoes, cucumbers, beans, and pumpkins, I saw a red calf with a long loose rope tied around its neck nipping the tops of the vegetables. I picked it up and with the help of a passing peasant hoisted it over the fence into the road.

Later, as I was eating breakfast, the old woman appeared at the door, her shrunken eyes bright with rage, and demanded hotly:

"Where is that rope? What have you done with it—and why did you put the calf out of the garden—our own precious little calf?"

She cried and scolded, and it took some time for us to soothe her.

"She's terribly upset these days," said the daughter-in-law, "because we've had no letter from my husband—her son; he's at the front. She thinks he has been killed and she weeps and weeps. But how she works! She's never worked so hard, never!"

How many thousands of women are there in Russia, old and lonely and worried over a son or grandson in the war, easily moved to tears yet trudging away, with inordinate zeal, at whatever task confronts them.

Newspapers and magazines feature with special respect the photographs of these old people. So do newsreels. So do war plays. One of the most stirring characters in *The Russian People,* by Konstantin Simonov, is the white-haired, sturdy ex-czarist officer, volunteer in the Red Army, who is more than the equal of younger men in vigilance and intrepidity. Hundreds of Cossacks in their sixties are riding side by side with their sons and grandsons and cleaving the heads off Germans with equal zest and even more skill.

I have before me as I write a finely printed little book entitled *Patriots of Trekhgorka,* which has just come off the press. Trekhgorka is the name of one of the oldest and largest textile factories in Moscow. The opening chapter bears the suggestive title "Four Generations." It is the story of the seventy-four-year-old Nikolay Nikolayevitch Kuzmin, who started working in this factory two generations ago when it was under private ownership.

For sixty years Nikolay Nikolayevitch spent most of his waking time in Trekhgorka. He still works in the printing department. He is never late; he is never in a hurry to leave. He is white-haired, and his muscles have not their former strength. Yet so marked is his perseverance that he has attained a wartime record of production which is the envy of many a younger worker. He has lifted his daily output by 200 and 250 per cent! He turns out daily twice and two and a half times as much as the program of the factory has set for a worker like him.

Nikolay Nikolayevitch has a large family—four generations. Three of his sons are in the Army, one is a colonel, another is a railroad engineer. One son-in-law, a musician in the orchestra of the Vachtangov Theater, has exchanged his violin for a rifle. His fourteen grandchildren are studying and working. He has been awarded a certificate of honor by the factory for the high quality of his work and also a gold watch with his name engraved inside.

Nikolay Nikolayevitch is a starik with a past and a present, and he is the pride of his factory. Young people salute him, gather around him, listen to his words of advice or to stories of his experiences as a textile worker for sixty years. He means much to Trekhgorka as a guide and inspirer of youth, and there is something almost old-fashioned (the Russians would laugh at the word) in the respect and adulation that is continually bestowed on this seventy-four-year-old textile worker. . . .

There is hardly a factory, a university, or any other institution where there are stariki, in which they do not command a confidence and enjoy a prestige which, in the light of the disrespect and the desecration leveled at age in the twenties and the early thirties, is as complete a reversal of attitude as it is unexpected.

Consider the case of Lubov Teplyakova in the village of Ovsyanniki, Yaroslavl Province. She is the *babka*—old woman—of the village. Nobody knows her age. To some people she admits being ninety-two, to others she says she is one hundred. Her home is older than she—the oldest in the village. A dwarfed cottage on the bank of the Volga with a roof of green moss, it is exposed on all sides to the winds of autumn and the blizzards of winter.

In 1926 a flood swept the village, and the babka's cottage was nearly sucked into the river. The local Soviet offered to move her into another house. But she refused. She had lived in the cottage since the day of her marriage. In it she had given birth to her children, had known the happiest and the saddest moments of her life. It was home—no other house however secure and comfortable could take its place. There she would remain to her dying breath. The village was proud of the babka, and though the years had enfeebled her body they had neither dimmed her mind nor dulled her tongue.

A day came when Ovsyanniki was confronted with the most momentous decision in its history, all the more so because the men were at war and the women had to make it and bear full responsibility in the years and years to come. Because the kolkhoz was badly mismanaged, the district Soviet offered Ovsyanniki an opportunity to merge with a flourishing kolkhoz in the near-by village of Rybino. Rybino had agreed to the merger, and Maria Sokolova, a Party member and chairman of the Ovsyanniki kolkhoz, called a mass meeting to act on the proposal. It hurt her to think of the village losing its identity. Other women expressed themselves in a similar vein, particularly as the people, or rather the women, in Rybino—their men were also at war—were haughty and boastful, as though they were performing an act of mercy for the unfortunate folk of Ovsyanniki.

Discussion raged late into the night, "until the cocks crowed." It was hard for these women to make a decision. So Maria Sokolova sent to the babka for advice. The babka's daughter acted as messenger. After hearing the question the babka laughed. To her it seemed absurd that anyone should give up his identity—forever. She wouldn't give up her home, old and dilapidated and wind-blown as it was, and here were these women earnestly thinking of giving up forever the identity of their kolkhoz and their village. . . . The babka said:

"Let these young women write to their husbands and say to them, 'We are weak and cannot take care of ourselves in our own home, so we are seeking shelter under someone else's roof. . . .' Let them ask their husbands' advice, but let them also ask themselves whether sending such a message is worthy of the character and the sense of justice of their husbands!"

So ashamed did Maria Sokolova and the other women feel that they decided to reject the proposal of the district Soviet and to work harder than ever and make their village and their kolkhoz worthy of independence.

When I read this story I could hardly imagine that a Party secretary anywhere, city or village, would seek the advice of an illiterate old woman on any subject! Yet it did happen in the Volga village of Ovsyanniki. It is happening in other places.

Stariki may have no learning. But they have lived. They have endured. They have a pride and a wisdom of their own. There is earthiness in their thoughts, saltiness in their speech, glory in the very years they have attained, and—they are Russian!

Stalin's speech of November 7, 1941, invoked Russia's *predki*—ancestors—for inspiration; and Maria Sokolova, member of the Party, and chairman of the kolkhoz in a Volga village, in a moment of great trial invoked the advice of an illiterate babka.

Here is an astounding fact about the Russia of today with a lesson that calls for the earnest reflection of anyone seeking to understand Russia and the Russians.

CHAPTER 12: **THE GOSPEL OF HATE**

SERGEANT TRENTYEV, native of a village in the Bryansk forests, in a letter to a Russian author, writes:

"As a child I often went to the woods with my grandmother to pick raspberries. My hands used to get red with raspberry juice. Now I want my hands to be red with German blood."

Not Fascist, not Nazi, not Hitlerite, but German blood!

In a diary as moving as it is brilliant, found in the pouch of an eighteen-year-old girl named Lidya Khudiakova, who was killed on the Leningrad front, are these wrathful words:

"Well then, you German bandits, you shall pay dearly for everything—you shall swill yourself full of your pure 'Aryan' blood, and there will be plenty left for fleas and lice."

In an editorial of the *Komsomolskaya Pravda* (August 20, 1942), captioned, "Kill the German," there are these tempestuous words:

" 'Kill the German!' is the plea of your old mother.

" 'Kill the German!' is the plea of the girl you love.

" 'Kill the German!' whispers the very grass of our native land, now so drenched in blood."

In still another editorial in the same fiery mouthpiece of the youth of Russia there is the exclamation: "You and the Germans cannot live on the same earth . . . so kill the Germans!"

Note again the use not of the terms Nazi, Fascist or Hitlerite, but *German*.

Pravda (September 9, 1942) printed a story from the front signed by "Battalion Commissar Vershinin." A Russian gunner named Fedotov, writes Vershinin, together with a group of other soldiers, broke into a village that had been held by Germans. A violent battle ensued. Suddenly Fedotov found himself out of ammunition. Flaming with wrath, he flung a German over his shoulder, stunned him with a blow on the head, then, falling on top of him, bit his throat with his teeth.

The world has been wondering about Russian morale. It has been asking why Russians fight as bravely and savagely as they do. No matter how crushing the immediate defeat, how stupendous the immediate disaster, they draw back, reform their lines, rush again into battle, fight on until death. Countless books can and will be written on the subject in the years and centuries to come, not only by scholars and historians but by novelists, poets, playwrights. None of them in this writer's judgment will fail to point out that one of the chief ingredients of Russian morale in the present Russo-German War is hate not only of Fascists, Nazis, and Hitlerites, but of Germans. Did not Stalin in his speech of May 1, 1942, proclaim, "You cannot conquer the enemy without learning to hate him with all the power of your soul"? Never in all her history has Russia, all Russia—in Europe and in Asia, in the deep south and in the remote north, the Russia of the Russians and the Russia of the Kazaks, the Tadzhiks, the Ukrainians, and the other nationalities—hated a foreign enemy with such violence and wrath as they now hate Germany.

It was not so in the beginning of the war. Under the impact of their political philosophy Russians thought and talked of Germans in terms on the one hand of Nazis, Fascists, Hitlerites, and on the other of Germans—workers, peasants, intellectuals—Germans who in their hearts and minds opposed the policies, the ideas, the wars of the ruling groups. They could not dismiss, and wouldn't anyway, the fact that in the German elections of 1932 the combined vote of the Communists and the Socialists was more than thirteen million! They were convinced that despite the Hitlerization of the German youth there were in the German Army soldiers—masses of them—who, if given an opportunity and promised cordial treatment, would be only too happy to desert their ranks and surrender. By radio, by leaflet they deluged German soldiers with appeals to come over to the Russian lines. But—few of them heeded these appeals. They fought with a demoniac energy—the worker no less than the Junker, the peasant no less than the Nazi.

On my arrival in Russia I heard a story which may or may not be true but which gives a clue to the disenchantment that came over Russians with Germans as Germans. At the start of the war a German flier made a forced landing in Russian territory. A Russian flier who was near by quickly went over, and though he had reason to think the German was a Nazi, he was ready to conform with the amenities of war, so he reached out his hand. Instead of taking it the German whipped out his revolver and shot the Russian. Thereupon other Russians appeared on the scene and had a swift reckoning with the German. Afterward, according to this story, Russian fliers no longer made the least attempt to show friendliness to captured or stranded German aviators.

This story bears out the facts as the Russians observed and experienced them, namely, that the Germans were not fighting an ordinary war of conquest. In subsequent chapters I shall describe at length the conditions and the torments to which Germans subjected not only Russian war prisoners but the civilian population of towns, farms, villages, which they occupied. I visited a number of these villages after the Russians had reconquered them. Here I shall cite only several incidents.

In the village of Lotoshino, Moscow province, a group of women gathered about me in the yard of a blasted alcohol factory and spoke at length and with emotion of their experiences during the German occupation. What stirred their deepest wrath was the burning of their homes and all other buildings in the village on the day the Germans retreated. In pairs armed Germans went from house to house, drove out the population into the cold, the bitterest the country had known in years, and set their houses afire. One woman said, "When they came into my house I was bent over a dying child, and I said to them, 'Have mercy on me—don't you see my child is dying?' And they only shouted 'Raus! Raus [Get out of here].'" Pointing to a tree on a hill that was facing us, she resumed, "Under that tree in the snow I saw my child pass out of life."

There were hundreds and hundreds of Lotoshinos in Russia.

The homes of Tchaikovsky, Rimski-Korsakov, Tolstoy, Chekhov, Gogol, Shevchenko, Korolenko, and other outstanding men in Russian and Ukrainian science, art, and history, which the Russians had converted into national museums, have been deliberately desecrated or destroyed. Not a monument of cultural and historical importance in any of the territories which the Germans held or had been holding but has been profaned and annihilated. Not even in Poland have the Germans sought more systematically to obliterate "the inferior" culture of an "inferior" people than in Russia. No wonder the *Komsomolskaya Pravda* in an editorial of August 9, 1942, exclaims:

"Remember, brother, if you are a Russian, a Ukrainian, a White Russian, you are no human being. . . . What cares *a German* for the life of a Russian citizen, or for his human dignity?"

In a rural district which had been under the Germans for nearly a year and in which I traveled after they were driven out, I asked peasants if any of the Germans who had stayed there had ever come to them in secret and said they were workers or liberals and completely out of sympathy with the action of the Reichswehr but powerless to stop its depredations. Only once, they said, in a village removed from the main highways, did they know of such a German. He was an officer of aviation and had come to the village on a motorcycle to collect grain. He requested a committee of peasants to gather the grain voluntarily as he did not want to force them to do it, nor did he want the military commander to send a punitive expedition. Only once! There may have been more such men in other occupied regions, but they were the rare exception. In one village which the Germans had controlled for ten months there is an old barn built of timber, with a low-hanging thatch roof, which tells a shattering story. In the dead of winter, the Germans drove into this barn several hundred Russian soldiers who were war prisoners. Despite the bitter cold, there was no heat in the barn, not a single wood stove. Nor did the Russians have warm clothes. The Germans had stripped them of the felts, woolens, and sheepskins they wore when they were captured. Exposed to cold and wretchedly fed—one bowl of almost grainless oat soup and one slice of bread a day—they died by the score from pneumonia and other diseases. One night two Russians attempted to escape. They were captured and hanged in front of the barn. For three weeks, in wind and snow, the frozen bodies swung from the gallows. I was in this village, and eyewitnesses who told me this story said they had never seen anything more gruesome.

There were hundreds of such barns in German-held lands. War prisoners and others lived, hungered, and died in crowds. No one knows or can know until after the war how many millions of Russians perished prematurely in conquered territory. The behavior of the German Army in Russia quickly sobered the Russians, even the most hardheaded Bolsheviks, into a reappraisal of the meaning of the word German—at least for the duration of the war. Hence the outburst of hate—against Germans. *Red Star,* official mouthpiece of the Red Army, no longer carries at its masthead the slogan "proletarians of the world unite." Instead in bold letters it prints the pronouncement, "Death to the German invaders!" So does *Red Fleet,* official organ of the Soviet Navy. In an editorial of *Red Star* of November 14, 1942, there is this sentence: "The Germans of today are the vile offspring of vile fathers." The *Komsomolskaya Pravda* crackles with hate—of Germans. Nor do *Pravda* and *Izvestia* refrain from boiling over with such hate.

On June 22, 1942, I was in Kuibyshev. I had been in Russia only three weeks. Late in the afternoon a British correspondent entered excitedly into my room and laid before me a newly arrived copy of *Pravda.* Spreading it out and pointing to a closely printed full-page article, he said:

"Read every word of it—it is one of the most significant documents that has appeared since the war started, all the more so because it is appearing in *Pravda.*"

The article was Sholokhov's well-known essay story, "The Science of Hate." I read it at one sitting and when I finished I almost gasped. Here was the most moving sermon of hate that had yet been preached by anybody in Russia; and the author was a Russian, a Cossack, the greatest literary figure that Russia has known since the coming of the Soviets.

Sholokhov tells the story of a Siberian lieutenant whom he meets at the front. The lieutenant's name is Gerasimov. So deep is the man's loathing for Germans that the mere sight of them even as prisoners of war stirs a paroxysm of rage. On investigation Sholokhov learns that Gerasimov not only has fought against Germans but had been captured by them. His experiences as a war prisoner—the degradation and the torment he has endured—have made him such a fountain of hate that he can look at dead Germans only "with pleasure." "Violently," Gerasimov tells Sholokhov, "I hate the Germans for all they have done to our country and to me personally." Again, and most significantly, he says, "No one ever imagined we should have to fight against such shameless villains as the German Army."

It is not only Gerasimov, Sholokhov, and *Pravda,* but the Russian people who voice similar sentiments and execrations on the Reichswehr, branding them all, privates and officers, from the highest to the lowest, from corporals to generals, as "shameless villains," with not a shred of the military correctness and the formal decencies which international law demands of a fighting and conquering army.

Sholokhov's story, which has sold in millions of copies in Russia, and which has been read, analyzed, discussed in all schools and factories, derives special meaning from the fact that only once does he use the word Hitlerite. Not once does he speak of Germans as Fascists or Nazis. He lumps them all together under the words *German* and *Germans.*

About the time that Sholokhov's story appeared another outstanding Russian author, Alexey Tolstoy, printed an appeal to the Russian soldiers and the Russian people under the title of "Kill the Beast." Like Sholokhov's story it was reprinted in all the newspapers of the country, including the ones published by the Army and the Navy.

"The only emotion possible at this time," says Tolstoy, "the only passion with which man must flame, is and shall be hatred of the enemy. Man must wake with this stubborn hate, fight with this hate, fall asleep at night with the hate unslacked. . . . You love your wife and your child? Then turn your love inside out so that it hurts and oozes with blood. Your task is to kill the enemy branded with the swastika, sign of Cain. He is the enemy of all lovers." Tolstoy ends his exhortation with the words:

"Friend, comrade, dear companion, at the front and in the rear! If your hate cools, if you get accustomed to it, then stroke the warm head of your child if only in imagination. It will look at you candidly and innocently, and you will realize that you cannot get used to your hate. Let fury burn in you like a constant pain, as though you saw a black *German* tightening on the throat of your child!"

Literary people in Russia make no secret of their bursting hate not only for Fascists, Nazis, and Hitlerites, but for Germans!

"The Germans of today," says Nikolai Tikhonov, who despite bombs and shells and hunger refused to leave his beloved Leningrad throughout the months of siege, "are a race of automatons and executioners, a people whose servile passion to prostrate themselves before every scoundrel of a police officer and every titled nobody has become second nature. . . ."

And Konstantin Simonov, whose *The Russian People* is one more hymn of hate, makes Marfa Petrovna Safanova say to the German officers before her execution: "I should love to fly unseen to your country and take your mothers by the scruff of the neck and bring them here and show them from the air what their sons have perpetrated on us. And I should want to tell them, 'See, you bitches, the kind of sons you have borne?' And if after what they have seen they didn't pronounce a curse on their sons, I should want to kill them and their sons and you too."

During my stay in Kuibyshev I was invited by a schoolteacher, a woman of over forty, to her house to meet other schoolteachers. The language of hate against Germans on the Russian radio, in the Russian press, in private conversation, was still so new to me and so contradictory of basic Soviet teaching that I asked them about it. . . .

"Yes, it is too bad," said my hostess, "but what shall we do? For nearly a quarter of a century we have striven to educate our people, particularly the youth, out of national prejudices and racial hates. Our old literature is packed with plenty of hate of Germans. Read Herzen and Dostoevski if you want to know how Russians can really hate Germans. But we had laid all that aside—bygones were bygones. In 1914, at the outbreak of the first World War, German shops on the Kuznetsky Most in Moscow were set afire and Germans were violently attacked, that's how strong was the feeling against them.

"But nothing of the sort has happened this time. We have removed Germans, hundreds of thousands of them, to faraway places, but we did not pogromize any of them. Think of it, in all the years under the Soviets not one insulting word have we said or permitted to be said against Germans. We strove to eradicate the old racial and national antagonisms— we schoolteachers especially. Even after Hitler came to power and the German press, orators, and radio deluged us daily with the vilest abuse imaginable, we did not retaliate in kind. As long as Hitler and other Nazis were content to call us names we didn't mind. Now when he is seeking to exterminate us—because we are Russians—we would rather die than compromise with him—in word or in act."

The other teachers, as sad and angry as the speaker, heartily concurred in this opinion.

So now Russian journals and Russian books and Russian pamphlets print not only the expressions of hate against Germans by living authors but by those who have long since been dead—Ryleyev, Lermontov, Gogol, Tolstoy, Dostoevski, Saltykov-Shchedrin, Herzen, Gorki and others. The literature of hate during the seven months that I was in Russia mounted almost daily in magnitude as well as in fury. Often enough have I read Saltykov-Shchedrin's famous indictment of Germans in his *Little Boy without Trousers.*

"Who is the most heartless oppressor of the Russian workingman?" asked one of the characters. "The German. Who is the most ruthless teacher? The German. Who is the most dull-witted administrator? The German. Who inspires abuse of power? The German. . . . Your [German] civilization is second rate; only your greed and your envy are first rate. Arbitrarily you identify this greed with justice and you imagine it is your destiny to explode the world. That is why you are hated everywhere, not only here but everywhere."

Nor do Russian publications fail to reprint a passage on Germans from a letter which the late Gorki wrote in 1911 to the monarchist Breyev in Nizhnii Novgorod in reply to his call on Gorki to repent his political sins. "Why is it, my gentlemen 'patriots,'" wrote Gorki, "that your favorite heroes are the Gerschelmans, Stakelbergs, Rennenkampfs, and others of whom there is no end and who fought so poorly with the Japanese and who so cruelly and so assiduously abuse the Russian people? Why do these Germans and barons play so momentously the part of hired servants whose duty it is to hold the Russian by the throat?"

Not that the Russians, people or government, aim at the extermination of Germans because they are Germans. Not that they seek to derogate or annihilate Germany's great contribution to the world's civilization. Schiller is continually in the repertoire of the Russian theaters not only in Moscow but in Siberia. Goethe and the Manns, Thomas and Heinrich, are not only widely read but reverenced. Beethoven, Bach, Mozart are as popular as ever with the music-loving public.

Nor has the German language been discarded or discouraged in schools and colleges. On the contrary, German is being assiduously promoted. "What are you doing now?" I asked the young wife of a Soviet diplomat who was drowned in the Baltic Sea.

"I am studying German," was her answer.

I had not been long in Russia, and in the light of the language of hate which I was everywhere hearing and reading the answer seemed a little strange; so I said with more surprise than reason: "But why German?"

"Because," came the reply, "it is one of the most important languages in the world."

On my last day in Moscow, as I passed the windows of a leading bookshop on the Kuznetsky, I saw on display new editions in German of Goethe's *Werther* and *Wilhelm Meister*. I am certain that Goethe, Schiller, and Beethoven mean more to present-day Russia than to Nazi Germany.

But the Russians insist that the Germans of today, as they have revealed themselves in Russia in the present war, have nothing in common with the great Germans of yesterday.

These Germans have divested themselves, or have allowed themselves to be divested of, the decencies and amenities that are a part even of modern warfare. They and they alone in all Russian history, in all the wars the Russians have ever fought, are treating Russians as outcasts and outlaws fit solely for subjugation and annihilation. Even the journal *Comintern,* Number 3–4, 1942, is constrained to ask: "Where is the German working class? Where are the German workers? . . . In your handshake we felt proletarian power and solidarity. But today? With your hands Hitler is conducting war against the Socialist Government of workers and peasants. Your hands are producing armaments for the bloodiest murderers of workers." The Comintern spokesman asks these questions and makes these charges, and about all the comfort he allows himself is to declare that a process of "sobering" has started in Germany. Yet nowhere does he offer the least concrete evidence of any *sobering* in Germany either among workers or among other groups. He readily admits in the next issue that "among German soldiers there are not as yet any mass desertions; that barring individual exceptions they do not yet openly revolt; that though they grumble they fulfill the orders of the High Command!"

The *sobering* may come after crucial and disastrous defeats which make it clear to workers no less than to industrialists that German appetites for Russian lands, Russian forests, Russian rivers, Russian oil can never be satisfied.

Meanwhile the hate, not only of Nazis but of Germans in present-day Germany, flames and crackles in Russia. Even children are impregnated with it. "Come, Mother," I heard a little girl say in a park in Kuibyshev. "Let us go home."

"But why?" said the mother. "It is nice and cool in the park."

"Yes, but there are too many mosquitoes here and I have to scratch myself and people will think I am a German." She had heard of German war prisoners so infested with lice that they always scratch themselves, and so she associated scratching with Germans and did not want to be mistaken for one.

At the end of the war the vocabulary of hate against Germans is certain to disappear unless, in the process of reshuffling the political alignments of Europe, Germany remains a military menace to Russia. But in the minds of the Russian people, in the memory of future generations, in the novels, plays, and poems that will for hundreds and hundreds of years be written by Russians, this hate, like a Vesuvius, will never cease to cast up its lava and its fire.

PART THREE: RUSSIAN CITIES

CHAPTER 13: TULA

SEPTEMBER 1942 was a rarely beautiful month in Russia, especially in the Moscow region. Instead of the chilly winds, the overcast skies, the stinging drizzles I had known in former years in Moscow, there were brightness and sunshine almost every day. Muscovites spoke with divided emotions of this wonderful weather. The longer it lasted the better they could prepare for the hard winter that was coming, the more fuel they could save for the freezing months ahead; but also the more comfortable the enemy's armies felt in their trenches and dugouts and the more vigorously they could resist Russian attacks on the western front.

Yet it was a cheering month, and the automobile drive to the city of Tula, about one hundred miles from the capital, was an invigorating experience. There were color and fragrance in the countryside, with marigolds, daisies, cornflowers, red clover dotting the fields as with bridal wreaths. In quiet weather the woods were silent and, with the sunlight upon them, the birch leaves gleamed with a golden luster.

The kolkhoz fields teemed with life. The grain had been reaped and stacked and was now being threshed. Potatoes were being dug, cabbage gathered. Winter plowing was under way. With the Ukraine gone, White Russia lost, the North Caucasus largely in enemy hands, Russians were more than ever striving to whip every bit of growth and fertility out of the soil. They were plowing day and night, plowing deep, so that more and more of the land would drink in the snow of winter, the rains and thaws of spring, and nourish the wheat, rye, oats, buckwheat, and other crops that would be planted in the spring.

The day before I started on this trip, Wendell Willkie had invited me to breakfast. He had been in Russia only a few days, and one of his first and strongest impressions was of the part women were taking in the everyday work and life of the country. "These amazing Russian women," he said. Amazing indeed! Here, as all over Russia, they were on every hand, as far as the eye could see, young and old, with but few men among them. They were threshing rye and wheat, oats and barley. They were digging potatoes and harvesting cabbage. They were plowing with tractors, with teams of horses, with one horse. With the shepherd boys and girls already in school, they were pasturing sheep and cattle. They were hauling wood in wagons, in handcarts, for their own homes, for the village schoolhouse. They were building high brush fences along the side of the road to prevent blizzards from packing it with unpassable snow. On their backs they were carrying sacks of grain and potatoes. On their backs they also carried children to a nursery or a doctor's office. Dressed in soldier greatcoats, in knee-high leather boots, in soldier caps with their hair bobbed and wind-blown or twisted into a pearlike coil, and with rifles on their shoulders, they were guarding bridges, flour mills, storehouses. They were doing everything—these humble, hardy, indefatigable, invincible Russian women!

In a country as young as Russia in nationhood and in modern industry, Tula is an ancient city. Its origin dates back to the twelfth century, and few cities have played so eminent and heroic a part in Russia's struggle for her independence and expansion. It is the home of the Russian *vintovka* —rifle—and of an array of other fighting weapons, the best Russia has ever made. These weapons have won Russia many a victory on the battlefield. The very names of the streets in the industrial district are symbolic of Tula's historic military eminence—Gunbarrel Street, Bayonet Street, Powder Street. There is no other Russian city with such street names.

In the fourteenth century, when Grand Duke Dmitri Donskoy gave battle to the Tatars on the rivers Nepryavda and Krasnaya Metch, the swords and hatchets forged by Tula blacksmiths were in no small measure responsible for the triumph of Russian over Tatar armies. In the seventeenth century, when the Poles overran Russia, seized Moscow and threatened to conquer the country, the trader Minin of Nizhnii Novgorod and Prince Pozharsky summoned the people to arms. Again Tula rendered the nation an inestimable service. The muskets and hatchets that came out of its smithies did their work well; the foreign invaders were battered and driven from the land.

In the eighteenth century Charles the Twelfth of Sweden launched a crusade against Russia and drove deep into the Ukraine. At Poltava, Peter the First made a decisive stand against the Swedish king, and the mallets, guns, hatchets, and sabers that Tula supplied once more proved their worth. Poltava became the graveyard of the Swedish king's ambitions.

Afterward, appreciating the importance of Tula as the mightiest tool of national defense, Peter mustered the scattered blacksmith shops of the city into the first national armament factory, and since then Tula has remained one of Russia's leading arsenals.

Tolstoy, who loathed Napoleon, scoffs at his military genius because in his march on Moscow he neglected to seize Tula. It was Tula's guns and powder as much as Kutuzov's strategy that smashed the once all-conquering French armies.

During the civil war of 1918–22 Tula again made history. Without Tula's arms and the fighting spirit of the factory workers, the Bolsheviks never could have routed their enemies. When Denikin moved on the city the Tula gunsmiths remained in their shops day and night. They assembled rifles and field guns, forged bayonets, and passed them from hand to hand until they reached the fighting armies. Denikin never crossed the threshold of Tula.

Geography has helped make Tula the mighty fighting fortress it is and always has been. Near by is the renowned *zaseka*—barrier forest—so called because in the past it had served as a barrier against enemies in the south who sought to push their way to Moscow. Again and again the Tatars had striven to fight their way across this primeval wilderness. But the Tuliaks, as the people of Tula are called, with the help of the population of the surrounding countryside, invariably barred the way. They felled trees, built barricades, and with their weapons shot and slammed away at the Mongols and drove them into retreat.

The land around Tula is rich in ore which in ancient times, as now, has given Tuliaks the iron with which to fashion their fighting weapons. It is this iron that has brought into being the skilled gunsmiths and the skilled gunners who have made Tula famous.

The Tula country is also rich in timber, coal, clay, and many other raw materials. The three Five-Year Plans have doubled its population and made it one of the leading industrial centers of the country. As in the days of Peter, the heart of Tula's industry is the gun and all that the word implies in this day of advanced technological development. Throughout the years that the Red Army had been developing its strength Tula was one of its leading armament sources.

In the fall and winter of 1941 the German armies kept pushing deeper and deeper into eastern Russia, and the defense of Tula became one of the main concerns of Moscow. With Tula in the hands of Germans, the road to Moscow was open. Napoleon had made the mistake of by-passing Tula. The Germans did not intend to repeat this error. So Tula, unconquered and unconquerable, was menaced by the most formidable army the world has ever known, and the people started to prepare for the

grimmest fight the city had ever faced. In more than one way the battle of
Tula was a rehearsal for the battle of Stalingrad.

In October 1941 Tula became a front. The streets were torn with ditches
and heaped with barricades. As in Stalingrad nine months later, thousands
of people, young and old of both sexes, went out with spades, axes, and
crowbars to dig trenches and build fortifications. From all over the
countryside crowds of boys and girls poured into the city, and everyone
was given something to do. Trucks and carts loaded with guns and stacked
with ammunition manufactured in Tula's own factories clattered along
the cobbled streets and roared away into the outskirts. Tula had heard too
much of the fighting ferocity of the Germans to leave anything to chance.

Autumn was upon the land, a harsh, unrelenting autumn with cold rains,
fierce winds. Orel fell. Orel, in whose province Turgenev was born, in
whose fields and forests he had wandered and hunted and where he had
written some of the most exquisite prose in the Russian language. Yasnaya
Polyana, Tolstoy's home, fell, and the Germans settled in the novelist's
old home. To the Russians there, farm workers and housekeepers, they
kept boasting: "We've smashed the fortifications of France, Holland,
Belgium, Poland, Yugoslavia; we'll blow to pieces the defenses of Tula. In
a few days the city will be in our sack." From heavy field guns placed in
the yard, in the flower beds, and in the birch grove on the Tolstoy homestead they kept firing heavy shells into Tula, which was only seven miles
away.

"You should have been here on October 29, 1941," said the janitor of the
hotel in Tula.

Tuliaks remember that date as the most eventful in their life and in the
history of their city. General Gauderian's tanks were on the threshold of
Tula. Streets became fortresses, with lines of defense close to one another.
The city was ready for a hand-to-hand fight with the enemy and to contest
every span of its land. The volunteer regiment of workers that had been
formed was as ready for combat as were the regulars in the Red Army.
Thousands of girls begged to join and many of them did.

On that day General Gauderian's tanks assaulted the city three times.
Behind the tanks came motorized German infantry. Machine-gun fire and
trench-mortar shells pounded the city streets. Fires blazed up and Tula
is easy to burn, for it is known as the city of wooden cottages. But so well
were the fire brigades organized and so many were the people in them,
especially youngsters, that the incendiary bombs were quickly extinguished
and no serious conflagration swept any part of the city. Gauderian's tanks
did not pass.

The next day the Germans flung themselves once more on the city.
Six times their tanks sought to break through the defenses, but in vain.

Meanwhile in the towns and villages they had occupied and in others they had hoped to seize, the Germans started the rumor that Tula had fallen, or "Tula *kaput*," as they expressed themselves. Yet Tula remained unconquered. The factories never ceased to forge armaments. Some of the shops became repair stations. From all directions damaged and shattered tanks, field guns, and trucks were hauled for repairs. The mechanics, among the best in the country, remained at their posts day and night. Some of them were wounded, and nurses were on hand to attend to them or to take them to the hospital for surgical operations. But the work went on. Hammers clanged. Engines roared. The men in the trenches and on the barricades lacked neither armaments nor ammunition. Everybody who could shoot was given a gun, even fifteen- and sixteen-year-old boys. Some of the boys were crack marksmen, and, like the older people, they evinced neither fear nor panic.

Indeed they performed deeds of astounding valor. They learned to throw hand grenades and bottles of kerosene at tanks. They made their way into the rear of the enemy and scouted for information about his strength and the disposition of his troops. They joined guerrilla bands. Shura Chekalin was a member of one of these bands.

Girls defied bullets and trench-mortar fire, crawled toward wounded men and carried them on their backs from the field of battle. In village after village which the Red armies had found necessary to abandon and which were already occupied by Germans, these girls managed to hide wounded soldiers in dugouts, in the woods, in haystacks, in other safe places. A girl named Makarova saved forty-six wounded Russian soldiers.

Maria Zhukova, only seventeen, and a teacher in the lower grades in the village of Voznesenskoye, had a breath-taking experience with Germans. She saved several wounded Russian soldiers and hid them in the forest. She even managed to fetch a Russian physician to one soldier who was dangerously sick. Meanwhile Germans came to her village; three of them stopped at her home. She heard that a Red-Army unit was moving on the village and decided to play the part of a gracious hostess to the Germans and detain them until escape was impossible. She invited them to sit at the table. She cooked breakfast for them. She played cards with them. Pleased with the hospitality of this young and attractive Russian girl, they made themselves at home in her house. The longer they stayed the more friendly their hostess appeared. She knew Russian soldiers were coming, but she did not want the Germans to know it. Finally, when Russian soldiers stormed the village, Maria Zhukova turned over to them her three German "guests" and hastened to the woods to rescue the wounded soldiers she had hidden there.

For one month and seventeen days Tula was under assault, almost under siege. At first the Germans sought to take the city by direct attack. Failing in this tactic, they shot out fanlike into many small pincers, hoping to feel out a weak link somewhere in the city's ring of defenses. Again they failed. They next pushed out long and powerful pincers with the intention of encircling the city and choking its resistance. But Russians pounded so severely at these pincers that they never could close around the city. In consequence Tula was never completely shut off from Moscow.

The Germans then decided to lay siege to the city, starve it into submission. They kept up only desultory offensive action. Yet the siege never was completed because the Germans never could envelop the city in their ring of steel. Their strength was waning, and when General Belov struck at their flanks they rolled back.

Had Tula fallen Moscow might not have held out. Tuliaks are human enough to say so and feel proud of the stand they made against Gauderian's tanks and Von Bock's soldiers.

Now Tula is still a city of hardheaded warriors but not the gigantic arsenal it was before the war or in the early days of fighting. Some of the best factories and their workers have been evacuated to the east, and Tuliaks know they will never be moved back. Russia is building a powerful industry in Asia, and the government will keep the evacuated factories in the east. In Tula it will build new factories, and in the future Tula will become even more powerful and more eminent than it has been in the past. This is the hope and faith of the Tuliaks.

As I wandered around the streets of the city I was amazed at the speed with which the ravages had been repaired. Gone were the barricades and all other overland fortifications, though the hedgehog steel contraptions were still lying around in yards, ready, if emergency came, to be again strung across streets and roads to hold up the movement of tanks. The deep-dug ditches with the walls of earth over their edges and the pools of water inside them remained in the streets. There was scum over some of these pools.

But Tuliaks were not lulling themselves into false security and blithe forgetfulness. They knew only too well that as long as Germans were fighting on Russian soil their city was not safe from possible assaults and possible vengeance.

Assiduously they were preparing for a hard winter. Everybody in Russia was. Not a family in the city but cultivated its own garden. In the residential section street after street with its double row of one-story wooden cottages, its grass-grown sidewalks, its broad and open drainage ditches, its gardens in the rear of the houses, resembled a village more than a city. Let it be noted that Tula, like most Russian cities, illustrates the anachronism that the Russia of today presents—industries as modern as any in the world and homes as ancient and primitive as in pre-revolutionary days or ages ago. The contrast between the two is as unforgettably impressive as the stories the Tuliaks tell of their encounter with German armies.

The stamp of the village was as deep as it was open in the residential part of the city. Workers not only cultivated their own gardens but often kept their own cow, raised their own pig, kept their own chickens.

A glance at the gardens showed that the Tuliaks were as good at growing vegetables as they were at fashioning guns. As I passed along the streets women and children were working in the gardens—digging potatoes, pulling onions, cutting weeds and nettles for the pig or the chickens. The cabbage patches gleamed with freshness and solidity, though the fruit trees for which Tula is likewise famous were this year barren of fruit. The harsh frosts of the preceding winter had winter-killed many of them.

Tula winters are always severe. The mercury drops low, the blizzards are violent. When I was there Tuliaks knew that the coal from near-by mines would be needed by industry. They therefore could not count on any of it for their own use during the cold months ahead. So they were cutting wood and hauling it home. With trucks and horses busy in the fields or carrying supplies to the front, Tuliaks hauled their wood home on hand-drawn carts. The roads were crowded with such carts loaded with freshly cut chunks of birch and ash and other woods drawn by old men, women, and children.

Tula samovars are famous all over the world. There are many of them in America and England, and anyone in any foreign land in possession of a samovar bearing the Tula imprint can feel assured that no one anywhere has a better samovar.

Tula is also a city of accordions—it makes some of the best in the country. Contrary to the prevalent impression in America and other countries it is not the balalaika but the accordion that is Russia's national and most popular instrument. From the Black Sea to the Arctic, from the White Russian swamps and forest to the Pacific Ocean, at weddings and festivals, in summer and in winter, there is someone in the village, in every village, to liven up the daily life and the holiday spirit with the melodious tunes of the *garmoshka*.

It was not until my visit to Tula that I learned of the city's indirect contribution to one of the greatest novels in the Russian language. Tolstoy, whose home was only seven miles away, went one time to a party in Tula. Among the guests was a striking-looking young woman named Maria Alexandrovna Gartung. She was the oldest daughter of Russia's most celebrated poet, Alexander Pushkin. So deep was the impression the young woman left on the young author that seven years later, when he started to write *Anna Karenina,* he vividly remembered her and used her as a model for his famous character.

Alexander Werth, of the London *Sunday Times,* was with me on this journey, and one evening, when we had gone to our rooms, the cook of the hotel came to see us. He was a little man with broad shoulders, massive chest, and rocklike neck. He had come up to see if we were comfortably cared for, so he said. In reality, irrepressibly Russian, he had come up to talk to the guests from foreign lands.

All his life he had been a cook and there was nothing in the world he enjoyed more than cooking. He had cooked for American engineers in the coal mines and he knew something of the ways and tastes of foreigners. If there was anything special we wanted—he was at our service. . . . Tuliaks were hospitable folk and when visitors came from a faraway land—especially from America and England—well then, a cook in the hotel had a special duty to perform. He was a real Tuliak, too, he proudly informed us, and therefore a stalwart patriot. He had served in the Russian Navy under the Czar, and when the Germans threatened to storm the city he laid away his apron, went to military headquarters, and said, "I am not going to be evacuated. I want to fight." So they gave him a rifle, a Tula-made rifle, and he joined the defenders of the city. When the Germans were driven away he put away the rifle, again donned the apron, and returned to cooking.

But he was still helping the Red Army in his own way—he was preparing cooks for its kitchens. He had already graduated three hundred of them, boys and girls, and they were all good cooks. . . . The soldiers whom they will feed will not suffer from lack of appetizing soups, meats, and other dishes. . . . He had five sons too—five grown sons. He had taught them cooking, and before the war they were all excellent cooks. Now they were all soldiers, excellent soldiers, he hoped. One was a sailor in the North Sea, one in the tank corps, one in the infantry, one with the picked troops of the Commissariat of Internal Affairs, and one an aviator.

They were brave fighting men, all of them, especially the aviator. A first lieutenant and pilot of a bomber, he had been in many battles. Once his plane was badly smashed. The major, the radio operator, the navigator were killed. Only he and the gunner remained alive. He clung to the plane, shattered as it was, until he made a forced landing in enemy territory.

As soon as they came out of the cockpit a German chauffeur made them prisoners of war. But they didn't remain prisoners. The lieutenant shot the German chauffeur, quickly packed his gunner and himself into the German automobile, and made their way to the Russian lines. That was the kind of soldier his son was—a real Tuliak.

The little man grew silent, put his hand inside a pocket, drew out a letter and gave it to me. "Read it," he said, and from the solemn tone of his voice I knew that the letter held no happy tidings. It was from the political commissar of the squadron in which his aviator son had served and it conveyed the sad news that the boy had been killed in an accident. The commissar eulogized the pilot, offered his condolences to the father, and in some detail described the military honors that were conferred on the dead soldier at his burial.

I passed the letter to Alec Werth. After he had finished reading it there was silence. "Unpleasant news," said the cook as he put the letter back into his pocket. He didn't weep, didn't sigh, didn't say a word, but it was obvious he felt not only pride in his son but grief over his sudden death. To break the tension Alec said, *"Vypyom tovarishtsh* [Let's have a drink, comrade]."

"Vypyom," said the cook, and Alec filled the glasses.

CHAPTER 14: MOSCOW

THE PLANE from Kuibyshev to Moscow was so crowded with passengers and baggage that some of us could find no seats. So we crouched down on the floor. No one complained—everyone was only too glad to have an accommodation to Moscow. A factory director from Kuibyshev was so happy he told anecdotes and made the passengers roar with mirth.

I found a place on the floor beside this director and looked outside. We flew low, at times almost brushing treetops. The day being clear and sunny, we commanded a superb view of the countryside. Here none of it had been devastated by war, and the ripening fields of wheat and the immense meadows were a cheering sight. The closer we drew to the capital the more varied and rugged was the landscape, with richer grass in pasture and meadow, with more abundant streams and denser forests. The weather had grown cloudy but there was no fog, and only the faraway burst of lightning and the echoing rumble of thunder disturbed the peace and the calm of it all.

On my arrival in Moscow the superintendent of the airport telephoned to the hotel for a car. To while away the time I went into the buffet. A broad band of sunlight fell with a silvery shimmer through a corner window, missing the glass-covered counter and leaving it in the shadow. The dark-haired unaproned waitress in a dark dress and with a dismal expression only deepened the shadow and so did the slices of dark bread laid over with fat-fringed slices of dark ham and the row of dark beer bottles. There was nothing else in the buffet—the war had stripped it of the sweets and pastries, the meats and dairy foods that had once made it one of the best buffets in Moscow. Yet people were rushing in, mostly men in uniform, fliers and others, eagerly partaking of the fare on hand and enjoying it with no less relish than the sumptuous dishes of prewar days.

Watching them, captains, colonels, others of lesser or greater rank, and civilians, only re-emphasized the physical sturdiness of the Russians and their unrivaled gift of adaptation. In time of stress they can dispense with comfort as easily as they wipe the water off their faces. They can eat the coarsest food, wear the roughest clothes, stay out in the open by day and by night, in rain and wind, in mud and blizzard, sleep on the bare floor of a barn or the bare earth of a dugout as soundly as in a feather bed. Boundless is their gift of endurance!

The hotel car came and we drove off along the Leningrad Chaussé, Gorki Street, the Sverdlov Square, and to the Metropole. The street lured me; as soon as I washed and cleaned up, I rushed out and wandered about the city. I could hardly believe I was in Moscow—the capital of a nation fighting for its life. It was so clean and bright, so green and spacious, so noiselessly and cheerfully awake—not at all like the city which only a few months earlier had withstood the siege of the cruelest foe it had ever known. The front there was only one hundred miles away, martial law was the order of the hour, but in outward appearance the city and the people evinced no hint of war or battlefield—not at first glimpse. Gorki Street, the Leningrad Chaussé, the Sadovaya were beyond recognition. I had never seen streets so broad anywhere in the world, including Salt Lake City, famed for its broad and beautiful avenues. In the six years of my absence rows on rows of buildings have been taken down and rolled back on gigantic wheels, corners have been squared and scraped, lines have been cut and straightened; and now the avenues sweep on like shiny and majestic rivers. The trees that were mere saplings a dozen or so years ago rise high and straight with umbrella-like clusters of limbs and foliage, enhancing the brightness and the gaiety of the streets. Outwardly Moscow appeared more comely and more lively than I have ever known it. After Kuibyshev, one of the most unkempt and shapeless cities in Russia, its cleanliness was almost startling. The main streets gleamed with asphalt, and remote side streets, whose ancient and battered cobbles billowed up and down like waves of a choppy stream, looked scrubbed and washed.

Of course Moscow has no fat newspapers like New York, Chicago, and other large American cities with which the citizenry can at will litter sidewalks and pavements. In Moscow, as all over Russia, paper is precious—any paper, more particularly newsprint. Men use it for rolling cigarettes, women, too, when they smoke, and it is good for wrapping purposes or for keeping rain and wind out of a broken pane. Besides, after years of campaigning, of exhortation and denunciation, the Muscovite, child as well as adult, has become too disciplined and too civic-minded to sully the streets with anything, even with the cork of a cigarette. He throws refuse into the wooden or tin receptacles that, sentinel-like, line the sidewalks. Dull green or dull blue, and perched against rain pipes or walls, they greet the eye of the pedestrian every few steps, as if to remind him of the purpose of their existence and of his duty to abide by this purpose. If he forgets it someone may remind him of it, not always with politeness.

All the more extraordinary is the cleanliness of the streets because Moscow is an old city with old buildings which off, more than on, the main streets, cry for repairs and for paint. Plaster crumbles, old paint peels like the skin of a diseased person. Now there is no time for new plaster, for fresh paint—and scarcely any material is available. But no matter—on Gorki Street, on any main avenue, an aproned man or woman darts out of somewhere with broom and dustpan, with brush and hose, and debris, like snow under hot water, melts out of sight.

The Moscow public on first sight seemed as neat and bright as the streets. Unlike the public I had seen in Stalingrad, Baku, Kuibyshev, it disdained even on workdays to appear in old attire. It was too aware of the dignity of the capital, too proud of personal presence to go back even in these tragic times to the slovenliness of the early years of the Revolution.

"Don't you think," asked a Muscovite acquaintance, "the capital is a little frivolous?" He admitted that he thought so because in spite of war, personal privation, sorrow, young people were sporting no small amount of finery.

By comparison with their appearance on my last visit, young Muscovites did seem as if on parade and in a festive mood. I must remind the reader that I am comparing Moscow of the summer of 1942 not with New York or any other American city but with Moscow of 1936.

I was amazed at the show of color in the street. There was of course much khaki, which only accentuated and enlivened the other colors— white, blue, and red predominating. The shimmer of silk, almost unknown six years ago, was now not inconspicuous. There were more hats here than in other cities, more lipstick, more handbags, more permanent waves, more dyed hair.

Many girls were in uniform—young girls of college or high-school age. Some were college students, others were from the factory and the farm. Some dressed like men in trousers tucked inside knee-high leather boots, in men's tunics, men's caps. Others matched the khaki tunic with a khaki short skirt, or else with a skirt of blue serge or some gray fabric, and sport shoes. Seldom did these "army girls," as foreign correspondents speak of them, appear in khaki dresses. In wartime, I heard one of them explain, the dress is too impractical, has not enough pockets for one thing; and an abundance of pocket space, especially in the tunic, is not the least of the many merits of Russia's eminently practical military uniform, whether of private or of marshal.

There were army girls who affected boyish bobs. There were others who, peasant fashion, flaunted long braids. Still others coiled their hair in old-fashioned housewifely manner. Nurses, chauffeurs, snipers, members of anti-aircraft crews, doctors, dentists, propagandists—they could all shoot, wield a bayonet, and most of them could operate a machine gun. Some of them had seen action, others were still in training or on their way to one of the many fighting fronts. They saluted and were saluted like men. They were gay and stately. In their stride as they marched about the streets there was energy and briskness, as much as in the men, and not a little more grace. Together with the girl militiamen, they added to the cheer and the vividness of the city.

Amazingly enough there were many old people around—some very old, leaning on stout canes, chins trembling with feebleness. One wondered how they had escaped evacuation during the dread mid-October days of 1941, when the city was in the throes of direst uncertainty. Old Muscovites, perhaps, who refused to leave, preferring to share the fate of the city, perish with it if necessary. Never before had I known them to be so visible in Moscow. They were grandfathers and grandmothers. They stood in line for papers. They went shopping to the market place, to the stores. They carried letters to the post office. They went to the libraries for books. They played with grandchildren, those that were left in the city—in the parks, on the boulevards. They sat on the benches, read, watched the passing scene, reflected—perhaps remembered and dreamed.

Involuntarily one stopped to look at them with no little respect and reverence. How much history they had witnessed—these quiet and digni-fied stariki: The last czar at the crest of his glory and the depth of his degradation; the Bolshevik Revolution, the civil war, the Nep, the mortal feud between nationalism and internationalism; the crash of the old and the eruption of the new social order; the Five-Year Plans; the new factory system; the smashup of the old village; the stormy birth of the kolkhoz; the havoc and the energy, the doubt and the exaltation, the death and the life that it all spelled—they witnessed it all. How much drama they ex-perienced and how much explosion they survived—like the lone birches and oaks on the battlefields of which Russian authors write with wonder and love as of the triumph of life over death.

If old folk in Moscow were conspicuous by their presence, children were conspicuous by their absence. In Baku, when I was there early in June, in Stalingrad, in Kuibyshev, children livened up the streets with their marching and their singing. Playgrounds, parks, courtyards swarmed with them. But not in Moscow. From here children were evacuated by the tens of thousands. The Three Hill textile factory—one of the oldest and largest, employing five thousand workers—transferred to the Urals eight hundred children, all sons and daughters of workers. Other factories did as much or only a little less. Also tens of thousands of children went to the country to help in the harvesting.

Nor, unlike other cities, was Moscow obtrusively a city of women. The population had substantially thinned, perhaps it had been halved, so immense had been the movement away to more remote and more secure regions. Yet to judge by the people in the streets, the proportion of women was not, as a Russian writer had said, "dismally" preponderant over men.

Many of the men were in uniform, but astonishingly large numbers were in civilian clothes, young men, too, in the late teens or early twenties, of an age not noticeable in villages or cities behind the fighting lines. They were engineers, directors of factories. They were political emissaries, delegates to conferences, students. They were messengers from the endless guerrilla detachments that fought in the enemy's rear. So youth in this land not only fought but worked and planned and came to Moscow. More than ever was Moscow the capital of the country, the motivator of plans and methods, of ideas and fulfillments. Like the roots of a tree it supplied all manner of sustenance to all other parts of the land. And so while Moscow abounded in women barbers, women directors of factories, women drivers of trolley busses, women mechanics—in fact, with women workers in every profession and every occupation—women monopolized neither the work nor the scenery of the Soviet capital.

Moscow had lost much of the undisciplined primitiveness which had so long brooded over it. Gone were the straw market baskets. No one carried them any more, not even in the market places. They were supplanted by the loosely knitted bag. Muscovites are happy with the substitution. So light is the bag that it can be neatly folded like a napkin or handkerchief and tucked into the pocket with no more bother. That, in fact, was what Muscovites did in prewar times. With the hamper always at hand it was no trouble to pick up something extra in the shop or the market place. Some wit had even given it the affectionate name of *avoska*—meaning "at your service, sir." Now, because of the war shortages, it has been renamed, also affectionately, *napraska*—"I'm no good any more."

More than the rest of the country Moscow was shorn of beards. The once-renowned Moscow "whiskers" are no more—except on the stage and only in old plays. The smooth-shaven face is a mark neither of emancipation nor distinction. It is not only the fashion but the habit of the times, as in England and in America. The mustaches, too, are disappearing. Few Russians youths in or out of uniform cultivate beard or mustache.

Beggars were around but usually off the main avenues. Only on steps of the few remaining churches during hours of service did they rise up like ghosts of yesterday—old folk, old not only in years but in manner. The blind and the halt, the maimed and the humpbacked, chanting and wailing, bowing and begging, reminders of an age now as vanished as the luster of the churches before which they gathered.

Thus, outwardly, Moscow looked more decorative, more cultivated, more aware than ever of its dignity, its worldliness, and seemingly as much at peace with itself and the world as the rows and rows of freshly grown trees. Outwardly, in their appearance and behavior, the people matched only too well the brightness and the liveliness of the city.

Together with several other American correspondents I went to the park to attend a concert of the Red Army Choir. This is the finest choir in Russia and includes dancers as well as singers. The immense outdoor auditorium was crowded with Muscovites dressed in their best, more than eager for diversion and entertainment.

On the program was a group of comical songs. The audience roared with amusement, clapped for encores. The dancing was as joyfully appreciated as the songs. To some of the correspondents who never associated joviality with Russian entertainment or Russian temperament, the robust humor of Russian folklore—in song and dance—was as much a revelation as the gay laughter of the audience. But watching the crowd, hearing the loud appreciation of the singing and the dancing, we forgot the country was fighting a war not of life but of death.

Yet it was only outwardly that Moscow glittered and more on first than on subsequent contemplation. The ravages of war were scarcely visible—superb were the city's defenses during the days when the Germans were rushing to conquer it. Yet evidence of vigilance and ordeal was ever present. The boxes of sand, for example, and the barrels of water with which to extinguish incendiary bombs. They were all over the city on the sidewalks and spaced only a few steps apart. In the back streets children played in the sand but were careful not to spill any of it. They seldom dipped their fingers into the water, as though it were something too dangerous or too sacred to touch. The larger and more important shops were "blinded"—windows were hooded with wooden frames or were hung with heavy black paper. In apartment houses and offices windows were crisscrossed sometimes in fantastic shapes with bands of white or black cloth. Here and there from tops of tall buildings loomed evidence in one form or another of military vigilance. Hardly a house but cried for new paint, for new plaster. New buildings were at a standstill. The rising steel framework of the Palace of Soviets, which when finished was supposed to rival in height the Empire State Building in New York, loomed like a skeleton of a gigantic monster. The glassed new office building of the Commissary of Light Industries was as if frosted with thick red rust. Camouflage, massive and ugly, tarnished the color and transformed the lines of many an eminent building. Safety mattered more than good taste and was the law of the land. On closer scrutiny the brightness of Moscow, so surprising on first sight, faded into ever-deepening haze, like skies at twilight.

All the more meaningful was this haze when one became aware of the barrenness it enveloped. Gone were fruit stands. I arrived in Moscow at the height of the cherry and berry season. On previous visits at such a time of the year booths gleamed with black and wine-colored cherries, with gooseberries, raspberries, blackberries. Now the booths were shut or dragged away. So were the gay-painted carts that once dispensed ice-cream sandwiches and Eskimo pies. Except in some of the choicest restaurants ice cream was rarely available, and these restaurants, like all the others in the city or the country, served only their own designated clientele.

Before the war Moscow was fast becoming a city of quick-lunch counters, cafeterias, some with automatic devices. They were all closed. Food was planned, rationed, and distributed according to assignment. Cafés, too, were becoming as distinctive a feature of prewar social life as clubs in factories and other institutions. One could go to a café, plump oneself into a comfortable chair, and browse at leisure over a glass of tea, coffee, cocoa. One could meet friends there, read newspapers and magazines, listen to the broadcast of a violin player or a choral concert, or just sit and dream. Now not a café was open. The comforts and the pleasures which Moscow had begun to know and appreciate in prewar days became a memory.

The food rations were none too ample. Workers in war industries received two pounds of bread, black and white, daily. Monthly they received five and a half pounds of cereal, the same amount of meat, two pounds of butter, two and a half pounds of fish, one and a quarter pounds of sugar. Bread is always available in the amounts prescribed but not meat, fish, or butter. Eggs might be substituted for meat at the rate of fifteen for two and a half pounds of meat. American lard, which has made its appearance here of late, provided another acceptable substitute. It was doled out at the rate of half a pound for each two and a half pounds of meat. People liked it, often clamored for it, ate it like butter by putting it on their bread.

The rations of office employees were lower. They received one and a quarter pounds of bread a day, and monthly three and three quarter pounds of cereal, one pound of butter, three quarters of a pound of sugar, three pounds of meat, two pounds of fish. They, too, were offered the same substitutions as workers in war industries and others, especially when meat, butter, and fish were not available or could be had in extremely limited amounts.

Children were allowed two glasses of milk per day and pregnant women received special quantities of sugar and fats. Of course factories and other institutions, especially schools, were cultivating farms and gardens and were seeking to supplement the prescribed rations with fresh allotments, especially of vegetables. In Moscow about 200,000 householders were cultivating in their back yards or in the suburbs individual gardens. . . . "See?" said a woman, pointing to a growth of lettuce in a wooden box on the balcony. I couldn't help laughing. "Why do you laugh?" asked the woman. "Because," I said, "when the campaign for eating lettuce was started in this country I heard Russians say again and again that they'd never eat it because it was no better than grass—good for chickens but not for human beings." Solemnly the woman replied, "It's different now. We've all become vitamin conscious, and we'll eat anything green—if only it is eatable." Indeed they would. I've seen children and adults in crowds searching meadows and fields, for sorrel, for young nettles, for other edible weeds and grasses—all in the cause of vitamins.

Many people were away from Moscow—people I knew and wished to see. Writers were at the front in Central Asia, in Siberia. Actors were scattered all over the country and the front—entertaining civilians and soldiers. Teachers were out of the city—fighting, driving trucks for the Army, helping to gather the rich harvest on the state and collective farms. Workers were also gone—the ones that I knew. Many I shall never see. The musical Ukrainian youth, who had married a Jewish girl as musical as himself and in whose home I had spent many joyful evenings, was dead. His wife became a sniper, ". . . killing Fritzes and doing it well," said a neighboring woman. Many workers, very many—and others too. Office clerks, schoolteachers, college students who joined the volunteer militia regiments in the autumn of 1941 and marched out to battle never came back. Some of their wives were evacuated to eastern towns, others remained in Moscow and were carrying on as best they could—working, maintaining their homes, bringing up the children, and hoping for a speedy and victorious end of the war.

I thought of Anna Vladimirovna and her husband, Boris Nikolayevitch, an eminent chemical engineer. I had their address and went to see them. They lived in a new apartment house on a remote side street. I climbed the six flights of stairs, for the elevator was not running. The stairs were nicked and uneven, as is often the case in the hastily built Russian apartment houses. On reaching the apartment, I saw no name on the door, not even on the black letter box. A white bell button peered out of the murk and I pressed it. From the inside came a girl's voice: "Who is it?" I asked if Anna Vladimirovna was home. The door opened and I saw a young girl, barefooted, with sturdy legs and in a red dress. I knew it wasn't Yelena, Anna Vladimirovna's daughter, who was only fourteen when I had last seen her six years earlier. Hesitantly the girl said, "Anna Vladimirovna is ill —she's in bed." "Tell her—" I began, and instantly from an adjoining door came Anna Vladimirovna's ringing voice, "Maurice—you—please, please, Marusia, show him in——"

Smiling and blushing, Marusia showed me into Anna Vladimirovna's room. It faced the courtyard and was well lighted by two large windows, red now with the glow of the departing sun. Anna Vladimirovna was in bed covered with a light brown blanket. Her face was pale and smooth with scarcely a wrinkle despite her fifty-seven years, and her large dark eyes glowed with their one-time kindliness and animation. Her illness, she said, was not serious: nerve strain—that's what the doctor had said. He had ordered her to bed, but it was not anything worth talking about, she insisted.

I asked about her family, one of the liveliest I had known in Moscow. Only she and Boris Nikolayevitch were in the capital now. The older son was in the Army, an engineer, and they often received letters from him. No letter could be more precious than one from a man in the Army—it meant he was alive. Had I talked to many Muscovites? Not a family but had someone at the front—and they were all waiting and yearning for a letter. If a month or two passed and none came—they imagined the worst; yes, they did—the poor things.

Anna Vladimirovna put her hand on her breast and sighed. Then she resumed the story of her family. Natasha, her older daughter, had graduated from medical school and was at the front; so was her husband, also a physician. They were at the same front but not in the same hospital. Yelena, the younger daughter, was in a distant city teaching foreign languages in a military school and hoping someday to become a professor of Eastern languages in a university. Lubochka, their niece, whom they had adopted, had graduated from the agricultural academy and was a livestock expert on a large state farm in Siberia.

Many Muscovites had gone to Siberia and to the Urals. . . .

"Never were we so close to Siberia as now," she almost chanted; "hardly a family we know but's got someone there or in the Urals. Thank God for Siberia and the Urals—such a wonderful country . . . and it'll always be ours—always." There was pride in Anna Vladimirovna's voice—pride of possession and of reassurance.

"How d'you like Moscow?" she asked.

"It glitters," I replied.

"Doesn't it? Life is hard, harder than you can imagine, but you'd never know it by the looks of the city and the people, especially in summer. That's the way it should be. No use moaning." She talked as her younger daughter Yelena might have talked, the language not of sorrow or self-castigation but of cheer and battle.

She gave a short and loud laugh. "I often laugh at myself, my husband, our friends. We'd gotten so spoiled before the war. Once my husband asked for mandarin juice. He wasn't feeling well, and the doctor told him to drink it all the time. The maid said she couldn't find it in the store —they were out of it—and would you believe it my husband got angry and said it was nonsense that the store should be out of mandarin juice." She laughed again. Then, placing her hand over her breast as if to hold back her own excitement, she narrated at length the improvements that had come over the capital in the years of my absence. Housing was still a problem but not food. Meats, fish, milk, eggs—there was plenty of everything, though not always in the same shops. In winter it might be difficult to buy milk, but it could be bought, and the prices were reasonable. Vegetables, too, were available all winter. Muscovites were taking to tomato juice like a calf to milk. They drank it all the time. They even started eating canned corn—"We did in this house, and remember last time you were here we told you we couldn't stand it—it was good for cattle but not for civilized human beings? Boris Nikolayevitch and I often remembered these words and laughed at ourselves for being so stubborn in our dislike of corn."

The information about food in the prewar days was not only new but exciting. It indicated how eagerly and successfully Russia was Americanizing herself in her food habits. The process had started soon after Mikoyan, the young Armenian Commissary of Food, had made a brief visit to America. Russian prejudice against corn flakes, for example, was loud and voluble. I remember a clerk in a shop outside of Moscow telling Americans who were on a picnic and who asked for a box of corn flakes not to buy it because nobody was buying it!

But in time, to judge from the words of Anna Vladimirovna, the prejudices were more than overcome. More and more factory-prepared foods were finding their way to the Russian dining table. Moscow had even taken to "frozen foods"—frozen vegetables, frozen berries, frozen meat—which might be bought even in winter. Canned goods, dried foods, preserved foods—Moscow had had plenty of them. But now—well, it was war! If they have plenty of bread and cabbage, and a bit of meat and a bit of precious American lard, they can live and work and fight—indeed they can.

As we were talking Boris Nikolayevitch came. I hardly recognized him —he looked stooped and wrinkled, a little stern, and minus his shiny Dutch beard. "He's aged, hasn't he?" said the wife, and laughed as if to reassure the husband she didn't mean a word she had said. Boris Nikolayevitch didn't mind. "Not at all," he said; "I've grown younger; one has to be young nowadays, in this country. We are all young—no old people any more—gone, abolished, made young again," and he laughed with the baritone melodiousness which was always as much a mark of his person as his now-vanished little beard.

We sat down to tea, bread, fish, cheese. I was loath to partake of the food when it was so closely rationed and when at the Metropole I was served ample meals. But host and hostess insisted, and it is never any use to resist Russian hospitality.

As we "ate" our tea, as Muscovites say, we talked of America, of the Allies, of Germany, above all of Moscow in the days of the German advance on the capital.

"One day," said Anna Vladimirovna, "our maid came in and said, 'D'you know, mistress, there's talk in the streets the Germans will be in the city by evening.' Did I shiver!" and she shook as if with cold.

"In my laboratory we could hear the guns," said the husband, "a terrible sound. Have you read much German literature?"

"Not too much," I answered.

"Lucky man," said the wife. I glanced questioningly at her. Hating Germans as much as they did, Russians did not permit their emotions to thwart their appreciation of German literature, German music—any form of German art.

"If you haven't read much German literature, then it's easy to believe everything you want to believe about Germans nowadays," explained Anna Vladimirovna solemnly.

"We've read a lot of German literature," said the husband, "and often my wife and I ask ourselves what's happened to those German mothers of whom German poets, novelists, dramatists have written with such ecstasy and whom we, too, in our imagination had idealized. How can they bring up such sons as the German soldiers now are?"

"For the first time in my life," said the woman, "I was really frightened. No, not of death. Death is easy—you lie down, you fall down, you pass away and are finished. Nothing matters any more. But on October 16, 1941, I was frightened of life, of myself, of my thoughts, my emotions, my tastes, my enjoyments, my friends, my books—everything. I realized there was nothing about me, my blood, my language, my ancestry, my ancient Slav ancestry, dating to the time of the Moscow princes, that Germans could respect or feel bound to spare. Of a sudden I had the feeling that because I was what I was, what Russia had made me, old Russia, new Russia, always Russia, I was in the enemy's eye fit only for death. I was so terror-struck that I did something terrible."

"D'you know what she wanted to do?" asked the husband.

"I wanted to burn our books," said the wife.

"Think of that, eh?" The husband laughed loudly and expansively.

"I got several sacks and boxes and hastily flung books into them by the armful—all these books on our shelves—Thomas Mann, Feuchtwanger, Goethe, Gorki, Chekhov, Tolstoy, others—German books, Russian books. When I had the sacks and the boxes filled I picked up one sack and started down the stairway for the furnace."

"Imagine that—throwing into the furnace Gorki, Chekhov, Tolstoy?"

"I dreaded the thought of their doing it."

"Luckily I happened to be coming home at the time. I almost had to drag her back into the apartment."

"A sin to admit it but he did—that's how hysterical I was with fear."

"I lifted the sack off her shoulders, carried it back to the house, put the books on the shelves, and said, 'These books are sacred to me and you; our very souls are in them, have been all the years we have been married—thirty-seven years—and only over our dead bodies will German hooligans take them away from us—over our dead bodies.' "

"He embraced and kissed me and said, 'We'll stay here, darling, in our beloved Moscow, and if they come, we'll face them—in each other's arms, beside the very books we love and they hate—Russians to the end, Muscovites to the last.'

"Yes, that's what he said!" Anna Vladimirovna sighed, glanced tenderly at her white-haired husband, and continued, "Then I was afraid no more. I was ready for the worst—the very worst. We're like that now—all of us —unafraid of them, or anything they might do to us, aren't we, Boris Nikolayevitch?"

"Aye, we are," and with sudden harshness, though without raising his voice, he went on, "unafraid to die and unafraid to kill. We've killed plenty of them. We'll kill plenty more. We hate enough to do it—Lord have mercy how we hate—and we know how to do it. Even our children are learning—some have already learned only too well."

Later I left. Midnight was the curfew hour in Moscow, and I wanted to reach my hotel in time. The black-out was so dense that when I came into the street I felt as though the world were inked out. I saw nothing, not even the outline of the house I was leaving. For some minutes I remained standing until my eyes could penetrate the darkness. Then slowly

I groped my way out of the courtyard and walked ahead with arms out-stretched to avoid bumping into people or obstacles.

Because of the piles of brick and rails at the curb, I avoided the side-walk and walked in the middle of the street. At the turn of a corner a truck with not a hint of illumination in front or in the rear roared past me, and I barely darted out of the way. I then took to the sidewalk and stumbled into Gorki Street. Here were red and green traffic lights, the same as in other parts of the world. Cars roared up and down, piercing the darkness with glints of blue and yellow—like the eyes of unseen and ever-moving monsters. Now and then, like a noiseless firecracker, a ball of white light burst over the sidewalk, lingered for some moments, then vanished. It was the flash of the ever-alert militiamen or military guards. Rifle and bayonet on shoulder, case of ammunition on back, they halted passers-by for an examination of their identification papers. Men in uniform, these "hawks of the night" checked with special rigor to make certain they were Russians and not Germans in the disguise of Red Army men.

I reached my hotel. The square in front and the city beyond were swal-lowed by the darkness and sounded almost empty of traffic. Yet now and then, as if appearing out of nowhere, a trolley bus, as void of light as the buildings around, glided past and sank ghostlike into the dense blackness. Curfew time was approaching. People tore out of the hotel door and plunged into the night, their footsteps echoing lightly over the sidewalk. I marveled at the speed of their movements. The city might have been brilliantly lighted for all the concern they showed.

Then lightning flashed across the sky. For flitting moments the city was in a dazzle of white light, exposing to the eye the roofs of buildings, pavements, trees—and couples in close embrace.

Yet all the time the image of Anna Vladimirovna and Boris Nikolaye-vitch was before my eyes and their words rang burningly in my ears.

CHAPTER 15: STALINGRAD

"STALINGRAD, city of our youth! Amidst the flowering of its hopes our youth has built and beautified this city. Determined and inspired, seven thou-sand Komsomols put up the tractor plant. . . . Here on the bank of the Volga . . . they dreamed, hoped, loved. All they held dear and beautiful is associated with this sun-flooded city. It is the cradle of our youth, our life, our love."

These rhapsodic words from the pen of Korotyev and Levkin, leaders of the Komsomol in the Stalingrad province, appeared in the *Komsomolskaya Pravda* on October 24, 1942—a crucial day for the mighty industrial city on the broad and shimmering Volga, the most crucial in its history. Forty miles long, the longest city in Russia, and hugging the steep and winding banks of the river with the firmness of a child clinging to its mother's arms, Stalingrad was the object of all the might and all the wrath of Hitler's highly mechanized legions. Thousands of cannon, thousands of planes were daily hurling thousands of tons of dynamite and steel. Not a weapon of destruction that Germany's superb industry created, not a tool of terror that the Nazi passion for cruelty had unleashed, but was showered on the city. From great heights pieces of rail girders, empty gasoline tanks and all manner of metal were flung down, not only to kill but to shock and deafen the population with the explosive noises of their falling. Fire and blood vied with each other for supremacy, and nobody could tell whether the fire would burn up the blood or the blood would quench the fire. . . . The Russian war map offered no consolation and no assurance. It stirred only the gravest premonition, the keenest anxiety.

"Ruined, burned, wounded," reads further the story of Korotyev and Levkin, "this hero city is still our native hearth. Now that fierce battles rage in its streets, it has become more precious than ever. The flames of the endless fires burn the very soul."

The flames burned the souls not only of the thousands of youths who had toiled and sacrificed building Stalingrad and who were fighting and dying in its streets, its houses, its underground entrenchments, but of the many millions of people all over the land. Centuries of Russian history, all that the Russian people had achieved and believed, all that they were and were aiming to make of themselves, the Revolution, the Plans, everything that was associated with the word Russia, old and new, was facing the harshest and fiercest test it had ever known.

Stalingrad had become a synonym of catastrophe, also a symbol of glory.

"If there is death over our heads," wrote the gifted Konstantin Simonov, "there is glory by our side. Amidst the ruins of our homes the cries of our orphaned children, glory like a loving sister never leaves our side!" Beautiful words like those of Konstantin Simonov were in those dark days the sole consolation the people were offered.

Yet at two-fifteen in the morning of June 6, 1942, when our train rolled into the railroad station, Stalingrad was still a city, still a ring of factories and oil tanks, still a center of schools, parks, theaters, and homes—the homes of half a million people, among whom were the flower of Russia's pioneering youth. Dense was the black-out, and though dawn was already breaking across the sky the people who moved up and down the platform seemed no more than hurrying shadows. Here and there along the tracks of the immense railroad yard someone was swinging a half-covered blue or red light. As it flashed upward out of the blackness and fell into it again, it looked like a magic toy which someone was fondly juggling for the edification of newly arrived passengers. Only the answering whistles of unseen and puffing locomotives made one aware of the stern utility of those flashing and vanishing lights.

The train was an express bound for Moscow and stopped for only ten minutes. Passengers tumbled out of the coaches, loaded as always with sacks, baskets, and bundles. They searched and called for porters, and when none were in sight, they grumbled with indignation. So did I. I was traveling with two young American diplomats and two British diplomatic couriers. Between them they carried a truckload of diplomatic mail. The hotel in Baķu had telegraphed the stationmaster in Stalingrad to meet us with porters and a truck and drive us down to the pier. But search the darkness as keenly as we might, we saw nowhere the least sign of truck or porters, or anyone to meet us. Hastily and excitedly we carried the mail outside and stacked it on the platform. Leaving my companions with their precious baggage, I left in quest of the delinquent stationmaster. I could not find him, and a man who had finally unearthed him in some faraway office returned with the news that he had received the telegram but had forgotten about it. So forgetfulness even in time of war was still a feature of Russian services. Angrily the man muttered, "What a stationmaster! He ought to be brought to trial for criminal neglect of duty." But the telegraph operator, a portly girl with shining blue eyes and a pale face, remarked pensively, "He is so overworked these days—you can't expect him to do everything."

The explanation only incensed the man. "If he were at the front, they'd show him and men like him what it means to forget an urgent telegram."

The girl shook her head and said: "He is doing his part; don't get excited."

"Oh," sighed the man. Forgetting his rancor, he turned to me and said, "Have you a *papyros?*"

I offered him an American cigarette. Hungrily he inhaled several puffs. His demeanor instantly changed. He smiled at the telegraph operator and at me, and when I suggested that we find the NKVD (Commissary of Internal Affairs), chief of the station and ask him to help us out, he answered with Russian soldierlike completeness: "Yes, the NKVD chief." We found our man and he more than came to our aid.

Walking along the platform of the railroad station, I became more acutely aware of war than in any other place I had visited since my arrival from Teheran. Not only was the black-out more complete here than elsewhere, but the station and the railroad yards were lined with hospital trains. I passed some of the trains, and the white uniforms of the nurses cut the thinning darkness with a knifelike sharpness. Young girls, most of them, the nurses were busy tending the wounded men. Some they carried on stretchers from the train to the platform for an airing. Others, who were unable to walk, they helped hobble around. Still others, limping badly, leaned on their shoulders and stepped along with the slow and awkward gait of children learning to walk. The nurses talked and cheered the men, many of whom were puffing incessantly on their hand-rolled cigarettes. In no way did they consciously betray the experience out of which they had just emerged. Yet the white uniforms of the nurses, the stretchers, the crutches, the bandaged heads and faces, the casings on feet and arms, the heavy smell of disinfectants, now and then the deep groans of those in pain, told the story of this experience more poignantly than words.

Overcome by a desire to speak to them—men fresh from the fighting front, the first I had seen—I offered them American cigarettes. I forgot, or rather did not bother to remember, that the two packages in my pocket were not mine but were given to me in Teheran for someone in Kuibyshev. I had intended to dispose of only a few but I had not reckoned with the hunger of Russians for American cigarettes nor with the good fellowship which prompted the soldiers to call to their companions to come over. Within a few minutes the two packages were empty.

Yet I failed in my purpose. The head nurse appeared and did not want the wounded men to excite themselves with talk—not at so early an hour. So I wandered on to survey further the scene at the railroad station. Presently I heard someone calling me. Turning, I saw a soldier on a crutch hobbling toward me. I waited, and when he came over he held a Russian papyros with its long, holderlike cork tip beside the American cigarette I had given him.

"This one," he said, lifting the papyros, "I can smoke to the last strand of tobacco. But this one," lifting the American cigarette, "I must throw away when it still has a good half-dozen puffs. Do you call that American efficiency?"

It was not the first time I had heard Russians discuss the economic merits of their papyrosy and our cigarettes. Nor was it the first time that in reply I merely shrugged my shoulders. Other soldiers stopped and heartily agreed with their comrade. For a few moments there was lively chatter and laughter. Then the head nurse once more came over and put an end to the discussion.

The war posters at every hand were more numerous and more flaming than any I had yet seen and only accentuated the intimate awareness of war. There was no mistaking the wartime spirit and temper of Stalingrad.

"Death to the German invaders!" read several posters. Another, showing a giant Russian soldier with a tommy gun shooting flames at fat barrel-like men who were spattered with swastikas, carried another quotation from Stalin, "If the Germans want a war of annihilation, they shall have it." Germans, not merely Nazis or Fascists!

Still another poster bore a quotation from Lenin, "To insure victory the masses must be inoculated with contempt for death." No general could have enunciated a more militant slogan. The stories in the daily press of men and women hurling themselves at enemy machine guns so as to silence them and make possible a Russian advance demonstrated only too eloquently the earnestness with which Russians had taken to heart Lenin's admonition.

Still another poster showed peasants, factory workers, and intellectuals armed with rifles and hand grenades. "To bear arms," read the caption, "is the duty of every citizen." Every citizen! Contrary to the expectations of Hitler, contrary to the prophecies of professional Sovietophobes, professional soldiers and professional diplomats—who beheld only the blood and the horror of the Revolution and none or little of its plans and its creativeness—the citizenry used these arms not to shoot one another, or the officials of the Soviets and the functionaries of the Party, but to visit death on the invading enemy!

Still another poster of gigantic proportions arrested attention. It showed two murdered women with a child over them, weeping. "Avenge," read the caption, "the foul insults to your mothers and sisters."

From fences along the streets, bulletin posts at the squares, walls of buildings, banners in the air, Stalingrad flaunted its defiance to the invading enemy. It was a gigantic arsenal not only of weapons but of will. A Volga city, a Russian city, a Soviet city, it was obviously girded for war with all its vast human and mechanical resources.

Founded in the faraway and dismal year of 1589, the city, originally known as Tsaritsyn, was by its geographic position alone fated to become a battleground and storm center in all internal and external wars. It has ever been the key to the rich lower Volga basin, to the North Caucasus, to the Caucasus proper, to the black-earth steppes of the Kuban, to Saratov, Kazan, Nizhnii, and other Volga cities, as well as to Moscow and St. Petersburg (Leningrad). Without Stalingrad neither czar nor revolutionary, neither foreign foe nor native conspirator could ever hope to reach and to spear the ever-moving heart of Russia and make himself master of the nation and the people. Hence the dramatic history of this most wind-blown, most sand-swept, once provincial town on the Volga.

Erroneously, foreigners and often native Russians have assumed that

the name Tsaritsyn, by which it was for centuries known, is derived from
the word *czar* and is associated with the Romanov dynasty. It is not. The
word was borrowed from the name of a muddy river, *sary-soo,* in Tatar,
and means yellow water. It was transcribed as Tsaritsyn into Russian.
This river surrounded a little island on which the city was originally
founded.

Short-lived was the career of the original city. It met an ignominious
death in the peasant wars that broke out in the ensuing century. In 1615
the *voyevoda*—military chief—rebuilt it, this time not on the island but
on the mainland, where it now stands. A timbered village of no more
than three hundred and fifty population, composed entirely of officials,
servants, and military guards, it was one of the most southern outposts of
the czars—a military fortress more than a town or a village. It was shut to
Cossacks and other plundering wayfarers who at that time wandered on
foot and on horseback over the steppes and laid siege to the mercantile
caravans that journeyed from one part of the country to the other and from
Russia to Turkey and Persia and back again home.

"Thievish Cossacks," wrote the voyevoda to the Czar, "have designs on
Tsaritsyn. They want to burn it and beat up the local population because
government guards and officials are everywhere interfering with their loot-
ing expeditions."

But strong as the city was in 1670, the redoubtable Cossack rebel, Stenka
Razin, stormed its timber fortifications. He called on the voyevoda, who
bore the distinguished name of Turgenev, to surrender. The voyevoda refused, and barricaded himself with ten soldiers in the tower. Infuriated by
the refusal, Stenka in person led an advance on the tower, seized the
voyevoda and the next day executed him.

Many were the subsequent attempts of Cossacks and others to conquer
the city and wrest it from the rule of the czars. One hundred and four
years after its fall to Stenka Razin, another mighty Cossack rebel, Yemel-
yan Pugachev, assaulted its fortifications. But Pugachev failed to take the
city. It was too strongly fortified. Merchants and serfs were so elated with
their victory over Pugachev that they staged an elaborate and clamorous
celebration. Afterward they obtained complete control over its political,
social, and military life and made of it a stronghold of reaction.

Though the population grew slowly, the trade of Tsaritsyn rose rapidly.
Fish, lumber, bread, meat, hides, and wool were the chief source of its
mounting wealth. Then the machine age crept into its midst; a railroad
was built linking it by land with the other cities, including Moscow and
St. Petersburg. From year to year its industries expanded until in 1875
the French firm of Barre opened a new era—that of steel. Other shops
followed suit to swell the number of those already functioning. Brickyards,

lumber mills, tanneries, textile shops, ammunition works, soap factories, though small in size, were rapidly multiplying. Tsaritsyn became a leading trading and manufacturing center on the Volga, though never as renowned as Nizhnii Novgorod or as widely known over the country.

In 1916 its population leaped to 150,000, about one fourth of whom were factory workers. Definitely and decisively it became a proletarian city—which in pre-Soviet Russia meant a center of revolutionary propaganda and revolutionary conspiracy. Yet like nearly all Russian cities, especially in the provinces, in its internal structure it had achieved but slight change. It sprawled in primitive wildness over hills and lowlands, over sandbars and ravines. Streets were mostly unpaved. There was no sewage system. The gullies and ravines were often used as refuse dumps and gave rise to cholera and other epidemics. It was a city of a rising revolutionary ferment. There were four hundred taverns, scores of bawdy houses, only two high schools, and five grammar schools.

Then came the Revolution. The streets of Tsaritsyn turned into sanguinary battlefields. Stalin and Voroshilov were in command of the Red armies. Generals Kaledin, Krasnov, and Denikin failed to dislodge the revolutionary armies. But General Wrangel succeeded in compelling Stalin and Voroshilov to withdraw northward and to bide their time. On January 3, 1920, the late Kirov, one of Stalin's closest associates, moved up with an army from Astrakhan and together with the northern armies drove Wrangel into flight. Since then Tsaritsyn—subsequently renamed Stalingrad—has been one of Moscow's stanchest and most powerful supporters. It was Stalingrad in more than name. In all his controversies with political and military enemies, Stalin could always count on its undivided support and devotion. Not only his name but his outstanding military triumph was associated with the city.

I first knew Stalingrad, then still Tsaritsyn, in 1923, shortly after the end of the civil war and the famine that had swept the Volga in its wake. Provincial and primitive and, except for the factories, with but few marks of modern civilization, it still bore the ravages of the fighting. Building after building was perforated with bullet holes. Window frames were empty of glass; doors were split and unhinged and leaned in helpless disrepair against walls and porches. Foundations tottered and were on the edge of collapse. Paint was peeling off the once-gleaming church domes. Though it was deep summer people often walked in felts—they had no other footwear—and some strutted around in bare feet precisely as in a village. Swarms of homeless children darkened the streets, the bazaars, the pier, the railroad stations, the cemeteries. More than the vivid tales of the inhabitants, more than the ruin and desolation in some quarters, more than anything I had heard or seen, these children testified to the devastation and the sorrow that the years of strife and hunger had visited on the city and on the nation.

Yet even then no city in Russia, not even Kiev or Moscow, boasted more ample or better food. That was the chief reason for the presence of crowds of homeless children. Nowhere else had I seen so much bread, meat, fruit, vegetables, milk, butter, eggs. The immense sand-blown market place teemed with caravans of carts, drawn by ox, horse, or camel and loaded with all manner of produce, including beef, pork, and mutton—on the hoof. Nep had struck its stride, and trade was booming not only in the bazaars but in the shops. Some merchants I met, in their unreasoned hopefulness for a return to the old days of private trade, invited me to their homes and deluged me with inquiries about American textile and other manufacturers with whom they wanted to establish trade relations. "We'll open to them the richest market in the world," said an energetic and enterprising Armenian dry-goods merchant. The swiftness with which the Russian village and the Russian city were recovering from the death and the wreckage of the civil war and the famine testified to the inordinate fertility and riches of Tsaritsyn and the surrounding territory.

Futile and short-lived were the hopes and anticipations of the Tsaritsyn traders. With the fury of the winter gales that sweep periodically over its squares and streets, the Soviets only a few years later cleansed it of all private enterprise. I revisited the city in those days. Dark and empty were the rows of booths and shops in the bazaar. Dingy and barren were the rows of stores on the main avenues. Girded with physical power, enormous, will, which were in strange contrast to their puny experiences, the Soviets took over everything and launched one of the most ambitious projects of their young career—the erection of a monumental tractor plant, one of the largest in the world. The cries and groans of the disfranchised and the disinherited were drowned out by the roars and sirens of engines and scoop shovels as well as by the lusty chants of the thousands of new builders, youths mostly, boys and girls, raw and inexperienced, who were daily pouring into the ever-expanding squares and streets.

Stalingrad became a boom town not in the American but in the Russian sense, with the government and no one else sponsoring and boosting all the new construction and all the new production. Gone were most of the old taverns. Not one of the bawdy houses, save those that had dug deep underground, survived the swift onrush of the machine and of the new morality that it brought in its train. Food became scarce again—because much of it was now shipped abroad in payment for machinery. But the building went on. More and more acres of age-old steppe yielded to the encroachment of cement and brick, of iron and steel.

Americans by the score arrived to help in the new task—not communists or propagandists, but industrialists, architects, engineers, among the best that America had. None of them, not even the late Hugh Cooper who built the Dneprostroy dam, stirred the imagination of Russians in Stalingrad and elsewhere as did the bushy-browed and steely-eyed Jack Calder from Detroit, Michigan. He supervised the construction of the tractor plant. His disdain for red tape and procrastination, his good-humored brusqueness, his speed at work, his little pet camel that followed him around the grounds like a faithful dog, his readiness to leap into a trench, pick up a shovel and work side by side with ordinary laborers, so excited the Russians that Pogodin, a leading playwright, made him the hero of *Temp,* one of the most popular Soviet plays of those days.

On June 17, 1930, in the midst of the tempestuous campaign for the fulfillment of the initial Plan, the first tractor rolled off the assembly belt of the Stalingrad tractor factory. An inferior piece of machinery, with almost as many faults as virtues, no match yet, so Russia freely admitted, to the tractors that were imported from America, its arrival was greeted with clamorous joy. At last the Soviets were making their own tractors, and the very word tractor in those days had a magic ring and spelled promise and triumph.

In the years, indeed in the months, to come, the quality of the tractor kept improving. Weak parts were made strong. Clumsy construction gave way to smooth integration of wheels, springs, levers, and other parts. Fewer and fewer foreigners were employed. Finally hardly any remained, and Russians themselves, trained in the new or rather American technique of production, assumed supreme command. . . . Tractors by the thousands rolled out of the plant and were shipped to all parts of the country, even to faraway Kamchatka. More than ever Russia was becoming tractor-minded, engine-minded—and the glory of Stalingrad was high and nationwide.

By the end of 1932—which marked the termination of the First Five-Year Plan, fulfilled, according to official proclamation, in four years—the population of Stalingrad leaped to 400,000! The Second Plan came and passed. The population rose to half a million and, as formerly, was made up chiefly of young people from all over the land. The Third Plan arrived. Stalingrad had changed its very visage. . . . It became, as the Russians love to express themselves, "an industrial giant," one of the mightiest in the country, and "a citadel of culture." Shipbuilding, machine-building, the manufacture of highest grade steel and of precision tools, lumberyards —some as mechanized as any in America—slaughterhouses, textile mills, chemical plants, furniture factories, many other industries found anchorage and throve here. Located in the heart of a rich food country, Stalingrad also became a canning center, and, of course, a manufacturer of arms.

Tens of thousands of trees were set out on its streets, many of which were still unpaved. Many of the trees withered from drouth and from lack of care. Others were planted in their place. From year to year the chestnuts and the maples rose higher and higher and added color and freshness to the once-sandy, arid, and unadorned avenues.

Schools by the score sprang into life. Colleges too—in mechanics, in pedagogy, in medicine, in politics, above all in engineering. Theaters were built, as imposing if not as decorative as any in the country—a youth theater, a summer theater, a drama theater, a musical-comedy theater. Housing was neglected, often woefully. Not that new houses were not built. They were—thousands of them—but as everywhere else in Russia the construction was hasty, often faulty. There was no time and no material, not yet, for modern and comfortable housing. Other plans in the future were to solve this problem, so official spokesmen proclaimed. Housing could wait, modern plumbing could wait, electric refrigerators could wait, many another comfort and pleasure, so common in America and now and then mentioned in the Russian press, could wait. But not factories, not blast furnaces, not engines, not guns nor shells. Neither modern plumbing nor refrigerators nor air-conditioned apartments could frighten away and kill the foreign invader, especially the fascist and the Nazi, who had already proclaimed in *Mein Kampf* and in endless other public utterances an inevitable clash, an indispensable conquest of Russia. . . .

So the city boomed with a new life, a new machine, a new power, a new dispensation. All over Russia even children had heard of it. Youths in Ivanovo, in Nizhnii, in Novosibirsk hoped for a chance to visit it and to be stirred by the sight of the new factories, the new schools, the new theaters. In those days I annually visited Ivanovo, Russia's leading textile city, and young workers in the Melanzhevy Combinat, of which I was making a study for the novel *Moscow Skies,* on learning that I had been in Stalingrad, deluged me with questions about the city, the people, above all, comparisons with similar factories in other parts of Russia and in America. They were more than proud of this mighty city on the Volga—they felt reassured of their plans and of their future.

The climate has helped make Stalingrad the unique and powerful city it became. Few other cities boast of so bright yet so austere a climate. Scant is the cloudiness in the skies, profuse is the sunlight—no less than 2,273 hours a year—and the almost continuous winds are often wild and bitter. Northwesters, southeasters, they blow and rage over the far-flung streets and squares. No more than fifteen, at most twenty-five, days a year are without wind. On my first visit I was wandering around one evening in the market place watching the fruit and vegetable dealers cook supper over little fires or make themselves comfortable for the night in tents or under their wagons. A sandstorm blew up. So violent and blinding was it that I lost my way and was nearly run over by a Cossack ox team.

Winter in Stalingrad comes late, stays long. Snow usually falls in December, lingers until April. Dry and powdery, it is easily blown up by the gales that sweep unimpeded from the Urals, and is no less blinding than the sandstorms in summer. Frost sinks deep into the ground, often to nearly five feet. On windy days the cold is hard and biting—only those clad in sheepskins, in furs, in heavy woolens, can endure it with comfort. The Germans knew this only too well. Hence the reckless speed and violence with which they flung themselves on the city.

Here, then, was this "giant city" on Russia's most beloved river. It was new and it was old. It was stern and sturdy. Machine shop and food store, fortress and arsenal, workplace and home town, it was as much the pride and the hope of the Soviets as any city in the land. It was a town of youth and audacity, a Stalin town in the Russian sense, a town with which not only the name but the will and the triumph of the nation's dictator was associated. It was a Russian town built in part by American brains, but by Russian hands, Russian gold, Russian sacrifice, a town therefore to fight and die for, to blow to smoke and dust, sink into the bottom of the Volga if necessary rather than to yield to the enemy!

The day I was there, in June 1942, it was flooded with sunshine. When we drove down to the pier in a truck which the NKVD chief of the railroad station had, after much effort, unearthed for us, the river, about half a mile wide, was as bright and inviting as I had ever known it to be. On the other side—known as the meadowland—were some of the most superb bathing beaches in Russia, long and broad and shiny, with white and yellow sand.

On previous visits I had often gone swimming there. Every day and late into the night, every night, the beaches and parks resounded with the loud talk, the gay laughter, the choral chants of vacationists. Children and adults sought in summer the meadowland's cool and refreshing shelters. An amusement park was opened there with buffets, restaurants, dance pavilions, bookstalls, and rows on rows of little cottages and tents in which young people and families spent their vacations. The park—all of the meadowland—was the playground of Stalingrad. Young and old sought refuge there after a day's toil in the hot and roaring factories and machine shops.

Now, though it was summer, the meadowland was empty of life. Not a sound came over the silvery sweep of water that divided it from the pier. No craft made its way there or came from there. The blue haze that hung over it was like a stage curtain that was rung down over the life and the promise it had known in previous years.

"It's as dead as in deep winter," moaned the gray-eyed, talkative policeman at the pier.

"Don't people in Stalingrad play any more at all?" I asked.

"Of course, but nowhere nearly as much as in prewar times," he answered gloomily, and squinted his eyes to the opposite shore as though in longing for the life that the war had suspended. Then, as if remembering my question and the incomplete answer he had made, he added, "People here work twelve or more hours a day—then there is social work—military drill —yes, even for girls—so you see there is little time for play." Looking up at the cloudless sky and the hot sun, he added, "Such wonderful weather for play—the devil only knows why people have to fight wars when nature is so beautiful—so very beautiful!"

Despite war Stalingrad was given to play and diversion. The philharmonic orchestra was holding concerts. The theaters were crowded. Shakespeare and *Rose Marie*—America's great musical gift to Russia—were vying for popularity with theater-goers. The tunes of *Rose Marie* were sung and whistled and strummed on guitars and played on accordions as zealously as the endless cycles of Volga tunes. Jazz orchestras, when they came to the city, were accorded a loud and joyful welcome. If the concert was in a park and all seats were sold, people leaned on fences outside and climbed trees and got their fill of the music. Nor did women neglect purely feminine indulgences. Beauty parlors, of which there were more than a few in the city, were doing a thriving business. War or no war, the factory girls of the city would have their manicure and their permanent wave, especially when they went to the theater. They danced too—all kinds of dances, including the *Bostón,* which is the Russian equivalent for the American waltz. Of course the war had made a difference in the dancing. Not many young men came—they were at war. But the girls didn't mind dancing with each other, and when they danced they appeared as carefree and happy as girls anywhere. Yet, testified the policeman on the pier, hardly one but had learned to handle a rifle and a bayonet, and more and more were learning to master the machine gun! They were Stalingrad girls; worthy of the name and the fighting spirit of this city of steel and machines!

The pier swarmed with people. The spacious, well-kept waiting rooms were reserved for soldiers. Civilians by the hundreds had parked themselves all along the steep, sun-flooded embankment. Some of them had built little fires and were cooking porridge or fish soup. Others lay or sat around eating, mending clothes, reading newspapers or books, talking to neighbors. Still others, wrapped in coats, with their heads on bundles, and unmindful of sirens of passing boats and of the loud noises of neighbors, were fast asleep.

They were all waiting for boats. In one respect Russia had not changed. People were still on the move, still wandering from one end of the land to the other. But this time the movement was under government direction, people were journeying to places where new factories were set up, new lands were broken. Many of the waiting passengers were refugees from the Ukraine, from White Russia. They were on their way to a near-by province or to the faraway Urals and Siberia. With the few personal belongings they had managed to salvage out of the devastation that had rolled over them, they were bravely seeking a new place of abode, a new start in life. They didn't weep. They did not wail. They had done enough of both. They were steeled to hardship and privation, and their one thought now was to be put on a boat and to sail away to their new destination and their new destiny.

There was the usual hot-water room at the pier. Two gigantic copper kettles with large brass faucets were continually kept at a boil. Anyone who cared could come and fill a teakettle or any vessel with steaming hot water without cost. The room was small and bright, well supplied with long benches for those who wished to sit down and talk or listen to the talk of others. Talking and eating were still the chief diversions of Russians who happened to be *v puti*—that is, on the road. A hefty and energetic woman, little given to speech, was in charge of the room. Oblivious to the visitors, she bustled about sweeping, scrubbing, above all keeping the kettles full and boiling.

Because the hot water was especially welcome, endless and varied were the uses and satisfactions it made possible. In it one could wash one's shirt, handkerchief, socks. If soap was lacking, sand made an acceptable and cleansing substitute. In hot water one could boil eggs. With it one could make porridge, soup, all manner of hot drinks. Russians have never been coffee-drinking people; so the lack of coffee was no hardship. But the lack of tea was no small privation. As I strolled about the embankment I observed the use of a host of substitutes. Dried apples, for example, and dried apricots, dried mushrooms, dried raspberry leaves, raw carrot— anything that would give flavor to the water. I saw one old man steep onion in hot water.

"Tastes good, Grandfather?" I asked.

"It's got to, little son," was the prompt, decisive reply. I stood a few moments and watched him make a meal of the onion-flavored water and cold, puffy griddlecakes.

"Want a taste?" he said.

I shook my head. But he gulped it with the zest of a man partaking of a precious delicacy.

Once as I came into the hot-water room I became aware of a middle-aged woman who was bent over a copy of the magazine *Propagandist*. So absorbed was she in reading that she seemed not to have heard the angry colloquy over peasant profiteering in the market place. She was short, portly, with heavy dark brows and a sunburned face. I concluded that she was a person of some political importance, for only such was likely to be absorbed in a magazine like *Propagandist*. Only when the policeman entered and asked me about life in America did she lift her eyes with a jerk, as though the question had startled her. She joined in the questioning. In a soft, leisurely voice she asked about the second front—when would it come—when—when—when? She repeated the word three times, as if to emphasize the importance of the question. Then she talked of herself. She was a worker in the tractor plant, a welder—taking the place of her husband who had gone to the front. For three years prior to the war she had not worked—stayed home because of a weak heart and because of the five children in the family. Though a member of the Party, she had not relinquished her social and political work. Nor was she the only woman in Stalingrad who had given up work before the war and was now back in the factory. Women had to take the place of their husbands who had gone to war. Some of these husbands were already dead. Many more were sure to die in this terrible war—the most terrible the Russian people had ever known. She paused an instant, and there was silence, as though all solemnly agreed with her pronouncement. Then with a show of passion she said,

"But no matter—we shan't be conquered—we cannot be conquered. We're Russians, and Russians have never been conquered—not for long. The fathers of our children may die, the German cannibals may come here to our beautiful Stalingrad. If they do we shall all fight—most of us may die. But we shall teach the cannibals a lesson they won't forget—a lesson in steel and blood. Yes, we shall—so they won't think of doing to our children what they are doing to us."

Again she paused, and again there was silence. Obviously she was accustomed to making speeches, as her political phraseology and her declamatory tone of voice so abundantly demonstrated. Yet there was no doubting the earnestness and the import of her words. Even the caretaker was impressed. Her rough hands on her hips, her lips tightly pressed together, her eyes contracted as if sullen with resolve, she looked at the speaker, looked and looked, in the expectation, obviously, of more information and more enlightenment on this war of wars.

Little did I then realize how tragically prophetic and heroically true were the words of this middle-aged woman. She never told me her name. I never asked for it. We just talked—she with the ardor with which Russians everywhere were talking of the inevitable reckoning with Germans—a deadly reckoning that would forever drain out of them the passion and the power to hurl annihilation and death on anyone, least of all on Russia. The war of the centuries between Slav and Teuton, Russian and German, was reaching a gigantic climax; Russians knew it, even if in their arrogant self-confidence the Germans did not, as the words of this woman so solemnly and so eloquently testified.

In the ensuing months I remembered her words and I wondered what had become of her and her children. Not a few married women in Stalingrad, including mothers of large families, joined the fighting armies, sometimes with rifles and hand grenades, sometimes only as cooks, nurses, stretcher bearers—or as "mothers." Had this woman become such a mother?

These Stalingrad mothers only added to the drama and the heroism of the city. Defying machine-gun fire and bombs, they crawled through deep trenches and shattered walls to some ruin of a building or to a bomb-blown crater with thermos bottles of hot soup and other foods for Russian soldiers. . . . Not always did they reach their destination. Something often hit them on the way and they were heard of no more. The others carried on by day and by night. . . . They washed and mended clothes for their "sons." They cooked and baked for them. They listened to their tales of battle and cheered them with motherly blessings. They read the letters from the real mothers or wives, and again and again, when a son went off to fight and never returned, they wrote to these mothers or wives words of solace and courage and of undying faith in eventual triumph over the "cannibal enemy." Great will be the tributes that will someday find their way into poems, plays, and novels to these Stalingrad mothers.

The city was bright and cheerful when I was there some ten weeks before the Germans began their campaign of destruction. I could not help thinking of the contrast it presented to the Tsaritsyn I knew nineteen years earlier. Then it was a provincial town of dwarfed homes, battered pavements, blinding sandstorms, with camel caravans whipping up clouds of dust in the steppes on which later many a modern factory was to tower its way into the sky.

Now the main streets were shiny with asphalt, green with trees, greener than ever, and interspersed with little parks which, as in Baku, daily resounded with the joyful shouts of children. Miles and miles of steppe on the outskirts were planted with factories, some as large and modern as any in the world, overshadowing anything, including the French steel shop which Tsaritsyn had known. A mere glimpse at these factories from a boat on the Volga, all built during the preceding fourteen years, gave one a fresh appreciation of the stupendous achievements of the Plans. Ruthless was the will, implacable the energy, immeasurable the faith, enormous the sacrifice, that went into the making and the execution of the Plans. . . . But they remade the city. They armed it with a power which Russia had never known. They crowded it with young people—men and women who were as ready to build barricades as they had been to build factories—out of their very bodies if necessary. . . . No wonder they wrote of it as the city of their youth, their life, their love.

Now the city is no more. I shall never see it again. Not a vestige remains of Tsaritsyn or Stalingrad. All is rubble and ruin—the brick and the cement, the iron and the steel, the buildings and the machines. But the enemy has reaped no reward. The city is gone, but he, too, is gone—to his death or to a prison camp. Stalingrad is the grimmest graveyard Germany has yet known. Not so soon will its ghosts cease to haunt and torment the self-proclaimed superman—whether he be Hitler, a German barber, a German bricklayer, a German clerk, a German shopkeeper, who had dreamed of rising to riches and glory over the dead bodies of Russians.

Stalingrad is no more. Yet to Russia it has never been more alive and more mighty. It is a sanctuary of sanctuaries in which is forever enshrined and consecrated the will of a people to die rather than cease to be.

PART FOUR: RUSSIA'S NEW SOCIETY

CHAPTER 16: FACTORY OWNERSHIP

SOMEWHERE in the Urals is the town of N —, which is how Russians designate an industrial city whose identity, for military reasons, they do not
wish to divulge. To this particular town there came early in June 1942 an
army of construction workers. They were of many nationalities—Russians,
Ukrainians, Tatars, Chuvashes, Jews, Mordvinians. They were also of many
trades—carpenters, bricklayers, blacksmiths, plumbers, electricians, mechanics, welders, sewer builders, architects, engineers, and all other categories required for the erection of a gigantic pipe foundry.

These workers were under the direction of Soviet Building Trust No. 22
and under the immediate supervision of a young engineer named Schildkrot. Their assignment was to build the pipe foundry with wartime speed
—that is, within four months.

On their arrival in the Urals they found neither dormitories nor kitchens
awaiting them. Bare fields lay all around the place where they were to work.
For temporary living quarters they put up four tents. They had all been
on war-construction jobs before, and lack of ready conveniences was
neither new nor distressing.

They set to work immediately according to a definitely mapped out plan
for every crew, every team, every worker. Long hours did not matter. When
necessary they remained at their job twenty-four hours at a time. From
eighteen to twenty hours a day excavators kept lifting masses of earth for
the foundation. Instead of trucks, a rolling mechanized transporter was
used for the hauling and the disposal of this earth. Concrete was carried by
the belt system straight from the mixers to wherever it might be needed.
Machines heaved and roared, and human hands never tired of keeping

up with them. Schildkrot testifies that never before on any building he had supervised had men and women worked with such energy and speed. Thirty miles of pipe line were laid within an incredibly short time. Construction which formerly required six months was now finished in as many weeks. Before the war a mammoth pipe foundry was built in Mariupol, in the Ukraine. It took a year and a half to install the machinery inside the shops. Now the installation consumed no more than fifty days. By the end of October, a few days before the celebration of the twenty-fifth anniversary of the Soviet regime, the foundry was finished. A residence community was also ready. The builders turned over both to the producers and journeyed eastward to build a new factory.

It is a far cry from this Ural foundry to the time in the early years of the Soviet regime when a delegation of miners from Kuznetsk, Siberia, came all the way to Moscow to appeal in person to Lenin for a dynamo for their colliery. Russia was so worn out and wrecked by the first World War and the civil war that no factory was making dynamos. Lenin told the miners to search around and if they found a dynamo to take it with them. The delegates scoured the capital and finally stumbled on a generator in the Maly Theater, the oldest playhouse in Russia. The theater had kept the extra generator in reserve and did not want it lugged away to Siberia. But the miners were determined to take possession of it and in the end, with Lenin's help, they succeeded.

Russia in those days was almost a barren land industrially and not much better agriculturally. In 1913 the output of Russia's heavy industry was valued at eleven billion rubles; by 1917 the output had dropped to 1.7 million rubles. Steel production in 1917 was only 4.6 per cent of the 1913 output. Speaking of that period, Stalin once said: "Ruined by four years of imperialist war; ruined once more by three years of civil war, a land with a half-illiterate population on a low level of development, with separate oases of industry drowned by a sea of parcellized peasant holdings—such was the land we inherited from the past. The problem we faced was to transfer this land from the tracks of the middle ages and darkness to the tracks of modernized industry and mechanized agriculture."

Industrialization all along the line—that was the dream.

Now the dream is one of the great realities of our time; and the story of the N —— pipe foundry in the Urals testifies to the power and momentousness of this reality.

Stories of the speedy erection of new factories, some larger and more modern than any in Europe, appear in the Russian press with recurrent frequency. While Russia fights she also builds. But these stories give only a hint of the stupendous construction that has taken place east of the Volga since the outbreak of the war. Nobody knows how many new factories have been built by the traveling construction crews of the various building trusts. There must be hundreds of them. Neither the Germans nor the Allies had ever suspected the existence of efficient building crews willing and able to pull up stakes at home and wander from one part of the country to another, from forest to steppe, from valley to mountain, and back again, and apply themselves with energy and zeal to the one task that the country needed immediately—the erection of new factories.

In *Pravda* of August 19, 1942, at the very bottom of the first page, is a brief summary of the output of heavy industry during the first half of the month. The most interesting feature of this report for the foreign observer is the names of cities in which metallurgical plants are located. The average American or Englishman has heard of Magnitogorsk, Kuznetsk, Stalingrad as centers of steel production. But unless he is a professional geographer or a special student of Russian industry, it is not likely that he has heard of metallurgical towns given in the *Pravda* story. Here is a list of them: Novo-Tagil, Achinsk, Beloretsk, Nizhniiye Sergi, Maykor, Verkhne-Sinyachikhinsk, Kuvshinovo, Satka, Teplaya Gora, Nizhnaya Salda, Severskaya, Amur, Alapayevo, Ufaley, Vyksa, Chusovo, Zlatoust, Serovo, Chermoz, Dobryanka. This list of twenty metallurgical towns is supplemented by a few more which appeared as inconspicuously in *Pravda* of August 29 reporting on output during the first twenty-five days of August. Chernaya Kholunitza, Omutinsk, Guryevskaya, Yelizavetino are the new names. They are all east or north of the Volga, chiefly in the Urals and Siberia, which gives the reader some understanding of the sweep and geographic distribution of Russia's metallurgical industry.

Nor, it must be emphasized, does this list exhaust the number of Russia's metallurgical plants. There are others, built since the war started, and the Russians speak of them as factory N —— or as located in the town N ——. Russia's migratory construction crews not only have been building new factories but have taken apart, packed away, and transported to new ground hundreds of factories from regions that the Russian armies have had to abandon to the Germans. Central Asia and the southern Urals are now dotted with these factories. Again and again, in the blistering heat of Kazakstan or in the icy winds of the Urals, men and women have set up machines and engines and then put roofs over them, walls around them; and indispensable war goods have started to roll out of the transplanted factories. No figures, no concrete information of any kind is available regarding the evacuation of industry.

"Someday," said a Soviet official, "we shall publish the story, and the world will be astounded at what we have been able to achieve."

Occasionally a Soviet editor bursts into exalted language and reveals, perhaps unwittingly, the magnitude of this work. The editor of *Pravda* does so in the editorial of November 10, 1942, when he exclaims:

"History has no parallel to the immense magnitude of the industrial evacuation we have achieved within so brief a period. Hundreds of enterprises have been transferred from territories temporarily seized by the enemy deep into our rear, thousands of kilometers away from the battlefield. . . ." The tank factory from Kharkov which had been evacuated to the Urals is, according to this editor, "turning out more tanks than ever before." Most significant is the declaration, made after Stalingrad had been ruined, that "at the present time our war industry, including aviation, tanks, and ammunition, produces an amount of fighting equipment which it had never before turned out. The Red Army receives increasing numbers of KV T-34 tanks, planes of the type of IL, YAK, LAG, and others, likewise cannon of varied names and caliber, mortar guns, anti-tank guns, and other armaments."

Every time the Russians lost an important industrial center and the outside world began to speculate—in Berlin and Rome joyfully, in London and Washington anxiously—on Russia's continued power of resistance, the Russians in some obscure way released a report of new construction somewhere that more than replaced the lost or shattered productive energy in the abandoned territory. Germans rejoiced openly and loudly when they seized Zaporozhe on the Dnieper, one of the great manufacturing centers of the Ukraine. They were certain that the Russians would have little aluminum, indispensable in the manufacture of airplanes. They had not counted on the new deposits of bauxite which the Russians had uncovered in the Urals and on the new plant they rapidly built for the manufacture of aluminum.

It is well to emphasize that the construction of new factories and the evacuation of all old ones are being achieved by Russian engineers and Russian workers without the help of foreign experts.

During the feverish Five-Year Plans it was widely proclaimed in the outside world that Russians were so hopelessly inefficient and so lacking in mechanical talent that it was futile of Stalin or any Bolshevik to imagine that Five-Year Plans would convert the Russians into skilled engineers and mechanics. . . . Many American engineers who were helping in the rapid industrialization of Russia often enough got so distressed over the stubborn cocksureness and brazen distrust of younger Russian engineers, as well as the clumsiness of Russian workers freshly recruited from the villages, that they despaired of Russia's industrial future. The late Hugh Cooper, who built the Dneprostroy dam, was one of the very few American engineers who never grew discouraged at the rawness of Russian labor or the vagaries of Russian engineers. He never failed to tell journalists and friends that despite all her difficulties Russia was destined to become a great and efficient industrial power.

The Russians still have much to learn about modern industry, especially from America. They say so themselves—Stalin as much as anybody. But they have learned so much within a mere thirteen years that, with the help they have been receiving from America and England, they have been supplying their many-millioned army with fighting equipment for a war against Germany, which has at its disposal the oldest and most productive industries of western Europe.

The period 1928–41 will rank in Russian history as the most decisive, the most epochal the Russian people have ever known. For centuries Russians and others will be writing of these years with amazement, with pity, with ecstasy; also with disdain and wrath, especially if they emphasize at length and with eloquence the inordinate sacrifices of the people.

Consider the trials and the achievements of the years. The Russian people had been through a disastrous world war in which they had lost millions of lives. They had fought a sanguinary internecine war in which they lost more millions of lives. A devastating famine caused by drouth and aggravated by economic disorganization further swelled the toll of death by many millions. For a brief period between 1922 and 1928 they had their Nep, with its legalization of minor forms of private trade and manufacture, its mounting supplies of consumption goods. They began to eat well, though they were wearing rumpled clothes and badly made or dilapidated shoes. These years were a welcome, though none-too-restful, breathing spell because, like a clouded sky, they were fraught with uncertainties. People did not know whether the atmosphere would clear or break into a violent storm. While leaders quarreled and grew more and more estranged from each other and historic decisions remained in abeyance, the people waited and wondered and hoped for the best.

Then in 1928 came the thunderous announcement of the First Five-Year Plan and the call for fresh effort and fresh sacrifices. Food, clothing, housing rapidly deteriorated. There was nowhere nearly enough of these to provide adequate satisfaction. People toiled, sweated, grew thin and weary. There was no respite from the toil and the inadequacy of consumption goods. Nothing mattered but the Plan and the factories it aimed to build and the new farming system it sought to launch.

Physically, mentally, technically Russia was ill-prepared for this gigantic task. She had neither the engineers nor the skilled workers. She had neither the foreign money nor the gold with which to pay for foreign machinery and foreign technical services. So wild and fantastic did the Plan appear to the outside world and so discredited was Russia's capacity to meet long-term financial obligations, that the banking community of the industrialized nations in Europe and America deemed Moscow an uncertain financial risk. Except in small amounts and for short periods Russia could not obtain credits abroad. She had to squeeze out of herself all the substance she could and bring it as an offering to the all-embracing, all-devouring Plan.

The strain and sacrifice, except for those imbued with a new pioneering spirit, left a harsh imprint on the body and soul of the people.

I was in Russia at the time and saw the population divided into two clear-cut groups: those who believed in the Plan and endured the sacrifices with equanimity, even with joy, and did not mind the lack of soap, underwear, meat, and other equally urgent daily commodities; and those who did not believe in the Plan, or who found the sacrifices too painful and too harrowing. The first group was sustained and often inspired by the grandiose sweep and the momentous promise of the Plan. The second group wept and swore, often laughed with mockery and pain. If they evinced doubt and sorrow too openly or too loudly, they disappeared. Officially no misdemeanor was so great, no sin so loathsome as doubt in the ultimate fulfillment of the Plan or disrespect for its beautiful promises.

Eggs, meat, butter, wine, cheese, which Russians needed, were shipped abroad—to England, to Germany, to any country that would buy and pay in money which foreign engineers and foreign businessmen would accept. Wheat and rye, which were terribly wanted at home, went to Italy, Finland, Turkey, England, and other lands.

I recall only too vividly a journey during those turbulent days in White Russia. The train stopped at a station known as Stariya Dorogi—the old road. I had been there before and had enjoyed many a sumptuous meal at the station restaurant. It was one of the most bountiful restaurants in White Russia. Now it was empty of meat and butter, of rice and jams, of all the other excellent foods I had eaten there on earlier visits. Black bread and tea without sugar were all it could offer to passengers. Yet on a siding I saw several freight cars with eggs on their way to Königsberg, Prussia. Russian goose and wild fowl, Russian butter and caviar were cheap in Berlin, Hamburg, and Dresden; but in Moscow there was little of any of them.

The vast majority of young people believed in the Plan. But if their fathers did not believe or were members of the propertied classes of prerevolutionary days, the young people publicly denounced them in the press, at meetings, in the privacy of their homes and offices. They changed their names and abandoned their old homes. They cut themselves off from the family as completely as though they never had known either family or home, father or mother. Once at dusk I met an elderly acquaintance in the streets of Moscow. With tears coursing freely down his cheeks, he spoke of his son whom he had loved all the more because he was an only child and who, without word of warning, had denounced him as a parasite and a pariah whose shelter and whose name he never again would want to share.

Russia was torn with family tragedies.

It was a time of war on a monumental scale, a new kind of war fought not on barricades with guns and bullets but in the hearts and souls of men, with weapons that were invisible to the physical eye yet no less devastating, especially in the wounds they inflicted on their victims.

But the Plan did not halt. Day and night the work pressed forward. With tornado-like violence Russia was driven to a new destiny—the most unifying and most challenging she had ever known.

In 1932, when the First Plan was finished a year in advance of schedule, Russia was enriched by 1,500 new industrial enterprises. The Second Plan, 1933–37, also completed in four years, involved a capital investment of 53 billion rubles, or nearly three and a half times the amount that the First Plan had demanded. The Third Plan, which had been mapped out for the years 1938–42, came to an abrupt end in June 1941, when Germany invaded Russia. During the three years of its operation Russia had managed to build 2,900 new industrial enterprises, some very large, some quite small. In its blueprint the Third Plan called for an output in the leading industries of more than double the amount they had yielded in 1937.

Now, in the light of the war and Russia's fierce resistance, the Plans loom as one of the most farsighted and astounding achievements of all time. The very father who had wept lonely and anguished tears because his son had disowned him in the early days of the First Plan, now had only praise for the Plans and for the men who had relentlessly driven the people into their fulfillment. He invited me to his home, and as we were sitting in his unheated living room drinking hot tea with candy and hard biscuits instead of sugar, he kept repeating over and over like the text of a sermon or the refrain of a song, "What would we have done without our *pyatiletki* against a Germany that is fighting us with all the industry of western Europe?" I met other such fathers in Russia.

He and his son are now united, for with the coming of the Constitution there was a nationwide reconciliation of fathers and sons, of mothers and daughters. The son is in the army, an artillery officer, and the father is happy that thanks to the Plans his son and the men he commands have all the field guns and all the shells they need, indeed more and better artillery than the Germans.

At every step in Russia one hears the words: "What would we have done in this war without our pyatiletki?" It is the Plans, harrowing as has been their cost, that are saving Russia from subjugation and from large-scale extermination.

Russia's Plans signalize a new technique and a new philosophy of industrial development. In many telling ways these differ from the methods of industrialization which other nations, including England and America, have been pursuing. The rancor and the conflict which the Russian Revolution evoked in foreign lands, particularly in the more advanced countries, have their roots in some of the ideas underlying the Plans.

Like an army at war, the entire population was mobilized for the task. With the exception of only a small group they were hopelessly unprepared for the work. The blueprints were new, the machines were new, the construction was new. Their hands were clumsy, their co-ordination of thought and action was faulty, at times catastrophically so. Unlike other nations, Soviet Russia did not bother to develop competent staffs first and only then launch the building program. The leaders followed not the individualized but the mass system of application, not the safe evolutionary but the spectacular revolutionary process of development and construction.

They did not mind the immediate expense or the immediate failures. They were seeking not only to cover the country with a network of modern factories within the briefest possible time but to train millions of men and women, many of whom had never seen anything more advanced than a village blacksmith shop, in the successful operation of blast furnaces, rolling mills, and all other modern and highly complicated machinery, also within the briefest time. They had no time for individual instruction. They literally flung masses of people, millions of them, into the task with the purely businesslike calculation that the immediate losses would be more than compensated in the long run by the competence which these masses would acquire by being obliged to perform tasks they had never before known. They needed this competence for the operation of the new industries and to fight a war, should it be thrust on them, a modern war with tanks and machine guns, with all manner and all caliber of artillery. They were profligate in their expense because they wanted to achieve a proficiency of mechanical skill.

They paid for the Plans with Russian resources. This is another distinguishing feature of Russian industrialization. Foreign capital, as already indicated, could not be obtained—not in very large sums—nor was it especially wanted; the Soviets would not accept the terms on which they could have it. They chose to pay for everything out of their own pockets, at the expense of their daily bread and shelter. As a consequence, no foreigner holds a mortgage on a single Russian steel shop, power plant, railroad, or any other enterprise. No one received a single dollar in interest on investment in Russian industry. Until the coming of the war, Russia had no external debt, except the prerevolutionary debt which the Soviets have officially repudiated.

The basic aim of the Plans was not only to distribute industrialization all over the country, which the czarist government had singularly failed to do, but to prepare Russia for war, for a long war, with one powerful nation, with several, or perhaps with all the powerful nations of the world. The thought of war had haunted Soviet leaders since the day they supplanted the Czar as the ruler of Russia. They expected war. They never ceased to speak of it. They sought to imbue not only the older folk but the children with the thought of its inevitability. They were laying plans to fight it even if they should have to withdraw to the deep rear—beyond the Dnieper, beyond the Don, beyond the Volga, anywhere. They wanted to make certain their soldiers would have the armaments, the transportation, the food to fight the invader, no matter how disastrous the outcome of individual battles or how deep inland they would have to retreat.

In prerevolutionary days about three fourths of Russia's heavy industry was centered in the south, chiefly in the Ukraine. As much as 90 per cent of the coal came from the Donets Basin, which is also in the Ukraine. No less than two thirds of all industry was crowded into the central and southern parts of European Russia. Siberia was frowned upon as a possible place of industrial development. Central Asia was treated as a backward colony. The Caucasus was regarded as a gay, exciting summer resort. Even the region of the Urals, one of the richest strips of land in the world, literally glutted with raw materials, where there once had been an old and primitive industry, was left largely to itself and to its harsh primitiveness. Until the coming of the Plans there was not a single *modern* blast furnace east of the Volga.

One reason why the Russian Army met with such disaster in the first World War was that so much of Russia's industry was centered in territory which the Germans overran. Had the Soviets centered their industrial development in the old industrial regions, all European Russia, perhaps even the Urals, would by this time have been under complete German rule. It is because of the wide, far-flung distribution of Russian industry that, no matter how deep into the rear the Red armies retreated, they always had an industry to grind out for them armaments and ammunition. Deliberately and with farsightedness the Russian leaders prepared construction crews that could travel from place to place in the Urals, the Caucasus, Armenia, Siberia, Central Asia, and build new or rehabilitate and enlarge old factories. Without the concentrated training of the three Plans these crews would of course have been impossible.

Russian industry now stretches east all the way to the Pacific Ocean; north to the Arctic Circle and beyond; south to the border of Afghanistan and India. The Red Army can always have an industrial base on which to lean in moments of catastrophe; it can always command weapons with

which to fight even if it should be driven beyond the European borders.
. . . The Urals are now the heart of Russia's new industry and Siberia, as
well as Central Asia, is rapidly being dotted with towering chimney stacks.
Including the Urals, Russia is now the most powerful industrial nation
on the Asiatic continent and in heavy industry and machine-building leads
even Japan.

The thought of war and the problems of defense east and west, in
Europe and in Asia, governed not only the geographic distribution but the
nature of the new industry which the Plans had encompassed. In czarist
times, in war as well as in peace, Russia was essentially an exporter of raw
materials and foods, particularly wheat, and an importer of manufactured
goods. Much of the native industry was in the hands of foreign investors
and foreign concessionaires, chiefly British, French, and Belgian. According
to Soviet sources, at least three fifths of the machine-building plants, coal
mines, metal factories were owned by foreigners or were under the control
of foreign capital. In 1912 three fifths of the industrial machinery of Russia
was imported from abroad. Czarist Russia also purchased in foreign mar-
kets one fifth of the coal it consumed, 98 per cent of the lead, 37 per cent
of the superphosphates. Russia depended on alien manufactures for much
of her industrial machinery, her electrical equipment, her chemical products, her coal, her pig iron, her non-ferrous metals. She even imported
scythes, plows, needles. She did not manufacture a single tractor, only a
few automobiles, a few planes, none of the modern mechanized equipment
for the digging of gold, the boring of oil wells.

The Plans strove above all else to make Russia independent of foreign
capital and foreign manufactures. They aimed to give Russia industrial
self-sufficiency so that in time of peace she could, if necessary, live off
her own factories, and in time of war she could fight with her own armaments—from bayonet and rifle to plane and heavy field gun—and with her
own ammunition—from ordinary pistol bullets to two-ton bombs.

The Plans have eliminated all foreign economic holdings. The Lena gold
fields in Siberia, last of the foreign concessions, British in this instance,
were taken over by the Soviets before the inauguration of the First Plan.
A few small plants (an Austrian sweater factory, a foreign cosmetic works)
lingered for some time but in the end were absorbed by Soviet industry.
The Plans have made Russia as free from foreign investment as the Soviets
have made her free of foreign political influence and intrigue. Though
fathered by Bolsheviks who have their roots in international socialism, the
Plans have converted Russia in her economic ownership into the most na-
tionalistic country in the world.

The emphasis of the Plans was on heavy industry—steel and machine
building. With plenty of steel and abundant machine-building capacity at

their command the Soviets felt they could easily develop in time all the light industries. They could then build factories anywhere without the need of foreign importation. The war interrupted the process. Yet the magnitude of the achievement of the Plans can be judged by a few simple figures. In 1913, which marked the peak of Russia's pre-Soviet economic development, industry accounted for 42.1 per cent of the national income, agriculture for the remaining 57.9 per cent. At the beginning of the Third Plan, industry had leaped so far ahead of agriculture that it yielded three and a half times as much income. In 1913, according to Soviet sources, the industrial output of the United States was fourteen times as high as that of Russia; Germany's was six times as high, England's four and a half, and that of France two and a half times as high. In 1940, according to the same Soviet sources, only the United States surpassed Russia in volume of industrial output.

In no other country, not even in the United States, was machine building accorded such a crucial place. In 1940 it constituted one fourth of all manufacturing. That is why during the Third Plan Russia was manufacturing excellent trucks and tanks, tractors and combines, planes and locomotives, but inferior, often execrable, shoes, clothes, razors, razor blades, furniture, and plumbing fixtures. Consumption goods in quality and quantity could wait—such was the theory. If a foreign enemy pounced on the country he would be fought not with razor blades and plumbing fixtures, not with silk stockings and silk underwear, but with guns and shells, with planes and tanks and other modern armaments.

From year to year Russia was building more and more heavy industry. In 1913 the daily output of steel was 11,000 tons. In 1940 it mounted from 58,000 to 59,000 tons. In Europe only Germany surpassed Russia in the production of pig iron and steel. In 1913 the daily output of coal was 85,000 tons, of oil and gas 25,000. In 1940 the first rose to 417,000 tons, the second from 97,000 to 98,000 tons. In 1939 the chemical industry yielded fifteen times as much production as in 1913. Kuzbas, the coal basin in faraway Siberia, accounted in 1913 for only 0.8 million tons of coal. In 1937 the figure leaped to 17.8 million tons. At the beginning of the Third Plan at least three fourths of all industry and all output in Russia came from factories that were new or had been reconstructed and modernized. The industry of "machines to make machines," as the Russians speak of machine-tool manufacturing, was pushed ahead of all other branches of machine building. This has made possible the rapid construction of new factories and the ready introduction of new inventions and new improvements which in time of war are often of decisive importance. New fuel sources have been tapped and developed in the western territory. The Russians speak of the second Baku oil field, which embraces the district between the

Volga and the Urals, and the Emba fields east of the Caspian. In 1938 one sixth of the nation's oil came from this region.

The Plans have given Russia a new and large heavy industry—a machine-building industry; a chemical industry; a tank and aviation industry; a shell and artillery industry (one of the largest in the world); a tractor and combine industry; an automobile and truck industry; an agricultural implement industry which is second only to America in capacity of production; a new food industry with a mounting output of canned vegetables, fruits, fish, and vegetable juices; many other subsidiary industries. One reason for the widespread national consciousness of Russia is the new unity, physical and social, that steel, machines, and electric power have made possible.

Equally significant is the machine-mindedness the Plans have engendered in millions of men and women. From the blacksmith shop to the engine room is the distance in mechanics that tens of millions of Russian folk have traversed in the brief space of thirteen years! Without this journey on the road to modern mechanics Russia could no more have fought against the entire land army of Germany and some of her allies than without her industries in the Urals, Central Asia, the Caucasus, Siberia. In this writer's judgment the Plans have supplied the sole means of saving the Russian people from the most ruthless subjugation any nation has ever known and from large-scale physical extermination. Anyone who has visited villages and cities which have been recaptured from the Germans, and who has seen and heard as much of German policy and German practice in occupied Russian territory, as has this writer, never can doubt the earnestness of Rosenberg's prewar boasts that the Russians must be expelled from Europe, or of Hitler's proclamations that the Germans must crowd out of Europe the Slavs, and in the first place the Russians, so as to make room for the two hundred million Germans who, to him, are the rightful inheritors of the European continent.

But the most distinguishing feature of the Plans and of all Soviet industrialization is their political basis and the implications they presuppose. Bolsheviks speak of their economic system as socialism. Socialists who are not Bolsheviks deny this claim. They contend that Russia has no civil liberties and certain other forms of freedom which are as much a part of a socialist society as collective ownership of property. This is not the place to discuss whether or not the Bolsheviks or their socialist opponents are right. Yet in any account of the Plans and of Russian industrialization it is impossible to overemphasize the fact that private capital and individual enterprise have had no part in it. On the contrary, the Plans, especially the first one, squeezed and throttled to extinction all but a most insignificant number of private enterprises.

Russia is now the only land in the world in which private ownership of income-yielding property is as much at an end as czarism or landlordism. Not a new factory in Russia but was built by the state. Not a Russian gun, shell, hand grenade that goes to the front but comes from a government manufacturing plant.

Many of the innovations and interpretations of the Revolution have been found inadequate or dangerous and have been modified or discarded. Theory after theory in education, from the project method of study and lax class discipline to the substitution of the class-struggle saga for history has been tried, denounced, and finally scrapped. Prerevolutionary values in social life, in aesthetics, in music, in the other arts, in daily life, which in the first flush of rebellion against the old world had been mocked and cast away, have been welcomed back with appreciation and fervor. As already stressed so often in the preceding pages, Russia has been zealously rediscovering her own purely Russian past, as well as the past of the human race, and has eagerly and passionately been reappropriating and reincorporating into daily life usages, customs, and relationships which old Russia had honored and other peoples had for ages made their own.

There has been a significant process of sifting and clarification in social adaptation, individual orientation, in cultural appreciation. There has been a visible return to old and long-tried values. This has led to frequent and loud proclamations—not in Russia but outside—that Sovietism is undergoing a transformation which is pushing Russia closer and closer to the civilization, the economic usage, the human motivation prevailing in England, America, and other advanced modern nations. This is only partially true and, like all partial truths, lends itself to easy distortion of all truth and therefore to rank falsification.

Let it be noted that on the main issue of Sovietism—the ban on capitalism or income-yielding private property in whatever form—there has been neither change nor deviation. The Plans not only envisaged but ruthlessly enforced this ban. The war has not lightened its application. Because of a shortage in consumption goods there has been in the market places an increasing amount of commodity exchange, principally between city people and the peasantry. City people offer clothes and household goods in exchange for food. This has led to no small amount of underground petty speculation. But it is illegal. All purchase for profit is unlawful. Persons engaging in it must sooner or later expect to face the consequences, which are very severe.

Dealing in the market place or through private transaction implies no lessening of the ban on private manufacture and private trade. There has been no compromise on this issue, which is the very basis of Sovietism. There has been only entrenchment of the ban, not only in daily practice, as is evidenced by the law of 1940 sharply reducing the acreage of individual gardens on collective farms, but in the consciousness of the people. After a quarter of a century of violent, ceaseless propaganda on the sinfulness and wickedness of private business, Russia is so alienated from the institution that young people under thirty, and even more of them under twenty, in conversation with foreigners find it hard to imagine the system of private enterprise as it exists in the outside world.

During a journey in a rural region in Central Russia which had been under German occupation for ten months, I stayed several nights in a peasant cottage which was also the home of the district secretary of the Party. He was an inquisitive young man in his late twenties. One evening he began to question me about the New York *Herald Tribune,* for which I had come to gather information on the German "new order" in rural Russia. He had been editor of a rural semiweekly journal which, of course, was not private but state property, and he could not conceive, he said, of a large daily newspaper under private ownership. Question after question tumbled from his lips. Who appointed the editors? Who hired the reporters? Who drew up the plan of daily stories and daily editorials? Who owned the buildings and the presses? My replies only made him shrug his shoulders in mystification.

As he talked and as I observed his incomprehension, I could not help thinking of the Americans and Englishmen who had often questioned me on Russian collectivized ownership and operation of property. Russia's ban on private ownership was as much a riddle and an absurdity to them as the lack of such a ban in America and Britain was to this Russian party secretary. Each is permeated with the ideas of economics prevalent in his land, his school, his home—his everyday associations. Even if he is a socialist in England and America he is at every step made abundantly aware of private ownership, while at every step in Russia the citizen is made acutely conscious of its absence, its sinfulness. . . .

Had Russia's historical development paralleled that of America or England there is no reason to assume that she would have risen at the time she did in deadly revolt against private ownership of "the means of production." But serfdom lingered until 1861. After it was abolished feudal rule largely continued to harass the village. The peasant continued to doff his hat before the landlord and the official. Though since the opening of the century Russian industry was gaining rapid momentum, it had not developed the large and mighty middle class that grew up in England and even more in America. The vast Russian masses and intelligentsia never became as property-conscious as the people in the English-speaking countries, particularly in America, where free land and political democracy, with their extravagant social rewards, fostered the cult of individual enterprise.

Only by taking cognizance of Russia's past as contrasted with that of England and America can we appreciate the motives and the forces that have led to the annihilation of income-yielding individual property and to a mentality that regards private business as the most hateful institution and the most loathsome practice in the world. The very expression "private property" evokes in the new Russian, who is usually a young Russian, the fiercest contempt or the loudest ridicule.

The war, it must be emphasized, has not shaken this attitude. On the contrary, it has intensified it. To the man in the Army the food he has been receiving, the one hundred grams of vodka that is his daily ration during the winter months, the clothes he is wearing, the weapons with which he is fighting, all come from government shops, warehouses, factories; all have been manufactured under a system of collectivized control, with no middleman or contractor or politician deriving personal gain from any of the transactions.

There may be petty graft and larceny and cruel bureaucracy in the administration of Russian trade and industry. Again and again at meetings Russian workers and office employees hear of incidents of incompetence and dishonesty. Again and again they read of such incidents in the press, in the wall newspapers of the institution in which they work. They don't like them. They denounce them. They clamor for severe punishment of the guilty and the recalcitrant parties. Yet with all the shortcomings and inadequacies of the collectivized economic system, they not only approve of it, they will fight to the end to maintain it. Let there be no misconception on this point.

One of the many reasons for the volcanic hate of Germans in the present war is that in occupied territories they establish, under German control, private ownership in industry, in trade, in land. No sooner do the Russians reconquer a community than the old institutions, with no sign of opposition even on the part of older folk, reappear and resume their interrupted functions.

The ban on private enterprise is the mainspring of Russian ideology and therefore also the mainspring of Russia's new nationalism. To view this nationalism solely or chiefly in terms of the nationalism of prerevolutionary Russia or of countries that rest their economy on private ownership in "the means of production" is to misread the intent and the nature of the changes that time and the war have wrought in the thinking and the living of the Russian people.

In 1913, according to the Russian *Little Encyclopedia,* the so-called bourgeoisie, the businessmen of the country, constituted 15.9 per cent of the population. In 1928, the first year of the Plan, the percentage dropped to 5. In 1937 it was zero. It has been zero since then, and there is nothing on the Russian horizon even now, red as it is with the fire and blood of war, to warrant the assumption that it is likely to be changed.

Since the private owner has vanished, who are the leading figures in the Soviet plant, and how do they administer their functions?

CHAPTER 17: FACTORY MANAGEMENT

As I WENT AROUND the yards, the offices, the shops, the warehouses of the Moscow Trekhgorka (Three Hills), one of Russia's oldest and largest textile factories, I could not help thinking of similar factories I had visited in America. I could observe nowhere, hardly even in the work clothes of the people, any external differences. The buildings were old and clumsy, but so are the buildings of old factories in America. The machines were as noisy as in America, the workers as busy, the secretaries as probing, the foremen as vigilant, the smells in the dyeing rooms as pungent, the samples in the showrooms, even now in wartime, as neatly and handsomely displayed.

A visit to this or any other Russian factory would swiftly disenchant anyone of the illusion that because it is not privately owned and intimately supervised by individual owners or their personally appointed executives, it is lacking in order, vigilance, discipline. The director, the foremen, the engineers, and the other executives have hardly less power and certainly no less responsibility, even with regard to the amount of annual profits (in this instance not for the investors but for the state), than have similar executives in England or America or in any country with individual ownership of property. The basic function of management is no different in Russia from what it is in America or England.

There was a time when Russian offices looked as untidy and rumpled as the people who presided over them. That day is gone. The three Plans have, among other things, chastened executives, especially in the highest positions, of neglect of personal appearance and have brushed out, as with a broom of steel, the disorder of their offices. The *kabinet* (office) of Victor Yeliseyevitch Dodonkin, the thirty-five-year-old director of the Trekhgorka, glistens with tidiness and comfort. Instead of the modern business suit, Dodonkin, like so many Russians, including artists and college professors, wears a loose-fitting tunic with spacious pockets. Instead of low shoes, he prefers knee-high leather boots, as does Stalin, whom no one has ever seen in a photograph or in life in low shoes. But he is as clean-shaven, as properly groomed as any executive in England or America. Because he is the highest executive he must set an example to others in everything, including personal appearance.

Yet despite these and numerous other similarities, the Trekhgorka, like any other Russian factory, unlike factories in other countries, is more than a manufacturing concern. It is a political institute of the first magnitude. It cannot be otherwise in Russia, for it was out of the factory that the Revolution sprang into the street. It was there that its physical power was born, grew, matured, and burst out of bounds. It was there that the Red Army was formed, clothed, armed, blessed, stirred into action. It was there that the main issues of the Revolution were fought out. It was there that the mighty battle between Stalin and Trotsky was decided. Had Trotsky succeeded in rallying and galvanizing the factory to his support, he never would have ended his spectacular career as an exile in an alien and faraway land.

Throughout the years of toil and sacrifice, of discouragement and sorrow, that the Plans, especially in the early stages, had imposed on the people, it was again the factory that flung itself, soldierlike, into the campaign of courage and faith, of patience and hope for the bright and rewarding tomorrow.

In Russia a factory produces but it also educates, entertains workers—in shops and in offices. It helps the family look after children. It trains youth not only to work but to fight. It supports the Kremlin in all policies, it espouses all the burdens the Kremlin imposes. Because of its historical importance, its political functions, its intimate relationships with the everyday life of the people within its walls and their families, its capacity to endure and impose discipline, to assume and evoke in time of crisis the greatest and gravest self-sacrifice, it is one of the main cohesive bonds of the nation, one of its mightiest sources of social and military strength. If, despite the ruin of Stalingrad, Russia, as Stalin assured the people on November 6, 1942, "has never yet commanded such an organized and powerful rear," it is because, among other things, of the many-sided and immense contribution the factory has made to national defense and national morale.

Yet the problem of production has been its main concern and its main task. The key to the material well-being of the people, to the national defense, production has been the main source and the main cause of all discussions and all conflicts within the ranks of the Party and especially of its leaders. For years no other subject has received such vehement attention in the press and on the lecture platform.

The Communist party publishes many journals as guides to lecturers and propagandists. Now, as before the war, the chief preoccupation of these journals is production. Not an issue of *Guide for Lecturers, Propagandist,* and the small, closely printed, widely disseminated pocket edition of *Notes for Agitators* but is "packed to the brim," as one Russian engineer has expressed himself, with the theme of production. All the inviting promises of the future reward for the Russian people, of an eventual system of society in which the slogan, "From each one according to his ability, to each according to his needs," will become a daily reality, are linked, indeed rooted, in production. The Soviet slogan, "to catch up and surpass" capitalist nations, applies first and foremost to production. Again and again Lenin had emphasized that Russia can achieve a higher system of civilization than any capitalist country can hope to attain only if and when the Soviet output per unit of human and mechanical energy exceeds that of the highest known in any capitalist country. In the last analysis the underlying and all-consuming challenge of Sovietism to capitalism, and of capitalism to Sovietism is not formula, rhetoric, blueprints, but achievement in the blast furnaces, rolling mills, coal mines, and all other agencies of production. It is there and not on the barricades that this monumental conflict of competition will be decided.

In the coefficients which Russians continually use in their discussions of factory output, of productivity of labor, they invariably use American figures as a basis of comparison, because to them America represents the very zenith of industrial achievement in the capitalist world. It is American figures that provide the guideposts on their road to what they never cease to repeat as their ultimate and final triumph over capitalism, which is the attainment of a higher coefficient of production.

The problem of production is therefore as much the preoccupation of the administrative staff of Trekhgorka as of any well-run factory in a capitalist nation. The director is the manager. He carries all the burdens that devolve on the shoulders of a manager in a textile factory in America or England. He is personally accountable for everything—raw materials, machines, division of labor, relations with labor, output of goods, above all for the discipline and the efficiency of other executives and workers. He must observe all Soviet labor laws, but he must be ruthless in enforcing discipline and all other requirements for efficient production.

For years the outside world has heard of Russian factories as meeting places rather than agencies of production. In the early years of Sovietism there was some justification for the charge. The Soviets had enthroned a new philosophy of life and work. But the leaders had formulas and blueprints but no experience to guide them. Gladkov's novel, *Cement,* one of the best portraying that period in Russia, is almost an amusing document when read in the light of the Plans.

Let it be noted that in the early years of Sovietism the Russian proletariat had not sloughed off the unwillingness to do more than was required or could be forced out of them by prerevolutionary employers. They talked during their working hours not only about their jobs but about their homes, their families, their girls, the cinemas they had seen or wanted to see. They spent much time smoking cigarettes. They found enough ways and excuses to shirk immediate tasks. They might have been violent revolutionaries with much glory to their credit during the civil war. At an instant's notice they were ready to shoulder a rifle in defense of the Soviets. But with only some exceptions they had no positive, enthusiastic attitude toward work. The fact that the standard of living was low, that there were not enough goods to satisfy immediate needs, and that with the wreckage incurred during the civil war and the general backwardness in Russian industrial development there was scarcely any hope of a rapid increase in available consumption goods, only fed the laxity and idleness.

But labor was the very heart of the ever-pressing, ever-present problem of production. It was essential to remold the worker's attitude toward his daily job, to inculcate in the mind of the millions of peasants, fresh from the village, a salutary conception of labor. To achieve this aim a propaganda campaign was launched through every available agency—the press, lecture platform, motion pictures, plays, cartoons. Labor was glorified as it never had been in the prerevolutionary days. All the socialist theories of the nobility of labor were galvanized into flaming rhetoric and dramatized into heroic sagas.

Simultaneous with this tumultuous propaganda there began a no-less-tumultuous campaign of practical measures to whip out of the worker all the output that he was capable of yielding. It was well enough to think of socialism as a paradise and of capitalism as a hell. But unless labor was made to do its utmost in production, there would be only stagnation and weakness, futility and collapse. Even before the coming of the Plans measures were put into force to lift productivity and to end the turnover of labor which for a long time had been the chief bane of Russian industry.

The Plans struck hard and violently at all laxity and indifference to labor discipline. Grumbling and protests were disregarded. "Once no private owner derives any benefits from your work," was the reply to malcontents, "you have no reason to object to measures that will make you give the best of yourself to your daily task." These and similar words constituted the text and thesis of a never-ending series of speeches and of as roaring a torrent of oratory as Russia had known during the civil war.

Severe disciplinary measures preceded or followed the oratory and the inspiration it was intended to excite.

From time to time these measures grew increasingly severe. In August 1940 they were made into so rigid and punitive a code that enemies of the Soviets outside of Russia denounced them as a violation of every socialist tenet that has ever been preached. Neither Stalin nor the others were moved by the denunciations. They proceeded on the theory that with war threatening they could not be too severe in the demands on workers and on all others, for if the war was lost all would perish, not only socialism but the Russian nation. It would be torn to pieces, mangled, subjugated, and in the end perhaps wiped out. Never had Soviet leaders been more impassioned in their belief that "the end justifies the means" as in the application of new and stern discipline in factories and in offices. The Trekhgorka, of course, was no exception to the rule.

A worker, whether in shop or office, has to be in his place ready for the day's job the minute the shift begins. If he is ten minutes late through a cause which is within human control, he receives an individual and a public warning the first time. The superintendent, the foreman, or the director, or all of them, say some unpleasant words to him. They also post a bulletin in all shops naming him and informing all workers of his dereliction.

Of course, if he lives a long distance away and there was an accident on the bus, the trolley, or the subway, the warning is withheld. But if he lives close to the factory and has overslept or has stopped to talk to someone or has lost valuable minutes in some other avoidable manner, the warning is issued in vigorous language.

If he commits the same offense a second time within a month, the worker receives not a warning but a reprimand which likewise finds its way to the bulletin board of all shops. If he is late a third time in one month he faces trial in a people's court. The sentence is usually a period of "redeeming labor," for from three to four months, in his regular place of work but not on full pay. A percentage varying from 15 to 25 per cent is deducted during the time that the sentence is in effect.

If a worker is late twenty-one minutes, even if it is his first offense, and he has no satisfying explanation to offer, he also faces trial. Again there is no jail sentence. There is only a more severe term of "redeeming labor" with the attendant deduction of from 15 to 25 per cent of wages or salary.

"Supposing," I asked the chairman of the trade union, "the worker's mother, father, or child is ill or dying?"

"We'd prefer even then for the worker to communicate with us, and of course we'd grant him permission to stay at home."

If a worker does not show up during the first half hour a messenger is sent to call on him at his home. Usually the messenger is from the trade union which looks after the health and social insurance of all workers. The absent worker may be critically ill in a dormitory or in his apartment with no one to offer or summon medical aid. The messenger then performs an errand of mercy. But if no legitimate cause is responsible for the absence, the law steps in with a vengeance. The factory and Soviet law are jealously vigilant of punctuality in everything, particularly in the time that the man at the machine or in the office must spend at his task. Never is the worker permitted to indulge his whim or his pleasure at the expense of his job.

In the Trekhgorka the staffs in all departments have become so habituated to punctuality that rarely is there any need of applying disciplinary measures to derelicts.

"We start from our homes so early," said a woman weaver, "that we get here in plenty of time for our shift."

A worker has no right to leave his place of employment without special permission. This is rarely granted by the factory management unless it is in the interests of the factory, the state, or the nation, and now, in time of war, of the Army. The worker's own wishes or advantages do not count. Not only does the law disallow leaving a place of employment but it bars one from securing work elsewhere. Unless the labor book, which a worker must always present when he applies for a job, bears written testimony that he is permitted to seek new work, he will receive short shrift from the place where he applies, unless the management is so hard pressed for new labor that it is willing to violate the law.

In wartime a change of work illegally achieved, or absence from work because of laziness, drunkenness, or other preventable causes, is regarded as desertion from "duty to the fatherland" and is treated on the same basis as desertion from the Army. Since the coming of war not a single person in the Trekhgorka has been branded a deserter. The men and women who work there are too acutely conscious of the war to indulge in offenses that will invoke such a charge.

Speed-up is another method vigorously and universally prosecuted to lift the output of factories. Stakhanovism, which in essence is rationalization of labor, is one of the means fostered to promote increase in production. The Stakhanovite is lauded, rewarded, set up as an exemplary citizen and worker whom others must never cease to emulate. Shock brigadiering is another speed-up method. The shock brigadier is a pace-setter and receives his reward in glory, in privilege, and in income.

By far the most widespread and most zealously espoused scheme of speed-up is known as "socialist competition." Factory challenges factory for increase in output, farm challenges farm, steel worker challenges miner, miner challenges sniper in the Army—not a production unit in the country, whether large or small, involving twenty-five thousand or only ten workers, but has literally been seething with socialist competition, especially since the war. Slogans, diagrams, teamwork, oratory, financial reward, all are the weapons or stimulants in this nationwide race for increased production.

After Stalin's speech of November 6, 1942, a new and more intensified wave of socialist competition swept the country. There is hardly a working person anywhere but is a part of it. The press literally storms with appeals and commands for more and more, better and better socialist competition. Nor are the appeals in vain. Foreigners in Russia often amuse themselves, sometimes frivolously, sometimes brashly, sometimes venomously, sometimes with innocent gaiety, at the expense of this widespread and violent campaign. But to Russians it is one of the most solemn and most momentous movements of the times. Victory or defeat on the battlefield is decisively linked with the work of the factory, and socialist competition does boost output—often enormously.

In one such campaign the Trekhgorka won second place. It was awarded the Red Banner of the Moscow Executive Committee of the Communist party. The victory not only carried glory and publicity but a handsome financial bonus—80,000 rubles for the improvement of living conditions in the factory, 110,000 rubles for distribution as a premium among the workers who had contributed most to the victory. The only one who is not eligible for a share in this premium is the director of the factory. All others with high records of production receive a share of the bonus.

In countries that maintain private ownership of industry, these Soviet methods of lifting the productivity of labor would hardly meet with the approval of labor leaders. Trade unions, for example, do not look with favor on efficiency experts who devise new methods of rationalization and speed-up schemes of production. Not only protest but now and then strikes have attended the introduction of such schemes. But in Russia the trade unions are among the chief sponsors of these speed-up campaigns. So is the Communist party. So is the Komsomol. So is the labor press. So is the rest of the local and national press. Socialist competition is considered not only a sport but a duty, almost a rite. The trade unions are the very soul of the movement.

"Don't workers protest?" I asked a foreman.

He laughed.

"Why should they? It is for their own good and for the good of the nation. Nobody derives any personal profit from it. They earn more money for themselves, and if the factory shows larger profits, these are used by the government as capital investment for new factories, as a source of national defense, and as a means of lifting the material and cultural standard of living of the community, which, in the last analysis, means the worker and his family. Why should any of our people object to our labor policy, when it is all not for the enrichment of an owner or an investor but for their own advancement. If they protest there is something wrong with them. You must remember that in our country there is not antagonism but identity of interest between state and factory, government and labor."

These words express the philosophy that underlies present-day Russian labor practices. All the more so now because the battle against the enemy is also primarily a battle for increased production.

"Will you always maintain such labor policies?" I asked. Again the man laughed.

"There is no such thing," he said, "as *always* in human history. Everything is subject to change, the process of manufacturing as well as the methods of lifting the productivity of labor. The one thing we'd never allow is the exploitation of labor for the sake of anyone's personal enrichment."

The management of the Trekhgorka follows a pattern of personal responsibility which is not much different from factory management in a capitalist country. As already stated, the director is the supreme command. He is analogous to the president-chairman of an American corporation. He is responsible to nobody in the factory, neither to the trade unions nor to the Party. His superior is the Commissariat of the Textile Industry. The chairman of the trade union or the Party secretary may make suggestions or criticize; they are called in for consultation, but they have no power of command over him. Not collective but personal responsibility, as in any capitalist country, is the very core of Soviet factory management.

At the Trekhgorka the director has three assistants or, as we would call them in America, three vice-presidents or vice-chairmen. The first is the chief engineer, the second is the treasurer, the third is the manager of the workers' supply stores. These four are in continuous consultation, but the director's word is as much law in the factory as is that of the owner in an American plant.

Since the Trekhgorka is what is known as a "combinat," it is divided into three mills—spinning, weaving, finishing and making the cloth ready for the market. Each mill, precisely as in America or in England, has a superintendent who is in complete charge of policy and production. Each shop in the mill has its own supervisor, who is responsible only to the superintendent. Each supervisor has several foremen who are responsible only to him. Each foreman has several assistants who are usually the toolmen and mechanics, and they are responsible solely to him. It is their task to keep machines in the best working condition. It must be emphasized that all these executives enjoy personal power and personal responsibility. They are accountable not to a committee or a mass meeting, but to their immediate superiors, precisely as in any well-managed establishment in a capitalist country.

Yet side by side with these similarities with American or English industry there are also differences. Present-day Russia does not know the closed shop; it is not regarded necessary, because of the absence of private ownership of "the means of production." The very expression "closed shop" is not part of the vocabulary of labor or political propaganda. Membership in a union is never compulsory. It is voluntary though it is continually urged. In the Trekhgorka everybody is a member of the union, including the pupils in the trade school. There is only one union. It is an industrial union and everybody is eligible, from director to janitress. Contrary to American practice, the executives are never barred from membership. They are welcomed. Indeed, since they are supposed to set an example to others, they are invariably members of the union. Dodonkin, the director of the Trekhgorka, is a loyal and active union man. So are the chief engineer and the two other assistants. So are all the superintendents, the foremen, the sub-foremen. The union offers so many advantages to the members, especially in social insurance, that most workers in Russian industry join—in 1938 the membership rose to twenty-three million.

One of the most surprising features of the Russian unions is that they do not know and do not demand the checkoff system of dues payment. The government collects by means of checkoff all income taxes, all payment for bonds, all other levies. The unions collect their own dues through their own agents. The amount is always proportionally the same—1 per cent of the wage or the salary.

For some years there has not been any collective bargaining in Russia. There is no need of it, say present-day leaders of the trade unions and factory directors.

I reminded the assistant director of the Trekhgorka, with whom I discussed the subject, that in the days before they achieved power the Bolsheviks were violently fighting any and all speed-up methods and were no less violently upholding the closed shop, collective bargaining, and many other commonly known demands of organized labor.

"Absolutely," he replied with emphasis. "We would still be doing it here in the Trekhgorka if it were privately owned. But now it is different, because labor and management are not enemies but partners, real partners." Not a labor leader or a factory manager in Russia I met but maintains a similar attitude. No doubt lack of conflict and antagonism between management and labor greatly facilitates the fulfillment of the plan of production by which factories in Russia are guided in all their daily activities.

Another striking feature of the Trekhgorka, as of all Russian industry, is that the director is not necessarily the highest-paid person. His salary is fixed by the Commissariat of Textiles. Dodonkin receives 1,400 rubles a month. He pays his income tax and meets all other levies of the government precisely as does any other worker. He has the use of the factory automobile. He does not personally own an automobile, nor has he an automobile for his exclusive use. Now that the Army has mobilized all available passenger automobiles, the Trekhgorka has only one car at its disposal, and it is at the service of all executives. Neither in the amount of rent he pays for his apartment nor in the price of the food he buys does he enjoy special concessions. There was a time when directors enjoyed such privileges, but not now.

What is more, the director is not eligible for any of the monetary awards that the factory may receive either from victory in socialist competition or from some other source. The Commissariat of Textiles may offer him a premium, but it does not come out of funds in which the Trekhgorka is involved. Unlike the director, the chief engineer, the treasurer, and the food-supply man, all other executives are eligible for factory premiums.

Again, strikingly enough, the highest-paid people in the Trekhgorka are usually the assistant foremen. They are the chief mechanics in the shop. It is their duty and to their material advantage to keep the machines continually at work. They are paid on the basis of the output of the machines assigned to them for repairs and supervision. The more production comes out of these machines, the more wages they earn. If the machines are kept in such a condition that they are seldom idle or only for brief periods, the assistant foremen enjoy all accruing benefits. Some of them earn as much as 2,000 rubles a month.

Some of the more energetic Stakhanovites, with their special systems of rationalization of labor, likewise earn more than the director or any of the high executives. All pay, except for executives, is on the basis of piecework. Naturally enough, workers who are fast with their hands and their minds earn more than others. As in the case of Stakhanovism, socialist competition, shock-brigadiering, the trade unions do not discourage or disparage piecework. It is one of their functions to lift productivity by whatever means. Therefore they not only uphold but foster piecework. Equalitarianism or equal pay for unequal work is considered as base a practice as lower rates of pay for women only because they are not men.

The right to hire new workers is vested in the hands of the factory management, which in the last analysis means the director. The Trekhgorka maintains an employment office, and when there is special need for workers it posts announcements on the bulletin boards in the neighborhood, on streetcars, and in busses. It may enlist the help of men and women in the factory. If they know someone who is without work—a boy or girl just out of school or a married woman who has been doing only housework—they make a personal call, explain the jobs that are open, the wages they command, and invite the unemployed person to make application at the employment office.

The employment office does not insist on union membership, and the union council interposes no objection. Its task, I must repeat, is to persuade the new worker to join the union *after* he has started to work.

The right to discharge workers is likewise vested in the management; that is, the director. Drunkenness, impudence to other workers, demoralization of other workers, above all incompetence and laxity, invite dismissal. The worker has a right to appeal to the controlling committee of the trade union. If he obtains no satisfaction from this committee, he may carry the appeal to the central committee of the trade union. If they uphold the director, the discharge usually remains in force, though the discharged person may seek reinstatement through court action.

Like all good executives, Director Dodonkin of the Trekhgorka does not remove himself from personal contact with workers. In fact, one of the chief duties of a director is to keep "close to the masses," to know their state of mind, their condition of living, their shortcomings, their virtues. Dodonkin sets aside three afternoons a week for personal conferences with workers. For at least two hours on each of these afternoons he receives them and listens to complaints, grievances, suggestions. The workers address him as "comrade director" or by the more formal Victor Yeliseyevitch. He cannot possibly know in person all the men and women in the shops, but he tries to meet as many of them as his work and time allow.

One of his obligations to his employer, which happens to be the Commissariat of Textiles, which in turn represents the state, is to show periodically and annually a generous balance sheet or a profit. Profit is as much an aim of Soviet as of capitalist industry, the difference being that in Russia all profit goes to the state for further capital investment and for social purposes. But Dodonkin cannot achieve profit at the expense of wages and salaries nor by personally boosting the price of the goods the Trekhgorka sells. He has a voice in the determination of both, but he cannot alter them at will. He must seek his profit principally through efficient management: economy in raw materials—cotton, metal, fuel, dyes; increase in productivity of labor; elimination of waste in office and shop; maintenance of machinery and other equipment in normal working condition; above all, by boosting volume of output.

The year before the war the Trekhgorka earned a profit of forty million rubles, which it turned over to the Commissariat of Textiles. Out of it Dodonkin received back not for himself but for what is known as the "directors' fund" 2 per cent of the sum. Part of it he used for improving the social and living conditions of the workers—helping the nursery, the Pioneers, the various youth clubs and organizations. Most of it he distributed as premiums. Beginning with the chief engineer and ending with the porters in the warehouses, and depending on the percentage of overfulfillment of their individual plan of work, they all received special financial rewards. Premiums out of the directors' fund are one of the outstanding material incentives in Russian factories. All workers know that the higher their output the higher the profits of the factory and the more substantial their share of the premium.

Today, the Russian's livelihood is integrally dependent upon his factory's production. More than that, the factory is the center of his life.

CHAPTER 18: **FACTORY LIFE**

ONLY A FEW YEARS AGO the neighborhood of the Trekhgorka was spoken of as the outskirts of Moscow. Sparsely settled, it abounded in acres of farmland, in fields of brushwood and weeds. After heavy showers there were mud puddles in the streets.

The Russian merchant Prohorov, who owned the factory, had built for his family and his workers a magnificent church, which was known as the Prohorov church. He had also built two mansions, one for himself, another for his son, many wooden cottages, and a few large brick buildings which served as dormitories and barracks for factory workers. But neither he nor the city administration had bothered to lay out the neighborhood by a plan. The result was that over winding side streets, dismal courtyards, straggling, half-sunken cottages, the factory towered like a mechanized giant.

Many of these courtyards have remained as primitive and unkempt as in czarist days. Many of the cottages have survived the Revolution, the Plans, the crash and the roar of the machine age. Their shingled roofs droop with decrepitude, their windows peep over winter banks of sand and battered sidewalks. They are reminders of a past that nestled close to an untamed earth.

Yet despite low ceilings, small rooms, unwieldy doors, these cottages gleam inside with neatness. The Trekhgorka women or the others who live in them have always been noted for their physical sturdiness. They do not mind rising a few hours early to scrub floors, wipe windows, then cook breakfast, look after the children, and go to the factory for the day's work. Electric light and running water, which the city of Moscow has installed, facilitate the tasks of cleanliness.

Now all around the Trekhgorka, looming as large and imposing as the factory, as red with burned brick, are the new houses which the management has built, particularly in the last twelve years. There are sixty-seven of them, four and five stories high, and they are carved into small apartments. They all have running water, electric lights; some have bathrooms. Three fifths of the 5,000 men and women employed in the factory live in these houses. The factory had mapped out an ambitious plan of further building to accommodate the remaining workers, but the war, which has put an end to all but factory construction, has reduced the plan to a paper document.

There are trees all around the Trekhgorka, but the neighborhood itself is still unkempt. There has been little landscaping, and village primitiveness broods over streets and courtyards.

Yet the factory has surcharged this community with a liveliness which in the old days it had never known. The labor discipline in the shops and offices is stern, the penalties for its violation harsh. But the social advantages and the cultural compensations that the factory makes possible arm the official propagandists with a formidable weapon against possible grumblers.

"Look around and see for yourself what is happening here," said one of the plant executives; "then you'll understand what we mean to make of our factories and why we are so hard on ourselves in our discipline and in our rush to lift production *higher and higher.*"

Frequently, as I visited public buildings, I was reminded, despite the ugly streets outside, of an American college. Sports, for example, and athletic teams are as much a part of the factory as its looms. Formerly the best choirs in Russia were in the churches. Not any more. Now the best choirs are in the factories and in the Army. The Trekhgorka has two choirs: one of adults, consisting of fifty voices; one of children, numbering over one hundred members. The choir of adults is so good that it is frequently invited for concerts in other factories and hospitals. The children's choir is invariably one of the most popular attractions at the celebrations of Soviet holidays.

There is scarcely a social need of the individual or the family that is not in some measure met by the factory. It has taken over many functions formerly vested in the church. Russia has no fraternal orders, no private charities, no socially exclusive clubs, no Y.M.C.A.s, Y.W.C.A.s, no settlement houses or town halls. Services and ministrations which these offer and which do not conflict with Soviet law and policy are in Russia entrusted to the factory.

"For us," said a twenty-two-year-old girl, foreman in the weaving shop, "the factory is a source of living as well as livelihood."

If it is as well organized, superbly managed, and highly profitable a factory as the Trekhgorka, it enriches the daily life of the worker in a multitude of significant ways.

At the time the Germans were making their drive on Moscow, the question of evacuating the children loomed as one of the most pressing of the moment. The management and workers of the Trekhgorka quickly organized themselves for the task. They gathered special clothes, food, and books. They dressed the children warmly and accompanied them to the railroad station. The government provided special facilities for their evacuation, and eight hundred of them, from three to sixteen years of age, including the little son of the director, started on the long journey to the Urals. They were accompanied by a committee of parents, attendants, and teachers. On arrival at their destination they were settled in specially prepared homes. The teachers started classes, and while Moscow schools were closed these Moscow children were able to resume their interrupted education.

During the summer the children who were old enough and strong enough worked on collective farms. They also cultivated their own gardens. They went berrying and mushroom picking. They hunted for medicinal plants and herbs. They collected metal and other raw material for factories. Their life, excepting for the fact that they were away from their homes, was normal, and—what counted most—they were beyond the reach of German bombers.

At regular intervals a committee of parents travels to the Urals to visit them. These mothers and fathers check on the living conditions, on the health, the play, the studies, the morale of the children. On their return to Moscow there is a mass meeting. Parents of all the evacuated children come to hear the report of the committee and to feel reassured of the well-being of their sons and daughters. The question of financing the Ural colony offers neither embarrassment nor torment to parents. The government and the trade unions bear most of it. Their own share is within their means. Nor do they have to worry about social and racial discrimination and the internal conflicts they may engender. The son of the director and the daughter of the janitress have been brought up to regard each other with a sense not of superiority or inferiority but of perfect equality.

In the village of Karusia, on the clear and cold Oka River, ninety-five miles from Moscow, the Trekhgorka maintains a summer camp for children of school age. As soon as vacation starts six hundred boys and girls, all sons and daughters of shop or office workers, go to Karusia for six weeks. Competent guides and attendants accompany them. These children bathe and play; they go on long hikes and study nature; they sing and dance; they observe insects and birds; they run races and have military drill; they sit around bonfires or listen to stories; they do all the other things that children like to do in a camp. Again no father and no mother need fear that the child will be insulted or snubbed because of social or racial origin. The least manifestation of snobbery is vigilantly fought. The expense of the camp is borne by the trade unions, the factories, and only in small part by the parents.

In Klyazma, twenty-five miles outside of Moscow, the Trekhgorka has its own summer home. It is nothing lavish or pretentious; but it can provide rest, comfort, outdoor play for three hundred persons at a time. The Trekhgorka also sends thirty-five workers annually to Sochi, the well-known Black Sea resort, and one hundred and fifty more to the Crimean summer homes. That is, it did this before the war. Now, of course, the children's camp and the summer home are closed. Workers are postponing their vacations until after the war. Children over twelve, if physically fit, spend at least part of their vacation on a kolkhoz or state farm. They pull weeds, hoe potatoes, gather berries and mushrooms, tend to the chickens, and do other chores.

Let no one underestimate the meaning of these institutions in a Russian factory and the ministrations they provide to workers and their families. Bureaucracy, which despite warning, denunciation, and purge continues to lay a parasitic hand on Russian services, may now and then subvert and degrade these ministrations. But not for long. In the factory, as in no organ of the government, the constant cry for efficiency and the access the workers have to some responsible executive inevitably restrains the recalcitrant bureaucrat.

The war has closed neither the factory nursery nor the kindergarten. It has disarranged their former schedule, though not seriously. Because of the bombings, nurseries and kindergartens in Moscow have been moved to safe quarters. The Trekhgorka nursery has consolidated with another in the neighborhood and sixty-four of its children, ranging from one month to four years, find daily accommodation there. The kindergarten has been transferred from the massive, timber-built house it occupied before the war to the spacious, concrete air-raid shelter underneath the Pioneer Home. Every morning at seven, one hundred and thirty boys and girls are brought there and they stay from twelve to fourteen hours. The shelter has been divided into many spacious rooms. It is hung with flags and decorations, furnished with low tables, low chairs and benches, and stocked with Teddy bears, toy horses, toy dogs, building blocks, crayons, drawing paper, and other play materials. A kitchen has been installed and it cooks four meals a day for the children. They are adequate meals even in these days of food shortages and include daily for each child three quarters of a glass of milk and thirty grams of sugar. Now that the factory farm has acquired eight milch cows the milk ration will be increased.

Dusk had already settled over Moscow when I visited the kindergarten. It was warmer and better lighted than the Metropole Hotel, in which foreign correspondents live. As I walked from room to room—the children are separated according to age, excepting those of six and seven—I was greeted by a salute and an outburst of merry jabber and loud laughter.

"When do you think the war will end?" I asked a group of the older children.

"This winter," came the clamorous response.

"Why?"

"Because we want our fathers to come home."

"My father," said a dark-haired, blue-eyed girl, "is wounded and he's got to come home so I can see him."

Most of the fathers of these children were at war.

Despite their cheerfulness, the older children were keenly war conscious. From blocks of wood and heavy rolling pins they built tanks, cannon, planes. They were playing war games, and there was always trouble because nobody wanted to take the part of the fascist. To solve the difficulty the older children now and then coaxed or "kidnaped" younger ones from the neighboring room and compelled them to be "fascists."

Once, in summer, a group of older children went out into the yard to hunt for fascists. They hid behind a building and when they saw an old woman pass by they surrounded her, pronounced her a fascist and their prisoner. For an instant the woman was terrified and swore she was no fascist but an honest Soviet citizen. Then the boys told her they were only jesting.

Less than half of the Trekhgorka children of kindergarten age in Moscow are accommodated in the air-raid shelter. The factory is making ready another building for one hundred and seventy more boys and girls. When it is opened there will be few factory children outside of the kindergarten.

The war has heaped on the civilian population many stern hardships, none of which has necessitated so much planning and scheming as the distribution of available food stocks. At the Trekhgorka, as in all other industrial communities, the factory attends to the problem. It has opened its own food shops. The service is so well organized that workers rarely have to form a queue. It had always maintained its own restaurants. Now it has made them more efficient. There are one main restaurant and three subsidiary ones. Breakfast—consisting of porridge, thin compote, and bread—costs one ruble, or the equivalent in American money, at the official rate of exchange, of twenty cents. Dinner or lunch of *shtshi* (thick vegetable soup), a small portion of meat, potato, and possibly another vegetable, costs two and a half rubles, or fifty cents. Supper—consisting of potatoes, herring, compote, and bread—costs two rubles twenty-five kopecks, or forty-three cents. Now and then for dessert there is a glass of milk costing the equivalent of four cents in American money; or a bottle of thick, creamy chocolate milk-shake weighing two hundred and fifty grams. This is the most luxurious and most expensive dish in the restaurant and costs three rubles ninety-five kopecks, or seventy-nine cents. Meat is served at most three times a week. Russian *kasha*—thick porridge—made of barley, millet, or buckwheat, takes its place on other days. Some workers, particularly those who are unmarried, eat most of their meals in the factory restaurants and charge them against their ration cards.

Those who eat at home buy their food in factory shops. The director's third assistant devotes all his time and attention to these shops. He negotiates contracts with collective farms for berries and vegetables. He enters into agreements with government food organizations. He operates a factory farm. The farm is not large, only two hundred and fifty acres, and is within easy reach of Moscow. The summer of 1942 was the first season it was in operation. Not all the land was tilled; there was not enough time or enough labor. Potatoes were the leading crop and there were fifty acres of them. Twelve acres were in cabbage. Cucumbers, beets, and carrots were planted on a much smaller scale.

The farm keeps livestock. It started with fifty sheep, fifty pigs, one hundred and fifty chickens. In October it acquired eighty more sheep, thirty-five more pigs, eight cows. One of the sows gave birth to a litter of twelve pigs, all of which remained alive. The pigs are the particular joy of the officials in charge of food supplies, if only because they grow fat quickly, and animal fat is the one food of which there is the greatest shortage. Eventually the officials hope to have a dairy of fifty cows, which will further ease the fat shortage.

The farm, the restaurants, the shops do not end the efforts of the factory aid to employees in the solution of food difficulties. There is a laboratory in Moscow that makes a vitamin preparation out of the juice of evergreen needles. Workers or their children who, on the advice of factory physicians, are in particular need of vitamins, receive adequate supplies of this preparation. In the summer of 1942 the factory also carried on a campaign for individual gardens and helped workers with advice and with seeds to start them. As many as four hundred and twenty-three families responded to the campaign. The courtyards in the Trekhgorka neighborhood were sliced up into beds and rows of onions, cabbage, cucumbers, and—quite unexpectedly for a Russian garden even in Moscow—summer radishes and lettuce. Because of the rich vitamin content of vegetables, the factory has had to urge and coax the workers into planting them, because the Russians overwhelmingly still show a marked aversion to such food.

Were it not for the factory—its organization, its spirit, its enterprise, its rigid discipline, its stern exactions from food-supply officials—the workers and their families could not possibly have escaped dire consequences during the war.

The factory supplies free medical service, free dental care, though the patients must pay for their own medical prescriptions.

There was a time when it also operated its own secondary schools, which workers attended evenings after work hours. In these schools adults could make up for lost educational opportunities. If they chose to do so, they could prepare for college or the university. Now these schools have been disbanded. When I asked the director why, he answered that there was no longer any need for them. The older workers who wanted a secondary education have already had it and the young workers obtain such education in the public schools. The vast majority of them who now enter the factory—or did in the prewar days—completed from seven to ten years of study in public schools.

With great vigor the factory emphasizes technical education. It offers a multitude of courses, varied in range, all relating to production in the textile industry. No worker need remain at an unskilled job if he has any wish to promote himself to the more interesting and more highly paid skilled jobs. The trade union, the Party, the Komsomol, which participate in all educational political industrial campaigns—though always with the agreement of the management and never at the expense of production—spend no little time in urging men and women to attend these schools so that they can better their own positions and simultaneously raise the level of the technical skill and increase the industrial output of the factory. Tuition in these courses is free.

For young people the factory maintains a trade school. A few years before the war the question of new staffs of industrial workers was becoming increasingly acute. A way had to be found to supply the ever-growing need of men and women in the hundreds of new factories that were continually being built. To answer this need hundreds of new trade schools were opened. The Trekhgorka always had had such a school, but now it has expanded it and made it more professional. Boys and girls between fourteen and sixteen are eligible for admission. They pay no tuition. They study at the expense of the Commissariat of Textiles, which also pays for their food, their lodging if they are from out of town, and half of the cost of their uniforms. The other half is deducted from their wages.

In the Trekhgorka there are five hundred and eighty students in the trade school. Two hours a day they study in the classroom. From four to six hours if they are under sixteen and eight hours if they are sixteen or over they spend in the shops. For the work in the shops they are paid from 125 to 175 rubles a month. Some of them work out high norms and draw as much as 200 and 250 rubles a month.

When they graduate they may remain in the factory or attend higher courses and finally reach the university and become engineers. The trade union, the Party, the Komsomol always encourage them to continue their education. Such schools lift from parents the burden of preparing their children for careers or professions.

The factory maintains three libraries—one on technical subjects, one for juveniles, one for adults. In January 1942 there were over 18,000 volumes in the library for adults. Many books have been carelessly carried away by readers who were evacuated from Moscow. Most of the volumes now on the shelves are fiction, which readers prefer to political studies. Tolstoy is the favorite Russian author. Dickens leads in popularity all foreign authors. There is some Dos Passos in the library, more than a little of Hemingway, who has an enthusiastic following among the more literary young people. Balzac, Mark Twain, Schiller, and Shakespeare are among the other foreign authors who are widely read. Gogol, Lermontov, Pushkin, Turgenev, Saltykov-Shchedrin are among the Russian classics in constant demand.

Before the war the library maintained thirteen portable units in the factory and anyone wanting a book could obtain it without making a trip to the library building. The portable libraries are still in operation but on a diminished scale. There is much less need for them. People have not the time for as much reading as before the war. The workday is not eight but eleven hours. Then there is always a call for social work, for military duty, for donating the rest day to work in a military hospital or to some community purpose such as unloading peat or wood.

By far the richest contribution the Trekhgorka is making to its employees is the system of many-sided diversions and recreations it offers them. The home of the former owner of the factory—a gigantic establishment of thirty-four rooms with numerous hallways and Dutch ovens—has been turned into a clubhouse. After complete renovation the building afforded quarters for any conceivable study and recreation circles—literature, politics, radio, military science, music, dancing, dramatics, painting, athletics. The former Prohorov drawing room was rebuilt into an auditorium and motion-picture theater. Day and night the place teemed with activity. There was a radio in the club, and evenings young people danced modern and folk dances to its music. Dancing was exceptionally popular among the young people of Trekhgorka.

Now the clubhouse is in the hands of the Army. Only the library remains open to factory employees. All the other facilities are used for military purposes. Men in uniform around the hallways scrutinize sharply the visitors in civilian dress, even though they know that without proper credentials no one could pass the armed sentries outside.

As I came out of the building, which towers on a height, I could see in the dusk rising above the snow-blanketed earth the full outlines of the factory. It was as shrouded in blackness as was the clubhouse. Not a glint of light showed through the drapes and curtains which at evening were drawn over the doors and windows. Yet I felt that within the walls of these buildings as nowhere else in the neighborhood the lights and the fires of the new Russia glowed bright and powerful.

The Trekhgorka fosters athletics on as huge a scale as it does technical education. It has its own stadium, a small one with no more than three thousand seats, but it hopes someday to build a large and imposing stadium for intercity and international athletic carnivals. The athletic directors have heard of baseball but have never seen the game and do not know how to play it. They are enthusiasts of volley ball, basketball, Rugby football, hockey, boxing, wrestling, skating, skiing. When snow first appeared in Moscow, Madam Zhukova, the trade union chairman with whom I was talking, exclaimed joyfully, "Do you know we shall soon get four hundred pairs of skis!" Only a little earlier I had heard foreign correspondents complain that they could not buy any—the shops of the city did not yet sell skis. But the Trekhgorka was getting four hundred pairs of them!

There are many athletic teams in the Trekhgorka, and they compete with one another. These teams are the only organizations in the Soviet Union that I know of in which the sexes are separated. Women have their own teams. Out of the best players on the male teams, a Trekhgorka team is formed which engages in contests with other factory teams. It is not colleges but factories in Russia which boast the best athletic teams and that engage in the most exciting athletic contests. Some of the games in Moscow draw crowds as huge as does a Yale-Harvard, Army-Navy football game in America. Nor is the air of festivity in the stadium any less marked, though the Russian crowds as yet are not quite as explosive in their expressions of disappointment and enthusiasm.

When freezing weather sweeps Moscow the Trekhgorka stadium is flooded and the factory has its own skating rink. Skating was one of the most popular sports before the war. Often an orchestra was hired to play lilting melodies and waltzes. Evenings, if the weather permitted, the rinks swarmed with skaters and echoed with loud gaiety.

By far the most popular cultural diversions are music and the theater. The Prohorov factory kitchen has been rebuilt inside and outside and made into a large theater with all the conceivable stage accessories. Beginning in September and continuing into May twice or three times a week its twelve hundred seats are crowded. No amateur musicians or actors appear on the stage of this theater. They would not draw an audience, as it is within half an hour's train ride from the best theaters and concert halls in Moscow. Recognized soloists—violinists, pianists, singers, dancers, string quartets, jazz orchestras—are brought to the factory theater. Professional companies from the leading theaters of Moscow and other cities present operettas, dramas, satires, and—Russians always insist on this—Shakespeare.

The entertainments are sponsored by the trade unions, who are in charge of all education and recreation. They are open to the public, though outsiders pay full price for tickets while factory workers enjoy a 50 per cent discount. No season tickets are sold. Each concert or performance is handled separately, but there is never any difficulty in disposing of all available tickets. In 1939 the Leningrad Drama Theater was engaged for a season of six weeks. It played every night and there never were enough seats to accommodate the public. The most popular play was Tolstoy's *Anna Karenina*.

The profits from the theatrical ventures are sufficient to pay all the expenses of the clubhouse.

During the bombing of Moscow the theater was struck by an incendiary bomb which burned the roof and the stage. Now the theater is empty, for neither material nor labor is available for repairs. When the war ends it will be renovated and refurnished and will resume offering plays and concerts to the public.

The young people of the factory invited me to *vecher*—an evening of entertainment. Before the war they often gave such parties, which invariably begin and end with social dancing. Now with the long hours of work in the factory, the demands for social work, the solemn mood of the country, a vecher is rare. But now and then, the Komsomol secretary explained, young people must have an evening of fun! So they arranged this vecher in honor of wounded soldiers and sailors who had won decorations in the war, some of them from their own factory.

Since the clubhouse was in the hands of the Army, the vecher was held in the Pioneer Home which was formerly the Prohorov church. Rebuilt and modernized, it is now the most attractive and imposing building in the Trekhgorka neighborhood, the only building with any architectural impressiveness. The war has shifted to its halls all the social life of the young people.

Since the curfew hour in Moscow was midnight the vecher started before sundown. Most of the young people were girls in their late teens. They were neatly dressed and groomed, though with none of the formality or the splendor of American college girls at a junior prom or a similar occasion. Yet their modest stylishness was a marked contrast to the comparative stylelessness of earlier generations of Russia's factory and college girls.

They gathered in groups and talked and waited. Then a military band in the gallery started to play. Instantly the floor swarmed with dancers. The war had drained the factory of young men and not even the invited soldiers and sailors could make up the disproportion of the sexes. But the men who were there sought to make up in camaraderie and generosity for their lack in numbers.

One sailor in particular outdid himself in chivalric action. He seemed the embodiment of the proverbial sailor whose heart spills over with emotion for all girls wherever he sees them. He loved to dance and the girls loved to dance with him. The floor shook with the rhythmic leaps of his feet, the violent sway of his body. He not only danced, he made others dance. He drew two girls together, put their arms around one another and saying *"davai, davai"*—get going—he whirled away with a third girl. Quickly he organized a group dance and no less quickly disorganized it. Not even during intermissions did he sit down. Sweat rolled from his face and deepened the glow of his large dark eyes. He enjoyed movement for its own sake. He radiated gaiety and mischief. Merry quips dripped off his lips as he recited verses of poetry; he hummed a line or two of a lively melody; he talked to any girl he saw, whether he knew her or not, and not a girl he talked to but shook with laughter. He did not invite a girl to dance, but, smiling, he put his arm around her waist and whirled away. He was so boyishly good-humored about it that no girl protested. He seemed only too happy to be with girls, look at them, talk to them, make them laugh. He filled the hall with his presence as amply as a good actor fills a stage. He was the soul of goodness and jollity—this unforgettable and lovable sailor.

There was a serious part to the vecher—a salute to the flag, greetings to the war heroes, speeches, and recitations. The crowd moved into an auditorium hung with flags and bunting and rapidly filled the seats. On the stage in rows sat the wounded soldiers and sailors—the guests of honor.

During the first speech there were shifting of bodies and audible whispering. There always is in a Russian audience if the speaker fails to catch its attention. Dictatorship or no dictatorship, a Russian crowd refuses to be bored by the man or woman on the platform or the stage.

Some years ago the late Ralph Barnes and I, while visiting the city of Poltava in the Ukraine, attended a celebration in honor of Gorki's birthday. An expectant audience packed the auditorium, but one speaker was a college professor who read from a manuscript not a speech but a philosophical dissertation. He hardly had read a paragraph when a buzz swept the auditorium and drowned out his voice. The chairman rose and appealed for order. The moment the professor resumed reading his dissertation the murmur arose once more. No appeals and tirades of the chairman

did any good. The college professor never finished his speech. . . . But
when a local poet, an athletic youth with wavy hair and glowing dark eyes,
was presented and read a poem he had written for the occasion, the
audience was instantly transformed. All whispering ceased and the people
were breathless with attention.

The audience in the Pioneer Home was equally intolerant of boredom.
It whispered when the first speaker in his opening words failed to capture
its interest. The chairman appealed for order, but the demand went
unheeded. It was different with the other speakers. One was a sailor who
told of his experience in a sea battle. Every youth in the audience leaned
forward and drank in each word of the speech.

The sensation of the evening was two poets who read their own poetry
from paper-bound books. They were not particularly accomplished recital-
ists though they had good voices and superb diction. The audience was
rapt with attention. At the end of each poem they clapped joyfully,
tumultuously, and shouted for more and more. Here were glory and
rapture for rather obscure Russian poets. Russians love to hear poetry read
even more than plays or stories. I have observed it in villages as well as
in cities. They like the rhythm of words and the stir of heroic emotion
they evoke. Pushkin, Lermontov, Byron, Nekrasov, Koltsov, Mayakovsky,
whoever the poet and however unskilled the reader, if only his voice and
diction are good, he is certain to elicit the warmest appreciation. Readers
of poetry are in great demand and so are poets who declaim their own
compositions. In Russia, poets can handsomely augment their income by
making public appearances before factory audiences.

When the speeches and recitations were over the audience flocked back
to the dance hall. They sang and danced. They threw confetti at each other,
and their hair and dresses gleamed with strips and flakes of variegated
paper. They listened to their own jazz bands, their own readers, their own
story tellers, their own singers, and watched their own tap dancers. There
was no drinking. There were no refreshments as in peacetime, not even
sandwiches. But nobody spoke of it. These young people were glad to
forget the food shortages. The battle for the plan of production was out
of their minds. The hum of looms and the smell of dyes had vanished.
The war was far away. The sorrows and horrors of the times were as if
exorcised from their consciousness. It had been a long time since they had
had a vecher. It would be a long time before they would have another.
They were young and confident and eager, and the world was now theirs
and they would strain out of it all the glory and gaiety they could. So they
sang and danced and shouted with glee and laughed and broke up into
groups and entertained each other and felt as lithe and free as the birds in
the forest.

I left close to curfew time. The moment I was out of the door I plunged into the blackness of the night. No glimpse of light anywhere. It was worse than in the heart of the city, where red and green traffic lights offer some relief from the dense black-out. Madam Zhukova, chairman of the trade union in the factory, led me down the stairs, and following her I groped my way to the automobile that had come for me.

Amid the ceaseless grumbling and swearing of the chauffeur we prowled away through side streets and squares and finally reached the traffic lights. My mind teemed with thoughts of the vecher I had attended. It was by far the liveliest and gayest gathering I had yet witnessed in the months I had been in Russia. There were health, stability, and camaraderie in these sons and daughters of factory workers. They danced and laughed and flirted and amused each other with as much energy and sprightliness as American high-school and college youths.

Yet they were different, a new generation, a new humanity in the world with an aim and dedication all their own, separated by vistas not of space but of thought and sentiment from those of the youth of other lands. . . . They had come from the factory, and all their hopes and ambitions were molded and galvanized by the factory. They might journey to Siberia, to Central Asia, to the Arctic, but always the factory would be uppermost in their minds—the factory which had been the source of all livelihood and all living, all travail and all glory. Wars may come and go, trials, purges, executions may shake the land, darken the hearts of the people in Moscow, of young people like themselves, perhaps of their own close friends. But like the sun above them and the earth under them the factory will always remain. The looms will hum, the engines will roar, the spindles will whirl, the smoke will float out of the gigantic brick chimney—and life will go on —with its shortages, its discipline, its plans, its competitions, its never-ending and ever-shattering demands for obedience, for toil, for sacrifice, but also with its ever-present hope and its never-failing promise of ultimate and irrevocable reward.

CHAPTER 19: INCENTIVES

NIKOLAY RATAYEV is the son of a Tula gunmaker. Bored by studies, he quit high school before graduating and started to work in the factory in which his father had been employed for twenty-one years. Within a short time he learned to operate a lathe, but, dissatisfied with his job, he went to work in another factory. He thought he could earn more money there, yet after eleven days he was discharged for laxity and indolence.

He became a shipping clerk in still another factory. After two months he sickened of the new job and went to work as a porter in a clothing shop. Here he lasted five days. He was dismissed for indolence. Tula is one of the busiest cities in Russia and, despite his adverse record, Nikolay had no difficulty in finding work in a new place. Once more he failed to satisfy the management and was obliged to seek work elsewhere.

Demidov, one of Russia's leading political writers, in his pamphlet, *Heroes of Socialist Labor,* says of Ratayev: "Why does this hunter after easy living bear the honored name of worker? Why do we bother with this violation of labor discipline?"

Demidov's pamphlet was published in 1940 in an edition of 100,000 copies. It went to leading libraries and bookshops. The stigmatization of Nikolay Ratayev is therefore known all over the country by people who have read the pamphlet and by those who have heard of the young man's misdemeanors. There is no doubt that in Tula, especially in the factory press, including the wall newspapers, Ratayev has been the subject of denunciation and vituperation.

I do not know where Nikolay Ratayev is now. He may have reformed and become a Stakhanovite. There have been many instances of workers who, under pressure of public obloquy, have changed their ways and have risen to eminence in industry or other pursuits. Nikolay Ratayev may now be an honored citizen of his community or he may be in jail. He may be fighting. He may have been decorated for valor on the battlefield. Not a few former factory slackers have become hero-soldiers. He may be dead. But to the reader of Demidov's pamphlet he remains to this day a moral and social reprobate.

Thousands of Ratayevs have suffered a similar fate in the press, in pamphlets, in speeches. The purpose of these campaigns is obviously to break the heart of the offender, ruin his standing in the community, incense and shame him into reform, as well as to warn others of the public scorn and obloquy in store for them if they permit themselves to descend to the level of the Ratayevs or fail to uncover within their own minds and bodies the powers to lift themselves above such levels.

Nor are the Ratayevs necessarily industrial workers. They may be engineers, directors, arctic explorers, farmers, actors, writers, editors, chairmen of Soviets, Party secretaries. The nature of the work they do is of no consequence. It is the way in which they do it that matters.

"What incentive," I heard an American businessman ask an American engineer who had done distinguished construction work in Russia, "have these people to work? What do they get out of it?"

Laughing, the engineer replied:

"A kick in the slats."

That was during the First Plan.

He told of instance after instance of engineers, directors, and other executives who had been publicly denounced and dismissed for laxity and incompetence. During the first two Plans the press seethed with stories of these dishonored men. They may have been good Bolsheviks who, during the civil war, had achieved distinction on the battlefield. They may have known by heart *The Communist Manifesto* and Stalin's *Leninism*. But if, whether because of indolence or lack of aptitude, they failed in the constructive tasks to which they were assigned, they were demoted, branded as "has-beens," loafers, slackers, scoundrels, and all the other vituperative epithets in which Soviet vocabulary is so volcanically rich.

In their drive for the enthronement of the machine, for lifting the country to habits of work comparable to those other nations had acquired through generations of measured, peaceful, evolutionary development, the Soviets have wielded the weapon of social obloquy with a passion and a vengeance which again and again have roused the scorn or amusement of foreign visitors and observers. To men trained in ways of business and manufacturing in individualized countries, "a kick in the slats," especially if administered in public, may seem an incentive not to good work but to demoralization and collapse.

But Russians, Bolshevik Russians, are fighting against time. They have no patience with mere evolutionary methods, whether in politics, in industry, or in education. Haunted by the dread of war, never forgetting that their success in Russia and in history, their very security, is conditioned by their ability first and foremost to win the war of production, "to catch up with and surpass capitalist nations"; above all, with such an advanced and technically fabulous country as America, they often enough seek to attain their aims by what they call "the revolutionary method." The essence of this method is speed. They do not mind hurting a person's feelings, making him feel low and degraded. They will stop at no means if it will result in the fulfillment of their plan. They cannot bother merely to teach and to wait patiently as can an American executive or industrialist. They have not his background of practical experience nor his access to all the skilled engineers and all the competent mechanics he may need; above all, his sense of security. They have "to forge" them into being, multitudes of them, millions and millions, and they have found social obloquy an invaluable ingredient in the process of forging.

"We have nerves of steel," I heard a Russian colonel say, "because of the drubbings we have gotten when we've made mistakes. That's why we fight as we do."

No doubt the weapon of social contumely has demoralized promising executives, stunted their creative powers, or ruined them forever. But, so

Russians avow, it has stirred or goaded multitudes into their best efforts and has in a significant measure been responsible for the swiftness with which the peoples of Russia have acquired mastery of the machine age.

But Russians do not content themselves with negative or punitive incentives. On men of meritorious achievement they bestow the emblems and rewards of social distinction as readily as they wield the weapon of social obloquy on those who fail or do not seek meritorious achievement.

I have before me the program of the performance of Korneichuk's play *Front* at the Moscow Art Theater. It is a significant play and a significant program, and indirectly it illustrates the two opposite sets of incentives the Russians are promulgating.

The play is an acid indictment of generals and commanders of armies who have soared to eminence through their triumphs in the civil war and who are too obtuse and too conceited to accommodate themselves to the demands of the age of technology. Written and performed while Russia was fighting for her life, it created a sensation inside and outside the Red Army, which had been immune from any and all public criticisms. For the first time the weapon of public denunciation was visited on men of the highest responsibility in its ranks. Though it is what the Russians call an *agitka*—a political sermon—on the stage of the incomparable Moscow Art Theater, *Front* becomes a drama of moving and epochal magnitude. The audience becomes excitingly, often angrily aware of the battle in the Army between youth and age, conceit and open-mindedness, tradition and audacity, the findings of the engineer's laboratory and the agitator's cocksureness. Free from the invective which usually accompanies Soviet denunciation, it is rich in proof as to the evil consequences of incompetence on the battlefield, and no less rich in indictment of such incompetence.

But the program is an exemplification of the incentive of social approbation that is visited on merit and on distinguished achievement. Here is Moskvin, one of the greatest actors of all time. According to the program he bears the title of "People's Artist of the Soviet Union," the very highest for men in his profession. Livanov, who plays the part of the young general, so moves the audience that it forgets the tradition of withholding applause until the end, not of the act but of the play; and in the middle of an act acclaims him with as tumultuous an enthusiasm as the theater has ever known. He bears two titles: "Artist of Merit" and "Stalin Laureate." The names of other well-known actors also bear titles.

The program of Chekhov's *Three Sisters* in the Moscow Art Theater, if only because of its distinguished cast, is even more revealing. M. N. Kedrov is spoken of as "Active Artist of Merit." N. K. Khmelev is "People's Artist of the Soviet Union" and "Laureate of the Stalin Premium." A. K. Tarosova bears two titles, "People's Artist of the Soviet Union" and "Stalin

Laureate." Three actors and actresses bear the title of "People's Artist of the R.S.F.S.R."—the Russian Republic—eleven others that of "Artist of Merit of the R.S.F.S.R."

Not only the Soviet Union as such or the R.S.F.S.R., but the other Soviet republics bestow similar titles on their distinguished artists, writers, scientists, and other intellectual workers.

These titles mean more than social distinction. Like the Nobel prizes they carry handsome and tax-free financial rewards. Since the war, in some instances even before the war, recipients have usually donated the money to the Red Army. They could have kept it with no loss of standing had they chosen to do so.

In 1941 the so-called "Stalin Premium" was bestowed on distinguished artists in music, painting, sculpture, architecture, theater, opera, dramatic writing, ballet dancing, cinema directing and acting, scenario writing, and also on writers of fiction, poetry, literary criticism. In each division from three to five persons receive a first award of 100,000 rubles each, and from three to ten a second award of half that sum. Of course no one ever receives such recognition if his political record is tainted with nonconformity. But the awards are widely publicized in the press, on the radio, at meetings. The men and women who receive them rise in the esteem of the public to the height of national heroes. They are spoken of with admiration, with fervor, often with reverence. To the people who receive them and to their families they mean much in personal satisfaction.

But these titles are not hereditary. They cannot be passed on to children. If the children wish to rise to the position of national heroes they have to earn their own titles or satisfy themselves with memories of the honors their fathers and mothers had achieved.

Titles and rewards of one kind or another are bestowed on workers in all walks of life. The Army, of course, has more titles and decorations than any other profession. But the factory has its share of both—the shock brigadier, who is usually the pace maker in the factory; the Stakhanovite, who is the rationalizer; the *otlichnik,* who stands out in socialist competition; the hero of labor, who may be any and all of these. Again and again one sees in Moscow on the coats of workers a row of decorations almost as rich as those on the tunics of soldiers.

I do not know of any other country which makes so much of the incentive of social approbation and of the purely egoistic pleasure which it implies as does Russia, which has banned the institution of income-yielding private property and all the individual reward and egoistic pleasures which it presupposes.

Yet, as in any capitalist society, the incentive of material possession and enjoyment ranks high in Soviet collectivized society. Excepting certain groups of executives and certain professions—such as teaching and acting, which do not lend themselves to any formula of rationalization—all compensation is based on piecework. In factories and on collective farms there are few jobs that are paid on any other basis. The principle or practice of equal pay for unequal work is alien and repugnant to Soviet usage. Quite the contrary: with respect to immediate individual reward Russia has outdone all capitalist nations. In no other country in the world, either now or at any time since the rise of the machine age, has the system of piecework and the inequality of pay that it ensures been in such universal use or so continuously and vehemently encouraged.

Eventually, so runs the theory, when Russia passes into communism, man will be so highly cultivated that he will need no speed-up schemes and no inducement of special material reward to make his greatest contribution to society. He will do his best at whatever productive task he chooses to pursue wholly because of an inner sense of duty and honor. . . . Then he will receive from society "according to his needs." But at the present, when Russia is still in a transitional stage, only "on the road to communism," personal material reward—or, as the Russians say, "to everyone according to his labor"—is the law of the land. Hence the sharply marked difference in financial rewards in the country, with the driver of a team in a Moscow warehouse receiving 375 rubles a month and a popular author like Sholokhov or Alexey Tolstoy earning in royalties perhaps half a million or more rubles a year. Of course these writers maintain a higher standard of living than does the teamster, though because of Russian emphasis not on production for consumption but on more production, the variety and supply of usable goods do not permit a luxurious life, in the English or American sense of the word, for anybody.

But even people in the higher earning brackets have not, under the Soviet system, the opportunity to build up what may be called "a family fortune." They must pay a high income tax. They pay it as does everybody in Russia, at the source—that is, the government collects it from their earnings, before these are paid out. In peacetime wages or salaries in the low brackets are almost exempt from taxation. If the sum is no higher than 1,800 rubles a year, the income tax is only 0.8 per cent. From this sum upward the tax rises rapidly, as is shown in the following table:

EARNINGS				TAX		SURTAX		
From	6,001 rubles to	8,400		168 rubles		5% above	6,000
"	8,401 " "	12,000		288 "		6% "	8,400
"	12,001 " "	20,000		1,064 "		8% "	12,000
"	70,000 " "	100,000		6,264 "		17% "	70,000
"	200,000 " "	300,000		41,364 "		45% "	200,000

If the payee is a member of a trade union, which he invariably is even if he is a writer, motion-picture director, or ballet dancer, he pays an additional levy of 1 per cent to the union. If he is a member of the Communist party, he pays 3 per cent more to the Party treasury. Always he must pay the government levy, known as the *Kultsbor* (cultural collection), which is used for building schools, theaters, libraries, and other educational and recreational institutions. This tax rises to about 5 per cent of the gross income.

The war has boosted the number and amount of the direct income taxes. By a law of July 1, 1941, workers and executives who earn from 300 to 500 rubles a month pay an additional tax of 50 per cent. Those who earn over 500 rubles monthly pay twice as much as they did before the war. If the payee happens to be a man of military age and is exempted from army service either because he is needed in industry or in some other peacetime pursuit or because of ill health, the rates are still higher. On earnings of 300 rubles monthly the tax has been raised 100 per cent; on sums between 300 and 500 the increase is 150 per cent, and on amounts over 500 monthly the increase is 200 per cent.

Because of the many orphans in the country, millions of them, the government has instituted a special tax for childless couples and for unmarried people—that is, for girls of eighteen and older and men of twenty and older who are single. This tax comes to about 5 per cent of the earnings. It has the indirect aim of encouraging childless couples or those who are unmarried to adopt orphans. Once they have a child to care for they no longer pay the special tax.

Owing to the war, there are further deductions for the defense fund which amount to one or two days' earnings a month, for government bonds, for occasional lotteries.

The peacetime and wartime taxes and levies devour from one third to one half of the average citizen's earnings, and of those in the higher brackets much more.

Before the war people in the higher earning brackets found it impossible to spend their earnings even after paying their heavy income taxes. They would have been only too glad to spend them on better housing, better furniture, better clothes, a car of their own. If they had a car, they might have wished to buy a better one or one of foreign make. The appetite for consumption goods is one of the things the Soviets *never* have denounced. On the contrary, the propaganda has been in the direction of developing it in a multitude of ways for the sake of hygiene and health; above all, for the sake of holding out the promise of eventual satisfaction. This promise by itself is one of the driving incentives in Russia.

Because of the heavy inheritance tax, people with surplus funds on their hands would be only too happy to spend them. Lacking the opportunity to do so or responding to the call and the pressure of the government to buy government bonds, and to deposit surpluses in savings accounts, they did both. The only form of legal unearned income in Russia was the interest on the sums invested in bonds or deposited in the savings accounts. This was the only "capitalist" concession the government allowed to people with surplus funds. It was a means of fighting inflation.

To encourage such disposal of money the government offered a still further concession: it exempted all bonds and all savings accounts from the inheritance tax. Neither America nor England has done so much.

Some foreign observers, noting Russia's system of piecework compensation and royalties to authors and inventors, as well as the unheard-of concessions to investors in government bonds and savings accounts, have been loud in their declaration that Russia is on the road to capitalism.

With this view the author vigorously disagrees. The Soviets have made no concessions to private enterprise. All ownership in land resources, "means of production," is vested in the state. Nowhere is there the least hint that changes are impending and that private enterprise, even in small-trading pursuits, is about to be legalized. There is therefore no valid proof of any real or decisive compromise with capitalism.

There has been a rediscovery and a recognition of the fact that human beings enjoy privacy even when they have been so little given to it as Russians. Nobody now thinks of building apartment houses without private kitchens for the family or private eating places. Individual enjoyment of consumption goods is continually emphasized. Before the war Muscovites who were earning high wages or salaries were buying more and better clothes, more and better furniture, more and better food. They were seeking to acquire frigidaires, high-priced radios, and automobiles. Some have actually done so—especially writers who earned royalties abroad. Many others would have been glad to make similar purchases had the goods been available.

Seven years ago I visited a village in the republic of Kabarda, in the North Caucasus, which was tearing down old houses and putting up what they spoke of as a "socialist town." Some of the new houses had already been built, and in the rear of each was a garage for a private automobile or a lot for the erection of such a garage. Building these garages at a time when the government manufactured only the small number of passenger cars that it needed for official purposes seemed a foolhardy, almost a ridiculous venture. On inquiry I learned that the people had no expectation of coming into an early possession of their own automobiles. But they were building a "socialist town," and the garages were merely a symbol of the prosperity and enjoyment that were to come.

By a special law of the Economic Council of April 1939, banks were ordered to make loans to individuals for building their own homes. Housing was one of the most neglected aspects of economic development, and to remedy the condition the government had embarked on a vigorous nationwide campaign of individual homebuilding. The project had no more than gotten started when the war broke out and put an end to it. It will be resumed on a larger scale than ever on the conclusion of the war to provide homes for the millions whom the Germans with their wanton incendiarism have made homeless, and to remedy the congestion that exists everywhere in the country.

Yet as for government bonds the average citizen in Russia does not regard them as an important source of income. The war came, and all over the country people donated their bonds to the government. People of the highest earning capacity set the example and pace in these donations. While no figures are available it is safe to assume that they have already canceled at least one half of the government's internal indebtedness. By the time the war ends, if it lasts much longer, there is no predicting how much of the internal debt will be wiped out by similar offerings.

The same is happening to savings, though at the time of this writing not yet on such a monumental scale as government bonds. Let the reader ponder on the meaning of the following telegram which a Moscow woman named Alexandra Smirnova, an executive in the textile trade, sent to Stalin in December 1942:

"In reply to your report of November 6, 1942, and following the happy example of the Tambov kolkhoz citizens, I want to appeal through the columns of *Pravda* to the Soviet women, the wives of soldiers at the front, to open a bank account for the construction of a tank column in their name. To open this account, I am depositing 1,000 rubles. . . . This sum of money I have saved from my salary."

Pravda never prints a message like this unless it has nationwide significance. It started a flood of contributions by wives of soldiers and others out of personal savings for the creation of this special tank column and other armaments. Not only the wives of soldiers, but soldiers themselves, especially officers, and civilians in every walk of life—factory workers, collective farms and farmers, teachers, and others—started to contribute savings for this purpose.

In all Russian history, whenever the Russian people were fighting a critical war, Russians made similar contributions to the war chest, but never on such a prodigious scale as now.

Between December 1942 and March 1, 1943, the amount of savings Russians donated to the Red Army Defense Fund reached 7,041,520,000 rubles, or about 1½ billion dollars!

But even if government bonds and savings accounts had not been donated to national defense, they could not have developed a leisure class which could live off income from investments. The moment there might be the danger that such a class would come into being, the basis of its rise would somehow be immediately and effectively undermined. The word *dialectics,* by which the Russians explain many of their changes and contradictions, would be invoked once more to justify a swift and drastic change in policy.

Besides, if ever the time comes when Russian production reaches a point where the government no longer needs to float internal loans, there will be no more government bonds. What is more, nothing will remain of the old bonds. They will be donated to the government and canceled! At a public meeting, in a factory or on a collective farm, someone will offer to make the government a gift of all his personal bonds. He will call on others to follow his example. The response will be overwhelming, as much so as the example of the Tambov collective farms and farmers with their contributions of forty million rubles for the construction of a special tank column. It is easy enough to say that the movement to cancel the internal debt can only be started under government pressure. Such pressure may play an important part, but one familiar with Russian conditions and Russian psychology must realize that pressure alone will not be sufficient.

"When we reach the point in our industrial development," explained a Soviet bank accountant, "where we can buy the things we want, we'll be so happy we'll be only too glad to donate our bonds to the government. Why not? Without any interest to pay us, the government can build better houses, lower the price of goods in shops, reduce the cost of travel, and we'll be better off. We don't have the kind of bookkeeping in our personal life that capitalist economy makes necessary."

This man may have been too optimistic in his explanation. But there can be no doubt that the Russians do not regard government bonds as a "nest egg" in the sense in which people in England and America do.

Nor will the government encourage people to put their savings in the banks. It will want them to spend their earnings on the goods it manufactures and puts on the market, whether clothes, food, houses, or pianos.

Of course Russia is a long way from the time when her production is so high she will no longer need internal loans or the funds of savings accounts. Yet not a Russian but is convinced that this time will come. In his speech at the Eighteenth Party Conference in April 1939, Stalin gave the people ample reason to feel that this condition would come as certainly as Russia would fight an invasion to her last drop of energy and blood.

With these conditions and expectations in mind, it is idle to talk of Russia's return to capitalism because there are income-yielding government bonds in the country and interest-paying savings accounts, both of which are now exempt from inheritance levies.

So while there is abundant material incentive and widespread inequality of income based on amount and quality of labor, family fortunes or accumulation for the sake of accumulation and the incentive it affords, whether in the feeling of power or superiority or any other purely egoistic satisfaction, is not of any consequence.

Indirect incentives, material and social, are likewise many and varied. Let the reader recall the account of life in the Trekhgorka textile factory. Not a large percentage of Russian factories can boast of so highly diversified and so ample a fund of outside satisfactions as the people of the Trekhgorka enjoy. It is an old factory, one of the best managed in Russia. Yet everywhere—in city and village—the aim and aspiration of the factory and the collective farm is to contribute more and more to the social life and the cultural enjoyment of the community.

One of the most powerful incentives in present-day Russia is the urge to achieve, to fulfill a plan, to pioneer for a new factory, a new city, a new university, a new farm, a new meteorological station in the Arctic or some other new institution. It is all a part of the pioneering spirit which the Plans have stirred and which Russian politics, economics, sociology are constantly and feverishly propagating. Russian literature, journalism and fiction, abounds in themes of this pioneering—the urge to create something new somewhere either on an assignment or on a voluntary but state-sanctioned mission.

"Why are you here on this collective farm?" I once asked a young woman, an architect and the mother of two children, who was supervising the erection of a new cow stable on a farm in White Russia.

"Because," she answered, "I made a special study of farm architecture and I want to build as many up-to-date cow stables as I can in the Second Five-Year Plan."

I have met Russian beemen, Russian livestock experts, Russian grain specialists who speak in a similar spirit of creativeness and adventure. Talk to a young Russian explorer, a Russian Arctic flier, a Siberian agronomist, and though his motivation may be different he speaks in words that often enough conjure forth the audacity, the spirit of adventure, the fighting energy of Jack London's characters. No wonder Jack London is so popular, far more now than he ever was in pre-Soviet times. His tales of man's triumphant struggle with nature evoke a quick and warm response in Russians, who in their quest for new treasures array their bodies and their minds against the steppes and the forests, the skies and the waters, the heat and the blizzards of Mother Russia.

True, a dense and unimaginative bureaucracy now and again besets the aims and the movements of pioneering men and women with endless and needless obstructions, which result in loud and venomous altercations between bureaucrat and pioneer. But these obstructions do not thwart the spirit or the will of pioneering folk except when they are oversensitive—which the new and toughened youth of the land most manifestly is not.

Another powerful incentive in the Russia of today is the faith in the righteousness of the principle of collective control of property and in the promise it so eloquently holds out for immeasurable future rewards. Not all, of course, share this faith. The campaigns to liquidate Nep, kulaks, and other official enemies of the Soviets or the Plans are grim testimony to the presence of multitudes who stood out in word, in deed, or in sullen silence against the new control of property or the manner of its administration. But their opposition is over. The new generations neither know nor care to know any other system of property control. This faith, far more than the severe discipline, enables one to understand the driving and surviving power of the Soviets, the extraordinary Plans with all the construction they have achieved despite initial lack of experience, low living standards, and incessant speed-up campaigns.

For those who are impregnated with it this faith is not only an incentive, it is a passion. Russians deny that it springs from idealism. Materialists, they abhor the very word idealism. They speak of their aspiration not as faith but as reality—the reality of today and tomorrow. Without this faith, neither Stalingrad nor Magnitogorsk, neither Kuzentsk nor Chelyabinsk, could have been built or rebuilt. Without it the Soviets would have long ago collapsed. With it they have made themselves invincible.

CHAPTER 20: THE KOLKHOZ

OUTSIDE the mountain country no villages in Russia overlook such beautiful scenery as those that rise over or near the banks of the Volga. Yet in outward appearance no other villages present so unappealing a picture.

Russkiya Lipyagi is no exception. Situated on a height not far from the Volga, it almost bends over a broad and translucent stream beyond which roll gorgeously grassed meadows, wave after wave of forest. Not even on the ocean in clear weather have I beheld such stirring sunsets as in this village. Not only the sky but the tops of trees blaze up with a crimson brilliance that lures and overpowers the eye. So clear and bright is the air after the sun has vanished that one almost expects it to reappear like a village bride who has slipped away to don a bridal wreath.

Never before in all my wanderings in Russian villages had I been so impressed by the grandeur of the Russian landscape and by its power to inspire love of land and country. In these days of battle for survival Russians speak much of this love, sing of it, too, not with heartbreak but with triumph, and their speech and singing have made them, as well as outsiders, more than ever aware of the glory of Russian nature. No wonder that Russian war correspondents—whatever the subject of which they write, whether the burning of Sevastopol or the flight of a bomber on a nocturnal mission to Budapest—never fail to speak fondly and at length of the Russian sky and stars, the Russian trees and the Russian earth; and that motion-picture producers linger long and lovingly over a willow on the edge of a Russian lake, the flowing gleam of a Russian river.

But in outward appearance Russkiya Lipyagi, like all the Volga villages I know, presents a discouraging spectacle. It is ancient and unkempt. The wooden cottages twist and bend and writhe. Neither roof nor walls nor windows bear the least vestige of decorativeness. There is no paint. There never has been any. The one whitewashed cottage in the village is the home of an outsider, a Ukrainian woman, whose love of ornateness has remained as much her Ukrainian heritage as her melodious speech.

Bearded little goats wander the streets or tug powerfully at posts to which they are tied. Chickens squawk in the courtyards. Dogs growl and bark on sight of a neighbor or a stranger. Children run races and shout with glee. The noises and the voices here are the same as in other villages, as in the blue-misted Caucasus, the gay-colored Ukraine.

On first sight the cottages speak of an unchanged and an unchanging way of life, of a people, whether happy or not, who seemingly cling to ancestral heritage with the immutable stubbornness of an oak to the earth in which it is rooted. As I walked along the village and looked around I had the feeling that it was older than time and as defiant of the dreams and the powers of man.

Yet this village is a kolkhoz—a collective farm—and the very sound of the word spells a social transformation more profound than any the world has known, transcending, in this writer's judgment, any and all changes that have swept the Russian city, if only because it originally embraced twice as many people, over one hundred million of them, once the most backward and the most disorganized in the land. I watched this transformation from the day of its inception. I was an eyewitness of the confusion and the terror, the agony and the devastation it had originally caused.

I heard the cries and the curses of men and women who beheld in the kolk-hoz an abyss into which they believed they were tumbling, and the chants and shouts of youths who proclaimed it the one and only deliverance of the muzhik from the evils and sorrows that had for centuries been devouring him, body and soul.

It was a time of stress and battle, the greatest the Russian village had ever known. In Slavenskaya, a Cossack village in the Kuban, the priest one Sunday morning mounted the dais and said, "There will be no services to-day. Go out and do your duty by yourselves and the community." The congregation took the admonition to heart, tumbled out of the church doors, women ahead, several rows deep, men following behind, fists clenched, hearts aflame with wrath, ready for battle and blood. They marched to the Soviet office and at the top of their voices shouted their hate and defiance of the kolkhoz. Some of the officials ran and hid, others stood up unshaken and unafraid and started to explain and harangue the crowd. But the crowd was in no mood for words. Blows fell hard and merciless, and blood flowed.

In a village in White Russia in a large thatched barn with two smoky lan-terns hanging on tall posts, a Jewish girl only seventeen years of age, standing on the back of a wagon, was passionately denouncing grumblers, disrupters, saboteurs. She was promising a crowd of angry, disheartened men and women all the bountifulness of the world if they would only for-get the immediate shortage of meat and sugar, of leather and textiles, of candles and kerosene, and apply themselves with all the energies of their bodies and all the fervor of their hearts to the work on the land and in the barns. There were outcries of protest and mockery; now and then there were bursts of bitter laughter. But the girl raced on with all the faith of an impassioned evangelist and in the end obtained a vote of confidence, though reluctantly given, in the new scheme of things in the village.

Afterward, while the mothers and fathers went back sullenly to the cot-tages, the young people gathered in the haymow in the barn. Nearly all night they argued and debated and laughed and sang and cheered and praised their red-haired, blue-eyed, seventeen-year-old leader from the city of Minsk and promised to fulfill all the assignments she would im-pose on them, including, if necessary, the use of their fists against some of the more excitable and more malevolent members of the new kolkhoz.

But the kolkhoz won. The modern machine and modern science triumphed over all protests, all grief, all sacrifice. Five years later, only five years, I was in Slavenskaya in the Kuban. Harvest was in full swing. A fleet of combines and tractors was reaping the collectivized wheatlands. The market place teemed with crowds who had gathered to buy and sell watermelons, milk, sour cream, apples, pears; and in the evenings in the village clubhouse a crowd of festively attired boys and girls had come to watch a dancing instructor, a good and loyal Komsomol from Krasnodar, demonstrate the steps of the *Bostón* (the Russian name for an American waltz) and the American fox trot.

Within a few years the modern machine had performed an even more stupendous transformation in the village than it had in the city. But in the village, as in the city, it was not the houses in which people lived but the fields and the factories in which they worked, and even more the hearts and the minds of men that bore the stamp of the epochal and violent upheaval. Houses could wait, are still waiting, but not fields, not factories, not with war threatening and with the battle for survival looming on the horizon. Such was the will of the Kremlin, such was the behest of the Plans.

Consider the sweep and the magnitude of this upheaval. In 1910, according to the census of that year, Russian villages were still in the throes of the age of wood. They boasted 10 million wooden plows of all makes, 4.2 million steel and iron plows, 17.7 million wooden harrows! Only a small fraction of these plows permitted the deep turning of the soil which is necessary to good farming. The light, stone-weighted wooden harrows made the preparation of the proper seedbed difficult or inadequate.

In 1918, at the end of the first World War, with import at a standstill, with Russian industry largely in ruins, with the new Soviet government facing outside boycott and internecine war, the age of wood remained supreme. Much of the machinery which some of the former landlords had acquired, as good as any manufactured in Europe, was out of repair or had been smashed or taken apart, with the parts carried away by peasants as pieces of metal and unavailable for repairs. It was a year of chaos and disintegration, in farming as well as in industry.

In 1928, the year of the first Plan, less than 1 per cent of the peasant families had joined the collective farms. They tilled an area of less than four million acres. The rest of the land, excepting the state farms which the government operated as its own exclusive enterprise, was divided into more than twenty million small holdings. These holdings were for the most part carved into endless narrow strips which were separated from one another by weed-festered dead furrows or ridges. Each family tilled its own allotment of strips with its own hands, its own implements, its own horse if it had any. Because of the small acreage, the vast majority of these families had no hope of acquiring the means to purchase mechanized equipment or adequate amounts of commercial fertilizer. They knew little and because of their conservatism welcomed even less the proper rotation of crops. The use of select seed they had grossly neglected. Spring-flooded lowlands needed drainage, but individually they were in no position to achieve it. Their age-old suspicion of new methods of tillage and new ways of living was a further impediment to the modernization of Russian agriculture. At best, under the old system, it could be achieved slowly and regionally. But the Revolution was impatient of delay; revolutions always are impatient—that is why they are revolutions.

How complete were the changed conditions in the Russian countryside in 1939! Nearly twenty million households, or 95.6 per cent of all, were banded into 241,000 collective farms, which had at their disposal for perpetual use over one billion acres of land. Including forest, swamp, and pasture, a kolkhoz averaged over 4,000 acres. Only an insignificant number of small farms had remained and they embraced only 0.4 per cent of the peasant lands. In 1940 the Russian farms were worked by 523,000 tractors, 182,000 combines, and all the needed accessory implements.

Within eleven years Russia had become a nation of large-scale mechanized farms. Hardly a shred had remained of the age of wood. The plows and the harrows of former years had been stored away in barns and garrets as museum pieces or for possible use in emergency such as a serious war might cause. From year to year the machine was deepening its hold on the Russian land and on the mind of the Russian peasant. A young generation was growing up knowing nothing of the strip system of farming and of the wooden tools. It knew only the machine—the tractor, the disc harrow, the combine. It knew only large fields of wheat and rye and oats and barley. It knew no individual but only collectivized tillage. It was acquiring a heritage such as its fathers and mothers never had known or imagined possible.

The transformation of Russian agriculture from individual ownership to collectivized holdings and tillage, because of the very speed of the process, was accompanied by inordinate sacrifice of comfort, substance, life. Yet great and tragic as have been these sacrifices they are small compared to what they would have been had the Red Army and the civilian population lacked the immense reserves of strength that the kolkhoz has made possible. In a subsequent chapter I shall speak at length of the "new order" which Germany has been seeking to impose on the Russian peasantry in occupied territory. Here I only wish to emphasize that the essence of the German program is subjugation and extermination. What would have been the cost in life had Germany won the war staggers the mind. Yet without the collective farm Russia could never have fought as she has been fighting. She would not have had the mechanical-mindedness, the organization, the discipline, above all the food. In this writer's judgment she would have lost the war.

A comparison between military and food conditions in the first World War and the second is especially illuminating. During the first World War—according to Stalin's speech on November 6, 1942—out of the two hundred divisions which Germany had mustered only eighty-five were thrown into the Russian campaign. To these were added thirty-seven Austro-Hungarian divisions, two Bulgarian, and three Turkish, making a total of one hundred and twenty-seven divisions fighting against Russia.

Germany was then fighting on two fronts. In the west she was facing powerful French and British forces. Nor did she have command of the entire industry of western Europe as she now has. France was France. Holland was Holland. Norway was Norway. Neither Poland nor Czecho-slovakia, both of which were under German rule, possessed the productive powers they now possess. Now—that is, at the time Stalin was making his speech—Germany, he said, held one hundred and seventy-nine of her two hundred and fifty-six divisions on the Russian fronts. Rumanians, Finns, Italians, Hungarians, Slovaks, and Spaniards swelled the number of divisions in Russia to two hundred and forty, and Germany had command of all the industry of western Europe. Also, out of the conquered countries she had mustered for German industry six million foreign workers. In numbers of men, in kind and amount of equipment, the Red Army was facing an infinitely more formidable task than had the old Russian armies.

The first World War had disorganized and shattered the Russian supply system. Russia had only a few tractors on the land. Most of the horses were mobilized for war. So were most of the men. Industry manufactured a small number of agricultural implements. In consequence, by the end of the war, according to Soviet sources (Moscow *Bolshevik*, November 4, 1942), the sowing of grain had been reduced by about twenty-five million acres. The output of grain had fallen by nearly one fourth, and so had the potato crop; fodder had been reduced by 43 per cent. Disorganized transport only aggravated the shortage of food and ammunition. Let it be noted that the Revolution against the Czar was initiated by hungry women in Petrograd who started to throw stones into the windows of closed bread shops.

Now, owing to the discipline, the organization, the planning, the science, the mechanization that the collectives make possible, the tilled acreage in Soviet-held territory has been continuously and substantially increasing. According to A. Benedictov, Commissar of Agriculture, in the autumn of 1941 in grain alone almost five million more acres were cultivated than in the preceding year. The Plan for 1943 calls for the cultivation of fifteen million acres more than did the Plan of 1942. Equally important is the care and the speed with which, despite the absence of man power, the crops have been gathered. In grain, in tobacco, in flax, in legumes, according to Benedictov, the harvesting was accomplished from ten to fifteen days earlier than in the preceding year. Never before had crops been gathered with such speed. Livestock in Soviet-held territories had likewise been increasing. The civilian population was on a basis of stiff food rationing. Yet the food was more or less evenly distributed. People got some food even in the far north, where little grain was grown. As for the Red Army, it knew none of the shortages which the Czar's armies had experienced. It had plenty of bread, meat, sugar, cabbage, and potatoes. Because it was so well fed the Army was fighting as well as it did against Germany and her allies. But let the story of the kolkhoz in Russkiya Lipyagi illustrate this point.

The kolkhoz in Russkiya Lipyagi is not one of the best in Russia nor in the Kuibyshev province. In more than one respect it is not quite the average. It has no pigs, for example. There is a quarantine against pigs in the district, which, to a peasant, is a serious disadvantage. If pigs are cared for with any degree of decency, no other animal on the farm, especially at a time of fat shortage, yields such swift and lavish rewards. Fed for six months or more, a pig may afford a family enough fat, if modestly rationed, to last all winter. So in Russkiya Lipyagi meat, especially in summer, is a luxury. But since nearly every family has a cow or a goat, chickens, ducks, sheep, the problem of fats is not so acute as in the city.

But the land is rich—deep black earth. It lies well—high enough and rolling to provide natural drainage. Without steep hills, with neither rock nor stumps, with the lowlands across the river devoted chiefly to meadow and pasture, it presents no special problems or difficulties in tillage. For these reasons it is as suitable for an appraisal of a kolkhoz as any other, however large in acreage or rich in material equipment. The institutions are the same, the laws are the same, the crops are more or less the same, the relations to the state and to the Army are the same. Also it demonstrates as abundantly as any other kolkhoz or as any factory that in Russia it is not the individual who rules or makes the community but the community that rules and makes the individual. Without the sacrifices of the individual the community never could have attained its present resources and powers and without these powers Russia never could have fought Germany so valiantly and effectively as she has. Let philosophers disentangle and decide the verities and the moralities such a system presents and symbolizes. The task of this writer is to record it as it exists and as it reflects itself in the life and in the personalities of the people.

As I walked the streets of the village I observed nowhere any sign or suggestion of the war. The scene was as peaceful and pastoral as it always is in a Russian village. But the moment I entered the kolkhoz office I became more than aware of the battlefields. The office was an ordinary peasant hut, no doubt the house of a former kulak—with high windows and a huge brick oven. The walls, doors, and posts were hung with vivid posters. "Blood for blood, death for death." "Avenge the savageries to our mothers and sisters, annihilate the child killers." "Be skilled in throwing the hand grenade—throw it fast and straight at the target." These are only a few of the captions I read on the posters. They were illustrated by pictures as expressive and combative as the words. In the headquarters of the kolkhoz the citizen was not permitted for a single second to forget the fight for national survival.

The day before, as the chairman of the kolkhoz and I were walking around, we met two wounded soldiers who had just come off the train. They had fought on the Leningrad front and had been sent to a hospital on the Volga. They were heavily bandaged, one on the head, the other on the feet. The one with wounded feet walked on crutches. Neither man any longer required hospital treatment, so both were sent to this kolkhoz for further recuperation. They needed rest, fresh air, simple food, and the kolkhoz was in a position to offer all these. The chairman immediately took them to the home of an elderly couple and arranged for their food and shelter.

Now the wounded men were in the office, and so was the woman of the house in which they had been stopping. The woman had been rude, she would not cook for the wounded men; so they came to see the chairman, and he sent for the woman. Gray-haired, with a dark kerchief on her head which only deepened the pallor of her face and the grayness of her eyes, she sat beside the chairman, calm, somber, hardly saying a word. "They fought in the war, *mamasha*," the chairman was saying, "way up in the Leningrad front. They were ready to give their lives for the fatherland, for you and for me, and yet—you wouldn't even cook for them." He paused, glanced at the wounded men, at the woman, and waited for her to speak. But she remained silent. "They don't even know where their families are—their wives, their children—they need attention. If we don't give it to them, who will?"

More and more people crowded into the office. But the chairman allowed them to wait, and they all listened—Russians love to listen as well as to talk—to the words of admonition that were slowly rolling out of his lips.

"Did you heat the bathhouse for them?"

"She didn't," interposed one of the wounded men.

"And I had requested you yesterday when I came to see you to heat it so they could wash up, and I asked you to be like a mother to them—and——" He paused, sighed, and there was more meaning in the sigh than in any word he could utter. Not until she had agreed to be like "a mother" to the wounded men did he end the conversation.

The chairman was a peasant under forty, with a deliberate manner, a drawling voice. But the very slowness of his ways, the calm depth of his finely rounded blue eyes spoke of will and strength and understanding. It was obvious that the woman found it too much of a burden to be "like a mother" to the strangers. But whatever her personal wishes, she had to submit to the will, not of course of the man who presided over the kolkhoz, but of the nation or the state.

As I listened to the conversation I knew that if the woman continued to ignore this will, the chairman would talk to her again. If necessary, at a regular meeting of the kolkhoz—at which, to make it legal, at least three fourths of the members would have to be present—he and others would lecture to her on her failures as a citizen, a patriot, a Russian, a woman. The pressure of public opinion is one of the most powerful weapons the Soviets wield against individuals who place personal wishes or interests above those of the community. In the end the individual submits. He may grumble and moan. Deep in his heart he may curse and wish he had lived in a world in which he could refuse to share his home with a wounded soldier, an evacuated woman with children, and exercise only as much social responsibility as he chose. But in Russia he is helpless. The individual submits to the community.

"How many wounded soldiers have you in the kolkhoz?" I asked the chairman.

"Twelve."

"Are other peasants as unfriendly as this woman?"

"Not all, and this woman means well—but I had to talk to her. Sometimes people are a bit forgetful of their obligations to the fatherland."

In addition to the twelve wounded soldiers in the kolkhoz there were fifty evacuated women and children. Most of them were from the cities and at first there was grumbling among the village folk who had to share their homes with strangers. But when the city women started to work and showed they could make their contribution to the daily task of the kolkhoz, the grumbling abated, then ceased. The kolkhoz had acquired much-needed help, and the *évacués* were enabled to resume a normal and productive life.

Supposing there had been no kolkhoz?

A few days earlier I had discussed this subject with an Allied diplomat who had been in Russia during the first World War. Refugees by the hundreds of thousands, he said, wandered all over the country. Seldom did they obtain accommodation unless they could pay for it, and ruined and battered as so many of them were they could pay neither with money nor with goods. Of course there were charity and government aid, but they were inadequate. So refugees wandered around, desolate and helpless, women and children and older folk, and often enough wounded soldiers. Many of them died for lack of care.

Now, and especially in the immense lands east of the Volga, the kolkhoz is absorbing millions of homeless people, even as the factory is in the city. With the Crimea in the summer of 1942 in the hands of the enemy, with many of the sea resorts under German occupation, wounded men who no longer needed hospital attention found food, shelter, and care on collective farms. If they got well enough to work they might join the kolkhoz. The need for workers was desperate, and once a soldier became a member of the kolkhoz he might bring his family and start life anew. . . . Seldom could he do it so easily or at all in the old days. He had to *buy* land. He had to *buy* a horse. He had to *buy* a cow. He had to *buy* implements. There was no community which he could join without funds and merely by the contribution of his labor resume his interrupted life as a citizen, a father, a husband, a son.

If there are no crowded or stranded and helpless men and women in Russia in these times of mass evacuations, if the roads and railway stations and public squares are not darkened by swarms of lost and homeless children, it is in large measure because of the numerous socialized institutions —not only the kolkhoz but the factories and the special state colonies which the government is sponsoring and which are somehow absorbing the multitudes of migrants.

The bookkeeper of the kolkhoz was a young girl of nineteen. She had graduated from high school. Her father was in the Army. Her mother had died a year earlier, leaving seven children, of which she was the oldest. In pre-Soviet days the children at best might be sent to an orphan asylum. Usually relatives and neighbors took them into their homes, treated them as "orphans," than which, with but rare exceptions, there was no more desolate fate for a child in the village. Now, at a special meeting, the kolkhoz passed a resolution taking upon itself complete guardianship over this large and helpless family.

It is easy enough to conceive that some members were reluctant to support the resolution. True, the kolkhoz was using 2 per cent of its gross income for the social-insurance fund—that is, for the support of the old, the disabled, the children's institutions. But this family was a special problem and needed special care. A woman would have to be hired to do the work in the house, and the kolkhoz would have to pay her as it would a field worker.

The expense involved would affect their individual income. They would get less for their labors. At the end of the year the balance sheet would be drawn up. Taxes and other government obligations would be deducted from the gross income, also the social insurance, the seed funds, other possible obligations. Out of the balance would come their individual compensation.The cost of maintaining the orphaned family would come out of this balance—that is, out of their income. It might not be much, but it would be something, and there were already plenty of deductions without inflicting a new one.

Such may have been the reasoning of certain or many individual members. But in the course of the discussions, led usually by the more political-minded members, they yielded to persuasion or to public pressure. In a conflict between personal interest and a public purpose, that is, between the individual and the community, the community invariably wins. But in this case a family of seven children without a mother, with the father at war, was enabled to maintain its home.

I was staying with a family in which, as in so many families, there were no men. They were at the front. This family, like others, had its own cow, its own ducks and chickens, a garden, a newly born calf, and several sheep. The law allowed as many as ten sheep per family, though only one cow. The garden was about three fourths of an acre and was superbly set out with rows of potatoes, onions, beets, cucumbers, cabbage, and a large patch of pumpkins. Because of the sugar shortage peasants were planting more and more pumpkins. Children liked it steamed, and the kolkhoz doctor encouraged them to eat more of it.

Out of this garden the family had to sell to the government, at its usual low prices, 15 per cent of the potatoes and a small amount of other vegetables. It was also obliged to sell annually, again at low state prices, one hundred and thirty quarts of milk, one hundred eggs, eighty-eight pounds of meat. Other families had to meet similar obligations to the government. These were not taxes. They were the so-called government collections.

No doubt there were peasants who would rather sell the produce in the open market, where the prices were several times higher than the government was paying, or who would rather eat it themselves. "But if they did," said the chairman of the kolkhoz, with whom I was discussing the subject, "where would we get food for the factories, for the city, for the Army? No, the individual must always be thinking of the community because it is the community that protects him. If the Red Army did not keep the Germans away from this kolkhoz, where would the individual peasant be, and what would his personal life be like? Under our Soviets the community is always higher than the individual because in the long run the community passes its accumulation to the individual. Consider, for example, our *contraktatsia* of livestock."

He explained that in order to replenish and increase the herds of live-stock in the country the government was contracting the purchase of young stock, especially heifers. "The peasant might want to kill the heifer or bull calf and sell the meat in the open market or eat it himself. That's what he did under the Czar during the first World War, and you know what hap-pened. . . . But now it is different. He has his own cow and some other livestock. He is in a position to raise the heifer and he sells it to the gov-ernment. By doing so he is building up the economy and the morale of the country, of the factory, of the people, of the Army. If he is a real patriot he is glad to do it. If he is not—law is law and he must abide by it. Otherwise, with our old and terrible backwardness which we haven't had time to overcome, what would happen to us? How could we fight the Germans and save ourselves, our children, our fatherland?" He was earnest and decisive and only laughed at the suggestion that the individual on the kolkhoz was bearing too heavy a share of the national burden. "What about the soldier who gets killed protecting this individual?" he said.

The day before I had heard an old man complaining that it was terrible for him at his age to go without meat because he would have to sell to the government the calf he was raising. Yet even as he was talking his wife brought a basketful of freshly cut grass for the sleek black calf in the courtyard. That was significant. She took good care of the heifer calf. No-where in the village did I observe the least manifestation of sabotage. The red calf of the family with which I was staying was the object of fondest attention of all the children. They petted it, ran after it, fed it potato peelings, freshly cut grass, other things.

By inducing the individual peasant to raise his calf for the state, the herds of the country are enlarged and maintained. The government is in a position to build up a reserve of cattle not only for food but for the collective farms in the occupied villages after their liberation from Ger-man rule. This is the plan and the practice. This is also the law. If the individual peasant, after selling his heifer, could spend his money on manufactured goods at the usual government prices, he would have no occasion to be displeased.

The government is doing all it can to enable him to do so. In Kuibyshev, for example, in every market place in the summer of 1942, there were spe-cial shops for the sale of shoes, slippers, household goods, sewing ma-terials to those peasants who disposed of their products at government counters. The government paid them not the fantastic "free" prices but its own, which are fixed by law. In return it offered them goods also at low fixed prices. There was no compulsion about it. Soviet law permitted the peasant to sell his surplus foods wherever he chose. But in order to enable some city people to supplement their regular rations at reasonable prices with foods that were hard to obtain in government shops, such as butter, milk, cheese, meat, it was seeking to induce the peasant to bring his products to its own counters. Komsomols and Party members circulated in the bazaars, talked to peasants, attempted to persuade them on grounds of self-interest as well as patriotism to bring their foods to the government counters. They even promised them textile goods in the city department stores if they made their products available to the government.

But war is war, and the country turned a multitude of shops, even little ones, into manufacturing armaments. There was not enough manufacturing energy and capacity to produce consumption goods and armaments, and armaments came first. Goods could wait, so could the appetite for goods, no matter how urgent. As long as the war lasted it remained the chief task and the chief burden of the community, the nation, and the individual.

It was not until the chairman drove me around the fields that I fully appreciated the enormity of the national strength which the kolkhoz was giving Russia. As I was sitting beside him on freshly cut grass in a jolting cart, he talked with his wonted drawl of the work and the people in the village. Of the one hundred and twenty-four men who had worked on the land before the war only fifteen remained. The youngest of them was over fifty, but most of them were over sixty. The others were mobilized for war. The chief burden of the work rested on the women. There were one hundred and forty-eight of them. Twelve were invalids, so that only one hundred and thirty-six worked all the time. They were helped by twenty-two girls and thirty-nine boys who were fourteen years old or over and were spoken of as adolescents. The younger children likewise worked at the lighter tasks on the farm. They were paid on the same basis as adults— depending on the type and amount of the work they did.

Despite the loss of so much man power, the kolkhoz sowed 500 more acres of wheat than the year before, planted 110 more acres of potatoes, many more acres of tomatoes and other vegetables.

"How did you do it?" I asked as the sprightly dark stallion turned into a grass-grown untraveled road that separated the fields of rye from the fields of wheat, the two basic crops of the kolkhoz.

"We had to do it, so we did it," was the laconic and telling reply. That they had done it well was testified to by the luxurious stand of wheat. The stalks were tall and fat and had already eared out. With bright sun on them, a breeze over them, they swept on like a rippling sea to the very horizon. I had never seen a better stand of wheat in America. "One of the best crops we've had," drawled the chairman, eying proudly the vast fields.

The deeper we drove into the wheat the heavier it was. Nowhere had it been lodged either by wind or rain. It was obvious that the land had been plowed and disked with painstaking care, over and over, so as to destroy weeds, and that the grain drill had followed an even line and had distributed the seed well. There were no bare or uneven patches such as I had often seen in kolkhoz grain fields on previous visits to Russia. Here the rows were as uniform as they were heavy, and the wheat was a glory to behold.

Large-scale mechanized farming on a new basis of ownership was no longer an experiment. It still was faced with the problem of striking a proper balance between self-interest and social advantage. But agriculturally and politically it was no longer an untried idea, a revolutionary dream. It was the most visible and the most momentous reality in the village, indeed in Russia. With no men to work the lands the women had acquired such extraordinary skill and competence in the new system of farming that they could grow wheat as well as any professional wheat farmers anywhere in the world. "Our women have performed a miracle this year," the chairman remarked. The sight of this wondrous field of wheat which we were passing emphasized once more the stupendous and heroic contribution that Russian women were making to the nation in this hour of battle and stress.

We came to a potato field. A crowd of children were pulling weeds. In their midst was a bespectacled man with stooping shoulders. He was one of the teachers of the school and supervised the work of the children. Eagerly the boys and girls gathered about us, glad, evidently, for a respite from labor. Their hands were black with dirt and so often were their cheeks and chins. They smiled and laughed and talked and danced around us as children will in a moment of playfulness.

The teacher came over, shook hands, asked questions. At once the children formed a circle around us, so eager were they to hear every word that was spoken. "Nichevo," said the teacher, "we'll have plenty of food for our glorious Red Army; won't we, children?" Nodding, laughing, they answered singly and in unison, "Aye, indeed." The rows they had been over were as clean as though a machine had plucked out of the ground all alien growth. As we left they waved and laughed and shouted joyfully not "farewell" but "come again"—the eternal Russian hospitality and friendliness which children imbibe with the very milk of their mothers.

The war had stamped itself deeply on the people and on the kolkhoz. The clubhouse was closed—it was needed for military purposes. Yet the school continued to provide entertainment, but only in winter. Before the war there had been a cinema in the village. Every evening, summer and winter, it had shown a picture, sometimes for entertainment, sometimes with a special message on farming, housekeeping, on health and education.

Now the cinema was closed. The Army had taken the projection machine for the front.

Everybody worked harder now than before the war. Sunday was seldom a rest day. Always there was work and more work. They ate less well, had less meat than formerly; but the children were well supplied with needed foods. Forty of them were in the nursery. Fifty more between the ages of four and seven were in the kindergarten. Parents paid little for service and meals in the nursery and kindergarten. The bulk of the expense came from the 2 per cent of the gross income that the kolkhoz was annually setting aside for social purposes.

Other children, who in previous years spent the summer in a Pioneer camp, now stayed home and worked. Their work was indispensable. They pulled weeds and bound sheaves. They gathered cucumbers and tomatoes. They did chores and everything else that their limited physical strength allowed. At sixteen they were regarded as adults and worked side by side with their mothers and their older sisters.

Many a home in the kolkhoz has been visited by tragedy. "Two of my sons are dead," said one woman to whom I stopped to talk in the street, "and my oldest daughter's husband is also dead. Her father"—she pointed to a lively little girl who was playing an extraordinary game all alone— "watch her," said the woman. I did. With her bare heel she pounded a little hole in the sand on the road and exclaimed, "There goes another, Grandma, and another and another."

"What is she doing?" I said.

"Her father is dead. She knows it, and she is taking revenge on Germans. Every time she pounds her heel into the sand she is 'killing' a German."

I watched the child. She was not yet five. Blue-eyed and round-faced, with soft, light brown tresses falling over her sunburned cheeks, she continued to stamp her heel in the sand. "There goes another and another— see, Grandma . . . see, Uncle?" she kept saying without even looking at us.

"Enough, darling; you'll be tired," said the grandmother.

"I am not tired . . . there—one more," she exclaimed. She pointed her finger at each little hole she had made and, talking to herself, exclaimed in triumph: "You are dead, and you are dead, and you, and you, and you, and you—you are all dead." The last hole she pounded with particular vehemence, as though to make sure of the effectiveness of her action.

In the Soviet office I saw the figures on births, deaths, divorces, marriages in the township of which Russkiya Lipyagi is a part. The figures tell a cheerful tale of family life in prewar days. There had been no divorces in 1939, 1940, 1941, and during the first half of 1942, the only years for which figures were available in the office.

"People here," I remarked lightly, "stay married."

"They had better," said the jovial Soviet chairman. "Our women, little brother, are nobody to fool with," and he laughed aloud.

The population of the township before the war was 3,243. In 1940 there were five marriages, 111 births against fifty-six deaths. In 1941, despite six months of war, there were forty-eight marriages, one hundred and twelve births, forty-six deaths. By July 1942 there were only two marriages, twenty-three births, thirty-one deaths! While church weddings have gone out of fashion, registration of marriages and home celebrations are the custom today. Rarely does a newly married couple fail to register in the Soviet office, though it is not obligatory.

There was a small crowd in the Soviet office and we talked at some length of the war, of world politics, of England, of America, and of the sudden drop in the local births and marriages. There was sadness in the speech of the women, and one of them, of middle age, whose husband was at war—she had not heard from him in five months—started to weep. The chairman would not tolerate sorrow in his office. He talked to her and solaced her and when she left she said: "God grant that my man is alive."

"He is, Auntie, of course he is; I say so, and you know I always tell the truth," said the chairman. The woman smiled and shook her head as she walked out of the door.

Turning and looking at the figures he had just given me on marriages, births, and deaths in the township, the chairman said, "They look pretty bad, don't they—especially marriages and births in 1942? But if you give us a second front in Europe and help us smash Hitler so our men can come home, it'll be no time before the figure'll jump up again. We Russians are past masters in this art—nobody in the world can equal us." Again there was robust laughter.

My hostess was a Ukrainian woman, and her house was typical of Ukrainian coziness—embroidered towels, window curtains, flowerpots, whitewash outside, paint inside, pictures and embroideries on walls and doors. There were two beds in the living room, mounted high with pillows and blankets. With customary Russian or Ukrainian hospitality she turned the beds over to a visiting Russian agronomist and myself while she and her children slept on benches and on the floor in the adjoining kitchen and vestibule.

She was manager of the village store and rose at dawn to open it before people went to work in the fields. Evenings she was home early to cook for us and to look after our comfort. Somehow she always managed to have freshly caught perch and pike, freshly laid eggs, butter, and cheese, and she fed us with an old-time Russian sumptuousness which I had not expected to find in wartime Russia. She was quite young, no more than thirty, a thin, tallish woman with a mass of coal-black hair and friendly dark eyes. She beamed whenever we told her how excellent was her cooking.

My last evening in the village the agronomist and I were sitting outside the house on a bench reveling in the coolness and fragrance of the Russian night. We talked endlessly of the kolkhoz—its past, its present, its future. The agronomist had been in the thick of the fight for collectivized farming and knew intimately the troubles and the sacrifices the country had endured in the early years of collectivization. He and others like him, he admitted, had committed numerous mistakes. They had had no experience because the kolkhoz was new not only in Russia but in the world.

He talked of the time when the first American tractors had come to Russia and of the blithe carelessness with which young people had raced them to the fields. Tractor after tractor had broken down or gotten stuck in a ditch, and neither the drivers nor the local blacksmiths could repair them. Those had been reckless, desperate times, and many an American tractor had been stupidly wrecked. Yet neither he nor men like him had wavered in their faith. Ruthlessly they had pressed on and on with the battle of the kolkhoz. Despite the prophecies of doom of older muzhiks and of the outside world, collectivization had deepened its roots in the rich Russian soil and in the consciousness of the Russian people. Millions of youth who had learned to operate a tractor were now among the best and bravest tank drivers in the Army. Had these youths known only the ox, the horse, the cow, as had their fathers and grandfathers, they never could have performed the feats of heroism which were thrilling the nation and assuring it of inevitable triumph over the cursed enemy.

Yes, the agronomist went on, the kolkhoz was a gigantic achievement. Now even women had learned to manage it. They were doing the work so well that the land was yielding bountiful crops. Of course there were difficulties. Not all the peasants had yet rid themselves of the individualist approach. There was still conflict between personal interest and the collective good, which to men like himself was the final source of individual good. When it was learned that the peasant was spending too much time on his own garden, his own cow, his own pigs, his own chickens, measures had to be enforced to oblige him to center his attention on kolkhoz production. The size of his garden was cut; so was the amount of his livestock. A new system of taxation was put into operation to spur the individual peasant and the kolkhoz into a more active and lively interest in the socialized lands. Now, in the grain areas, a peasant had to work from sixty to eighty days a year in the kolkhoz and in the cotton country as many as one hundred days a year.

Now, because of the war, people worked harder than ever. If they had no tractor, they went back to the horse. If there was no combine, they wielded sickles and cradles. If no engine was at hand for the threshing machine, they pushed out of the shed the old-fashioned wheel horse and hitched it to the thresher. If there was no threshing machine or if it did not arrive in time, they used flails. They did their work better than ever in spite of worsened living conditions. . . .

But after the war all would be different. People would be more patient, more understanding, more certain of themselves, of their own way of life, and of their own aims. Eventually the conflict between self-interest and social good would vanish. It was all a question of production. When the kolkhoz produced enough of everything for everybody, not only bread and potatoes but meat and dairy foods and vegetables and fruits, the peasant would have no inducement to cultivate as large a garden as he now does, or bother with his own cow, his own pigs, his own chickens. Why should he? He would rather spend his leisure in the clubhouse or read a book or do something else for fun and pleasure.

"Yes, we'll come to that—wait and see," the agronomist raced on. "Our Russian land is rich—none is richer. In time the machine, science, and collectivized labor will make it so productive that there will be no need for any individual garden or cow. The kolkhoz will be supreme. Come back ten years from now—ten years—you won't recognize this land."

CHAPTER 21: RELIGION

"CITIZEN, pass this candle to Nicholas the Miracle Worker."

The words startled me. Though I was in the Moscow cathedral attending the regular Saturday-evening services, the nature of the request and the words in which it was uttered mystified me. In all the years I had been traveling in Russia, I had rarely heard people speak of Nicholas the Miracle Worker, even in most remote villages, though in pre-Soviet days no saint was more highly revered, more fervently worshiped, more frequently petitioned. The speaker was a young man, tall, smooth-shaven, with a broad bony face and a nervous manner. Observing my surprise, he said:

"The ikon of Nicholas the Miracle Worker—please pass it on," and shoving a thin wax candle into my hand he added, "I am a stranger here. I don't know where the ikon is."

"I, too, am a stranger," I replied. Turning to an elderly man beside me, I said: "Maybe you can help this young man out."

Thereupon a kerchiefed woman who had been kneeling in front of us arose and said,

"I'll take it over."

I gave her the candle, and she walked away and disappeared in the crowd. The eager young man knelt down and was lost in prayer.

Nikolay, the Kiev Metropolitan, was conducting the services. He is a handsome man, over sixty, with smooth-shaven cheeks, a soft beard, and iron-gray hair cut shorter than in czarist days. He is one of the most scholarly, literary priests in Russia and loves to quote poetry in his sermons. He has the distinction of being the first churchman to be appointed to a Soviet committee of international importance—the committee which is to investigate German atrocities in Russia. Wearing his gilded miter and silvery vestments, he stood on the dais facing a large ikon of the Mother and the Son and, as if directly addressing both, intoned a prayer. His voice was soft, undistinguished, yet it rose with exalted informality all over the high-domed cathedral. There was rhythm and cadence in the intonation, and in the long-drawn-out finales it was not unlike the ancient folk melodies which Russian choirs feature nowadays in their concerts.

Though the audience was large the cathedral was not packed. I did not see many young people or a single man in a soldier's uniform, though as I entered one uniformed man came out of the door. Nor did I see many elderly people. In other churches on other occasions I saw more young people and also soldiers, though seldom any officers. Most of the congregation was made up of women of middle age or close to middle age. Earnest and devout, they followed the service with keenest attention. There was no talking, no whispering, no looking around to observe or identify or speculate about neighbors. Some of them were on their knees making the sign of the cross. Others were standing, eyes and ears fixed on the man in the gilded miter and silvery vestments.

There were no electric lights in the cathedral, but candles burned brightly in sand pots, in chandeliers, on the dais on which the Metropolitan and the deacons stood, and far away in the open inner room, into and out of which the senior deacon kept coming and going. The frames of the many ikons and of the many chandeliers gleamed with a golden glow which imparted to the cathedral a color and opulence in great contrast to the dark and gray winter garb of the worshipers.

The many lighted candles were a revelation. On several occasions I had attempted to buy a candle in Moscow shops. In an emergency such as might occur during an air raid or the shutting of a power plant, a candle is a priceless treasure. But not a single candle did I find in any of the Moscow shops I visited. Here in this cathedral hundreds of candles of all sizes, though mostly thin and little, burned brightly at every turn. Since there are no private enterprises in Russia, the candles could only have come from a government shop or some co-operative society. Obviously the government was keeping the church plentifully supplied with candles.

There was a choir of mixed voices in the cathedral. This was as much a departure from former practice in this renowned cathedral as the haircut of the Metropolitan. In the old days the women had their own choir, off by itself, on the opposite side from the men's choir; it sang responses or in unison with the men. Now the men and women sang together. The church was too poor to maintain separate choirs.

Some of the best singers have been engaged by the factories and the trade unions. They can pay higher salaries and offer singers privileges, including better food and lodging, than the church is able to provide. Only people with deep religious convictions or with a special love of church music, or both, remain in the church choirs. That there are such people in Moscow was testified to by the high quality of the singing in this splendid Greek Orthodox cathedral. True, the choir lacked the magnificent basses and tenors I had so often heard in the former Temple of the Savior in Moscow. The numerous secular choirs of the country have absorbed them. But enough good singers have remained loyal to the church to make the choral singing an enjoyable and edifying experience. Russian church music is so rich in melody and solemnity that those who love it render it with an exaltation that is beautiful and moving.

As I was leaving the cathedral I passed three former monks. In the lead was an elderly man with a pointed beard. His black cassock, cone-shaped skullcap, and the shiny silver cross on his breast spoke of former glory and prosperity. The two other men were shaved and wore old and bedraggled civilian clothes. They all had their hands stretched out for alms. I slipped a bill into the hand of the bearded man. The hand was clammy and it hardly moved, so stiff was it with cold or perhaps infirmity.

Suddenly out of the dark vestibule there poured a small crowd of men and women. They clustered about me, thrust out their hands, and intoned with an ancient and mournful cadence,

"Help a blind man for Christ's sake!"

"Help a sick man for Christ's sake!"

"Help a widow with little children for Christ's sake!"

Here was old Russia, unabashed in its humility, unregenerate in its abasement, pathetic in its rags and wretchedness yet sturdy in its defiance of the mighty storms that had swept the lands.

There are twenty-two churches open in Moscow. They function normally. They draw large crowds. Even now, in wartime, they have candles, incense, and in season all the flowers they need. There is less pomp in their services than in the old days but no less solemnity and a good deal more humility. On important holidays large queues gather outside the doors, and not always do the waiting multitudes manage to squeeze their way inside. The Easter services are particularly popular. Half of Moscow seems bent on seeing the dramatic ceremonial and hearing the exalted music. Since the war began the military commander of the capital on Easter night has lifted the curfew hour, which is midnight, so that worshipers and others might freely walk about the city on their way to and from the night services.

All of this testifies to a cordial relationship between the Soviets and the Greek Orthodox Church. Though Byzantine in origin the Church has always been vigorously nationalist, and its nationalism in time of war fits the spirit of the times and the policies of the government. Throughout the centuries the Church has earnestly supported the wars which Russia has fought. It is doing so now on as prodigal a scale as circumstances allow.

All over the country the Church has been collecting money for the Defense Fund and warm clothes for the soldiers. Moscow churches in 1941 collected one and a half million rubles. In Gorki the churches collected one million rubles in cash and several hundred thousand in winter clothes. Church women are sewing and knitting for the Army. Late one evening, as I was leaving the home of a friend in Moscow, I saw on the floor in the hallway a white-haired woman knitting mittens by the dim light of a smoky little lantern. She was a pious churchgoer and spent her evenings knitting in the hallway because it was warmer there than in the living room. She isn't the only such woman in Moscow.

In its publicity the Church has been not only permitted but encouraged to unfold to the world its patriotic record in all the wars Russia has fought. It upheld and after his death sanctified Alexander Nevsky, who, in the thirteenth century, smashed the Swedes at Lake Ladoga and the Germans at Lake Chudskoye. The Church, as already related, encouraged Dmitry Donskoy to wage war against the Tatars. In the sixteenth century, when, owing to internecine feuds, the Poles swept into Moscow and threatened to impose their rule on Russia, the Church again offered its wholehearted help to the trader Minin in Nizhnii Novgorod, and to Prince Pozharsky, who mobilized the Russian armies and ousted the Poles from the Kremlin and from Russia. During the Napoleonic Wars the Church stood shoulder to shoulder with the government. In the dramatization of Tolstoy's *War and Peace* at the Maly Theater in Moscow one of the most moving scenes is the high mass which Field Marshal Kutuzov attends. He kneels before the ikon, kisses it, and humbly accepts the blessing of the priest for a successful campaign against the ungodly French intruder. Proudly and eloquently the Russian Church is now publicizing its century-old record of devotion to the fatherland in all the critical wars it has had to fight.

I have already mentioned that the Kiev Metropolitan Nikolay is on the committee to investigate the German atrocities in the occupied lands. The other members are well-known scholars and communist leaders. On November 7, 1942, on the occasion of the anniversary of the Revolution, church leaders for the first time addressed letters of greeting to Stalin. Among other things Metropolitan Sergey said that the faithful of the Orthodox Church "heartily and prayerfully greet in your person the God-chosen leader of our military and cultural forces." Andurachman Rasulev, speaking for the Mohammedan clergy in Russia, wrote: "The Mohammedan world knows you as a fighter for the liberation of the oppressed people. May Allah help you to bring to a victorious conclusion your glorious efforts in behalf of the redemption of these people. Amen!"

Whatever one may think of the thoughts and the eulogies in these messages, or the motives that prompted them, the fact that they were sent and were printed in *Pravda,* the official organ of the Communist party, adds to the evidence of a new relationship between Church and State.

Since the coming of the war, atheist propaganda publications have been banned. Some years ago, when I visited the city of Ivanovo, I attended a performance of Molière's *Tartuffe,* which had been transformed into an anti-religious play. Such a performance is unthinkable now and has been unthinkable for some years. The change in attitude was first manifested rather unexpectedly, and, characteristically enough, it was in the theater.

In the autumn of 1936, after exciting advance publicity, the Kamermy Theater in Moscow presented a new operetta, *Bogatyri—The Knights.* The music had been composed by Borodin and had been recently discovered. The libretto was composed by Demyan Bedny, at that time the best-known communist poet, high in the esteem of the Kremlin.

The Art Committee, official censor of the theater, had approved the script, the dress rehearsal, the entire production. The newspapers had paid it glowing tributes; audiences applauded it with enthusiasm. Then with the violence of a hurricane came an official outburst against Demyan Bedny, against Tairov, director of the Kamermy Theater, against the opera —and it was banned. Higher powers objected to the portrayal of Russian folk heroes as vulgar louts and of the Christianization of Russia as a drunken debauch. The new theory was that Byzantine Christianity was a step forward for the Russian people, because of the contacts it made for them with the outside world and with more advanced civilizations. The official recognition by the Bolsheviks of the service that religion had rendered the Russian people was an unexpected and epochal concession to history.

There was a time when Russian schools spurned the study of history. They studied instead the class struggle. Now they emphasize history with a vengeance. In Pankratov's textbook, which is used in all secondary schools of the country, there is the following significant passage:

"The acceptance of Christianity had a great influence on the life of Russia in its Kiev days. By comparison with paganism Christianity was a step forward. . . . Christianity helped to spread among eastern Slavs the higher civilization of Byzantium. Under the influence of Byzantium, Kiev began to build stone houses, decorated them with painting and mosaics. . . . Education began to spread."

The reconsideration and the reinterpretation of the part religion and the Greek Orthodox Church played in Russian history paved the way for the cordiality that now prevails.

The Constitution, which followed in the steps of this reinterpretation, made possible salutary improvements in the life of the clergy. They were no longer pariahs but citizens with the right to vote and to be elected to office. Their children, who had been barred from higher institutions of learning, were now eligible on a basis of equality with all other children for admission to any school anywhere for which they qualified by the special entrance examinations. Russian priests lifted up their heads in relief and hope.

I happened to be at that time in the village of Reshitilovka in the Ukraine. I called on the priest. The year before the local kolkhoz had used part of his church as a storehouse for grain. He didn't like it, but the kolkhoz authorities told him that they had to store their grain somewhere, and with the small congregation in the village he could carry on in the rear part of the church. He acquiesced because he was helpless. Now he said he would insist on his "new rights." Besides, it was time, he said, that the kolkhoz had a real granary in which to keep surplus grain. The firmness with which he spoke was indicative of his growing awareness of individual rights and of a fresh spirit of self-expression. Turning to his twelve-year-old son, I asked, "What are you going to make of yourself?" Promptly the boy replied, "An aviation engineer." Proudly the father added, "Yes, now the son of a priest can go to the university."

The changing attitude of the Soviets was preceded and followed by an amazing shift in the ideology of the clergy.

"In our country," writes a village priest named Sergey Lavrov, "capital is vested not in the pockets of individuals but in the hands of our collectivized socialist society. Our system of life and government will never allow the concentration of earthly blessings in the hands of the minority."

This priest teaches and preaches economic Bolshevism. Nor is he an exception. All over the country the Russian clergy now propagandize against the sin of personal accumulation and of personal riches. They are for the kolkhoz. They uphold state-owned factories. They prove by holy writ, sometimes with great eloquence, that Christianity was first to enunciate the sinfulness of the individual accumulation of wealth and the exploitation of labor. The new political ideology of Russian churchmen is a complete reversal from the attitude they had once espoused.

In pre-Soviet days the Church had been practically harnessed to czarist state policy. In 1700, when the patriarch died, Peter the Great, who strove to lift Russia literally by the scruff of the neck into the age of science and the machine, forbade the selection of a successor.

The Ryazan Metropolitan Stefan Yavrosky sat on the patriarch's throne but was deprived of the patriarch's prerogatives. In 1721 Peter abolished the patriarchate and created in its place the Holy Synod. Peter did more— he appointed himself head of the Church. The czar became the court of last resort in all matters that came up before the Synod. In subsequent years, under later czars, the emperor, according to the code governing the Church, "is the supreme defender and protector of the dogmas and the ruling faith."

Under the pampered protection of the czars the Russian Church lost not only its political independence but its spiritual integrity. It became one more weapon in the hands of the czar to wield over the people. Of course there were churchmen who fought the new impositions on clergy and religion. Some of them were unfrocked and jailed. Others, while still in the seminary, became revolutionaries. Chernyshevsky and Stalin are outstanding examples of such men. But as an institution the Church remained at the supreme service of the old government and all its aims in politics, economics, and social thought.

In consequence it brought on itself the increasing disrespect and condemnation of the progressive intelligentsia. Memorable is the passage on the Russian Church which Vissarion Belinsky, Russia's foremost literary critic, wrote in his famous letter to Gogol. The author of *The Inspector General* and *Dead Souls* had made public an eloquent eulogy of the Russian autocracy and the Russian Church. Ill and dying with tuberculosis, Belinsky was at that time abroad in Salzbrunn, Germany. The year was 1847. Gogol's declaration aroused his wrath, and so he wrote Gogol an excoriating letter.

"The Russian Church," wrote Belinsky, "has always been the pillar of the rule of the knout and the abettor of despotism. . . . Why do you drag in Christ? What have you found in common between Him and any church, especially the Orthodox Church? He was first to enunciate to mankind the doctrine of liberty, equality, fraternity, and with His martyrdom He has set the seal on the verity of His teaching. . . . Is it possible that you, the author of *The Inspector General* and *Dead Souls* can with all your heart chant a hymn of praise to the heinous Russian clergy? Of whom do the Russian people tell ribald tales? Of the Orthodox priest, his wife, his daughter, his manservant. In the eyes of all Russians is not the priest the living symbol of gluttony, miserliness, sycophancy, shamelessness? As though you didn't know it all. How strange! According to you the Russian people are the most religious in the world. It is a lie. The basis of religion is piety, reverence, fear of the Lord. But Russian folk will pronounce the name of the Lord even while they are scratching themselves. . . . The Russian says about the ikon—'If it does you good pray before it; if not, use it as a cover for the cooking pots.' "

Violent as is this denunciation of the Russian clergy, it was not exceptional among the intelligentsia. It was shared by revolutionary groups, particularly by the Bolsheviks.

From their very first days in power, while they upheld the doctrine of freedom of conscience, the Bolsheviks perceived in the Russian Church one of their most formidable enemies. The defense of the Czar, of Kerensky, of the old order, which leading clergymen were at the time making, and their condemnation of the Soviets and the Bolsheviks, only sharpened the conflict until it became one of the real battles of the Revolution. Deprived of the financial, military, and political support it had been receiving from the czars, with no reserves of inner strength to weather the crisis, the Church found itself driven from position to position until it was utterly at the mercy of the new regime.

Now after a quarter of a century of Sovietism the political relations between the Soviets and the Church are more than cordial. But this cordiality does not, in this writer's judgment, indicate a revival of religion in Russia or an end to the conflict between religion and Bolshevism. Let it be noted that State and Church are one thing, and Bolshevism and religion another. The Constitution upholds the right of worship, and the Soviets punish local officials who try to interfere with this right. Since the war started, according to the testimony of some priests and some church-going civilians, the number of churchgoers has increased. More women, it is said, seek the solace of God and the Church in these days of tragedy and horror. But Russia is a large country and the attitude and behavior of individuals, or even of groups, do not in my opinion warrant the judgment that religion is rising in Russia. Churches in Moscow are crowded, especially on high days. But the population of Moscow before the war broke out was over four million, or more than twice what it was a quarter of a century ago. Yet the number of Orthodox churches has fallen from five hundred and sixty to twenty-two.

From certain villages come reports of religious revival or rather of the failure of the campaign of infidelity. In the Prokovsky, Marfinsky, Golitzin-sky townships in the Tambov province, for example, school children made the rounds of peasant homes on Christmas Day and "praised Christ." This happened in 1937, declares an official report of the Tambov Board of Education. In another township—Sukhonsky—in a northern province, school children wore crosses and went to confession. On Easter Day 1937, in the villages of the Pecherniki district, Moscow province, one hundred and fifty school children attended High Mass, and of this number forty had previously fasted. In still another rural district in the province of Voronezh children refused to sing during the period of the "great fast." They said "it is a sin to sing now." Other instances of village children's following the faith of their fathers or their grandfathers are often cited in official reports.

The fact that they are cited means that they exist. It also means that they are combated. That is all it means. In Russia grandmothers, in families that have them, are the real nurses of children. In the village and in the city, health permitting, they devote most of their time and attention to grandchildren. If they are religious, as nearly all of them are, they naturally seek to instill their own faith in their grandchildren. Often they succeed only too well. There is neither ban nor interference with the teaching of religion in the home. Mothers and fathers who were reared in a religious environment in pre-Soviet days likewise strive to inculcate their faith, whether Christian, Mohammedan, or any other, in their sons and daughters. With older folk religion is still a power, but if only because grandmothers and other older folk are dying off, this power is continually waning.

In matters of principle and doctrine, authority holds a supreme position in Russia. It is authority which makes and unmakes law and not the reverse. Ultimate, final, irrevocable authority is vested at present in four men: Marx, Engels, Lenin, and Stalin. Their utterances are the court of last resort in all doctrinal or ideological disputes. Other spokesmen of socialism and communism carry no weight in Russia and influence neither the thought nor the behavior of the population.

On the subject of religion the pronouncements of these four men are only too clear and pointed. Their speeches and writings are permeated with a passionate denial and an implacable repudiation of all religion in any and every form. Let the following brief quotations speak for themselves.

Marx: "It is not religion that creates man but man who creates religion. . . . Religion is the groan of the downtrodden creature. . . . It is the opium of the people."

Engels: "Every religion appears as nothing more than a fantastic reflection in the minds of the people of the external forces which rule them in their everyday life, a reflection in which natural forces assume the form of the supernatural."

Lenin: "Religion is the opium of the people—these words of Marx are the cornerstone of the whole philosophy of Marxism on the subject of religion."

Stalin: "The Party cannot be neutral toward religion and conduct anti-religious propaganda. The Party stands for science, whereas religious biases are opposed to science. Every religion is diametrically opposed to science."

I could fill pages with similar and more trenchant quotations from these four apostles of Marxism. Nowhere in any of the recent pronouncements of Stalin or of the Party or of any of its subordinate leaders has there been a retraction or a modification of this basic antagonism to religion. Nor is this antagonism a mere abstraction, an article of faith only. It is a living force. The ruling party of Russia will not admit a believing man or woman to membership. Education from the lowest to the highest grades in the public schools, and in the other institutions of learning, promulgates infidelity. I have yet to meet a college student or a college graduate in Russia who is a believer, except among the Protestant sects, and even they no longer exercise the hold on their young people they once did.

With authority, education, press, all organizations—Party, Komsomol, trade union, Pioneers, Soviets, kolkhoz, factory—with all these institutions arrayed against religion, the church of whichever denomination, however cordially treated by the government, has small chance of influencing a telling number of young people. Nor, it must be emphasized, will the Soviets under any circumstances permit foreign missionaries to Russia, any more than they will admit foreign capitalists to engage in any economic activity.

At present the Orthodox Church is no longer a danger to the Soviet regime. It has no property. It commands no power. It is not a member of any foreign body. In its economic teachings it has become bolshevized. It preaches the sinfulness of exploitation, the wickedness of accumulating wealth. Besides, it represents an aspect of national civilization which the Russians may want to preserve, not for religious but for purely historical and cultural purposes. Twice I was in the little town of Istra outside Moscow and saw the ruins of the ancient Cathedral of Jerusalem which the Germans have wrecked. It was one of the great monuments of church architecture in Russia and had been made into a national museum. On the expulsion of the Germans from the town the Soviet government announced it would restore the cathedral. Scientists and architects are already at work on the drawings for the restoration. It is not inconceivable that the Orthodox Church will be similarly fostered as a memorial of a Russia that is no more.

In the future it may become a national sanctuary which even non-believing Russians will respect and revere.

Also, it professes the same faith as does Bulgaria, Rumania, Yugoslavia, Greece, all neighbors of Russia and all destined to be of no small importance in the postwar settlement. If backed by the Soviet Government it might yet become a far more powerful force on the international scene than is now apparent.

Yet the basic and irreconcilable antagonism between religion and Bolshevism persists. It must persist until such a time, if it ever comes, when Russians feel the need of something mystical which is contained neither in their pioneering spirit nor in their new ideology and in the teachings of the four men who have enunciated it and given it living expression.

CHAPTER 22: MORALITY

LEGENDS DIE HARD, particularly when they are sensational or shocking. In the early years of Sovietism the story of the nationalization of women swept the world. Though nearly a quarter of a century has passed, this legend is not dead. True, no one with the least conception of the sturdy character of the Russian woman, the immense role she plays in all avenues of public life, now believes that laws could ever be passed putting her on the same level as coal, iron, or land merely to provide pleasure for men.

But——

On my arrival in Russia I was present at a luncheon at the Grand Hotel in Kuibyshev. Here one American correspondent who had not been in Russia for some years said to another:

"The longer I stay here the more impressed I am with the stabilization of morals and family in this country."

Thereupon the other with more indignation than humor exclaimed:

"Do you carry in your pocket an American or a Soviet passport?"

A Russian journalist who speaks English was also present at the luncheon and said afterward, "It's incredible what fantastic notions people have about our women and our morals."

It *is* incredible. The implications of the story of the nationalization of women linger in one form or another to this day in many a mind. Among many foreigners in Moscow who live in a colony by themselves, and who not always through their own fault are separated from the Russian community, with neither knowledge nor appreciation of the character of the Russian woman and of the momentous part she has always taken in all progressive movements, appraisal of her morals and her character is only slightly less fantastic than that of the above-mentioned correspondent.

To a large extent these foreigners base their judgment on the few women they know and to whom they themselves often refer as "floozies." Russians speak of these women as *shpana*—spawn. They are not necessarily prostitutes. Some of them speak foreign languages and find the acquaintance of foreigners with whom they can converse in a foreign language pleasant and edifying. None of them are averse to partaking of the special foods and drinks that foreigners command. Nor are any of them reticent in seeking for themselves silk stockings, a wrist watch, special shoes, or cloth for a dress which might be brought in by a foreigner on his return to Moscow after a vacation in Teheran, Cairo, or some other foreign city.

Now and then a "floozy" may actually be in love with the foreigner she cultivates. Varied and many are the motives that actuate these girls. Their morals are of their own making, to suit their own wishes and their own emotions or those of the men they know. I must repeat that they are not necessarily prostitutes, though cohabitation with a man who does favors for them or whom they may admire is to them neither an unwelcome nor a dishonorable experience.

Yet they represent the morals of Russian women even less than taxidance girls in America represent the morals of American women. The fact is that the Plans, which have built up a firm, new foundation of Russian industry and Russian agriculture, have also remade the foundation of Russian morals. The war, with the miseries, uncertainties, and confusions it has stirred, may here and there shake or shatter this foundation. But only here and there. In the light of the ingredients that have gone into its making, the new morality is, I feel, as indestructible as the ideas out of which the Plans have originated and which for years and years to come will determine and govern the direction of Russian social and political life and the flow of Russian history.

There was a time in the early years of the Revolution when, in the rebellion against the old world and all it represented, Russian youth, or rather certain groups of it, rose in clamorous, almost hilarious revolt against the established sex morality of czarist times. The peasant youth was not a participant in this rebellion, though the old morals had suffered visible setbacks even in the countryside. Chiefly it was the student youth that championed the new age of liberation from all the restraints and compulsions of the old days. This youth grew riotously lax and often stupefied itself with sex orgies. A vast literature grew out of and around this new "emancipation."

The most vivid literary document on the subject is Panteleimon Romanov's short story, "Without Flowers." Romanov pictures a young

man, a student, visiting a girl acquaintance. He has come for only one purpose, and he treats the relationship with the girl with no more show of
romance than if he were visiting a brothel. The girl is saddened and
infuriated. She craves romance and beauty. But to this youth she is only
an object of physiological gratification. He loathes the thought of romance, the least suggestion of exalted emotion.

Those were tense, irresponsible, exciting times and the "emancipated"
youth made the most of it. With no little scorn and sarcasm Lenin denounced its carefree behavior. To him undisciplined sex indulgence was no
more edifying than "drinking from a mud puddle." Other Party leaders
likewise raised their voices in protest against laxity and dissipation. But
the campaign was essentially of a verbal nature. Meanwhile the policy of
official *laissez faire* in sex led now and then to sensational forms of hooliganism. In Leningrad, for example, a crowd of boys collectively raped a girl.
The trial of these boys, which was given nationwide publicity, elicited the
startling fact that some of the offenders were Komsomols. Nor was this
the only instance of such a crime. In consequence a fresh campaign was
started against "hooliganism" in all its manifestations, including, of course,
sex laxity.

With the coming of the Plans, when there was need for disciplined
living and of concentrated effort and energy on industrialization and collectivization, the campaign assumed a more decisive and creative form.
Theoretically, the individual was still master of his own emotional destiny.
If a man and woman cared to indulge in a love life which did not interfere with their duties and did not involve serious social consequences, it
remained their private, unquestioned affair. Yet even such relationships
were in a multitude of ways frowned upon and discouraged.

Of course religion, except among the faithful, played, and to this day
plays, no part in the determination and formulation of sex morality or of
any other social relationships.

"To us," said Lenin, "morality derived from a power outside of human
society does not exist—is merely a deception."

Scriptural injunctions on the subject, whether from St. Paul or St.
Matthew, are never quoted, and, if they were, would not be accorded the
least respect, except among the faithful.

The motive and purpose of sex morality in a Soviet society was expressed by Lenin in one simple sentence.

"Morality," he said, "serves the purpose of helping human society to attain new heights of development and to rid itself of exploitation of labor."

This is not a theistic or deistic but a social approach to the problem. Its
solution is imbedded in the needs of the new society in which women
enjoy the same rights, socially and sexually, as men, and in which not the

economic advantage of the individual but the social good of the community is the determining force.

Yet the new morality that Russia has since elevated into a rule of living is in many respects a resurrection and a sanctification of old and universal sex decencies. There is no reference to adultery in the Russian legal code. Yet woe to the man who gives himself to an indiscriminate indulgence in adultery. Once I said to one of the leaders of the Komsomol of Moscow: "You have grown surprisingly puritanical." She laughed and replied: "No, we are not puritanical. We do not like the word *puritanism*. We do not condemn sex as a sin. We only advocate sex decency based on the individual's responsibility to himself and to society."

But however sovietized the rhetoric, the morals now in vogue in Russia have little to distinguish them from puritanism except that the element of sin is absent, as is the least suggestion that in and of itself sex is a degrading experience.

"If you had in your Komsomol organization a boy," I asked a factory Komsomol leader, "who was known to be enjoying or seeking cohabitation now with one girl, now with another, what would happen?"

"We'd expel him."

Another factory Komsomol leader in reply to the same question said: "We'd not only expel him, we'd disgrace him in public at a meeting and in the local press, perhaps even in the *Komsomolskaya Pravda*."

There was a time when Soviet Russia was ridden with prostitution. In part it was the result of unemployment, in part of despair, in part of the contumely visited in the early years of the Revolution on the children and the descendants of former owning and aristocratic classes. The coming of the Plans also meant an onslaught on prostitution—not only against women who were soliciting but against men who responded. I recall the picture in a Kiev daily newspaper of a man with the legend underneath that he was a perverter of morals and a renegade. He had been caught with a prostitute, and the fact was prominently publicized in the press. Public disgrace, one of the commonest weapons the Soviets have forged against evildoers of all types, was often invoked against men who disregarded the ban on prostitution.

But it was not the only weapon. In the battle against venereal disease in America and England, citizens are informed that they may apply to any public-health clinic for free treatment without running the danger of being questioned as to the source of the contamination. "No questions asked," is the way the announcements sometimes end or begin. Not so in Russia. Quite the contrary. In his *anketa*—question blank—the applicant for the treatment of a venereal disease, whether a man or a woman, must always disclose the source of infection—that is, give the name of the person with whom he or she had cohabited. He is refused treatment until he complies with the regulation.

"We make no ceremonies about it," said a Russian physician. "What's the use of curing a man of gonorrhea or syphilis if you don't stop the person who infected him from infecting others?"

Brothels are banned by the law. Were anyone to conduct such an institution secretly the penalty would be stern—most likely "the highest measure of social defense"—death by shooting. In the early years of the Plans there were instances when such action was invoked against violators of the law.

Not that prostitution has been completely wiped out. The *shpana* linger on the fringe of the best hotels, but only in the large cities, especially in the capital. Yet the entrenchment of the Plans, the elimination of unemployment, the vigorous propaganda against frivolity in sex, the rigid community control of apartment houses, all of which are state property, the fierce social condemnation of anyone patronizing a prostitute—all these have practically driven and shamed prostitution as a trade or profession out of existence.

With millions of men in the Russian Army there is little venereal disease among the soldiers. I have yet to meet a military physician, an officer or political leader in the Red Army who has been faced with the problem. Since the outbreak of the war not one word on the subject has appeared in the press. Had there been the least hint of a problem it would have been accorded frank, violent rhetoric in the press. With Russia fighting a war of survival and seeking to muster all the energies and capacities of the nation, the press never would have been silent on a condition which might mean impairment of the health and fighting capacity of men in the Red Army. The fact that it has not done so only confirms the pronouncements and opinions I have heard from officers, soldiers, and political leaders.

Besides, the economic and social life of Russia renders impossible the practice of prostitution in the vicinity of army camps. Brothels are banned. Hotels and taverns are owned by the government. The profit incentive to turn hotels into bawdy houses is absent, and there is no chance of recruiting girls for the purpose. A manager or waiter who might yield to a bribe or to some other inducement and secretly provide facilities for the traffic, if only to one woman, would soon be caught. With the many checks on individual behavior by the Party, by the Komsomol, by the trade unions and other organizations, he could not hide his unlawful action for any length of time.

There is also in the Red Army the unwritten code of honor, for recruits as well as for seasoned generals, which forbids the patronage of prostitution. One of the many reasons which have fanned such flaming hate against Germans in Russia is their practice of establishing for their soldiers brothels with Russian girls in the towns and cities they occupy. Now and then a woman of easy virtue in one guise or another may penetrate the ranks of the Russian Army, but if discovered the man who patronizes her must face public denunciation and condemnation.

There have been a few instances of the violation of this code among Russian officers in Persia. The penalties imposed on them were swift and vigorous. They were not demoted. They were not fined or put under irksome army discipline. At a meeting of officers they were publicly disgraced and denounced. It is not likely that any other officers will dare readily to repeat the offense.

Folk tradition is still a powerful restraining force in Russia, especially in the village. In one form or another the cult of chastity has always been the basis of this tradition. In the writer's native village in White Russia it was as deeply entrenched as the belief in the house goblin. Not that the everyday relations between boys and girls suffered from any special restrictions. Unmarried folk always felt free to visit each other, in their homes, in the street, in the churchyard, at the bazaar. In summer they rode their horses together to the communal pasture. Guards of both sexes remained awake all night before an open fire to watch for wolves and to prevent horses from straying into someone's unfenced field or meadow. The others went to sleep on the grass, and many of the boys and girls slept together snugly wrapped in home-woven, blanket-like hemp cloths. But rarely did these associations result in cohabitation. The girl knew the stern social penalties that might follow. If she became pregnant and gave birth to a child, she was stigmatized by the community for the rest of her life. The law did not compel the boy to marry her. Nor did fathers or mothers, as sometimes happens among American mountaineers, descend on the boy and at the point of a rifle compel him to take the girl as his wife. The girl was the sole sufferer.

But even when there was no child she no longer enjoyed the standing with the attractive boys of the village that she did previous to the relationship. With her chastity gone—and the boy usually told his friends of his achievement—she was no longer sought after as a wife as she might otherwise have been. If she gave birth to a child she usually had to marry an elderly man or a widower with children.

The coming of the Soviets, of the new enlightenment, and of the machine age to the village has loosened the hold of former folk traditions on the community. Theoretically the cult of chastity has ceased to exist. For a long time it was mocked and cursed. But in practice it largely continues to govern the morals of young people in the village. The girl who cohabits with a man before her marriage remains in the eyes of the eligible boys less desirable as a wife. She is deemed frivolous and incapable of marital loyalty. If she gives birth to a child out of wedlock she is likely to forfeit the chance of marrying the man she may want as a husband. Knowing the consequences of careless behavior, the village girl will protect her chastity almost as stubbornly as in the pre-Soviet days.

Only to a slightly lesser degree does this usage obtain in the factory. Since the passing of the law of June 27, 1936, banning abortions, the city girl almost as much as the girl in the village, as a matter of self-protection, and with the wish not to jeopardize her feminine prestige and her eligibility as a wife with the men she knows, will exercise all the self-control she can muster to hold aloof from compromising behavior.

Since the passage of the new law against abortions, sex has been relegated more and more into the framework of marriage and motherhood. Most enlightening was the discussion I had with Fyodorova, Komsomol secretary of the Moscow province, and therefore guide and teacher of the youth of the province in all their social relationships. Fyodorova is a handsome young woman of twenty-five, with large dark eyes, and heavy dark hair. She is so tender and feminine in voice and manner that on first acquaintance one would not associate her with the severity of disciplinarian action against anyone.

"Supposing," I said, "one of your best girl friends in a moment of frivolity or ecstasy submitted to a man she liked and in consequence found herself pregnant. Would you help her avert childbirth so as not to spoil her chances of marrying the man she might love?"

Swift and vigorous was the reply: "Never!"

"But the girl," I persisted, "might be heartbroken and see herself doomed to marry someone she never could love or even to remain unwed."

"It makes no difference," said Fyodorova. "If she had an abortion I would regard her as the worst enemy of herself and of society. We would throw her out of the Komsomol."

"What would you do to help her out in her despair?" I further asked.

"I would surround her with all the care and love I could. I would make her feel that no greater glory can come to a woman than the birth of a child. I would lift her out of any sordid thoughts about herself. I would make her feel that she had committed neither a crime nor an offense, that she was not in any way forfeiting her chances of marital happiness. But never, never would I encourage an abortion."

To me this was a new voice in Russia, decisive and compelling and as relentless in opposition to abortions as the voice and the word of the Roman Catholic Church. I said so to Fyodorova, and she smiled and answered,

"That's the way we feel and act."

Contrary to the opinion of those foreigners in Russia, especially in Moscow, who derive their knowledge of Russian morals and Russian women from "floozies" they meet, Russia since the coming of the Plan has set in motion an array of forces, social and legal, which militate against frivolity and promiscuity in sex. Basically, it must be repeated, sex is associated with marriage and with all its consequences. The new morality, the new or rather the resurrected and chastened folk tradition—particularly strong in the villages, where the majority of the population is still living—the new marriage law with its discouragement of birth control and its ban on abortions, the banishment of prostitution—all combine to promulgate no other attitude and no other result.

Of course there are exceptions. There are illegal abortions. They are expensive but not impossible. There are transgressions of the established code. I heard of a community in the Urals which has been but slightly stirred by the new laws or the new enlightenment. As in the pre-Soviet days men and women neither hide nor feel ashamed of promiscuity. They know no birth control and do not care for abortions. The women have many children, some of whom live, some of whom die. The death of a child is no particular calamity, though at its burial the mother may feel prostrate and may wail aloud. Other children come into the world to take the place of those who have passed away.

There may be other communities or scattered individuals within any community who are as undisciplined in their sex life. Russia is an immense country with diverse peoples, climates, and heritages. In faraway places the erection of a steel factory may not necessarily mean the inauguration of a new morality. But sooner or later even the most backward community is destined to yield to the demands and persuasions of the new social order.

Now and then there appear stories in youth publications which illumine the subject of Soviet morality with the glow and the freshness of a stirring experience. Such a story, in the form of a letter to the editor, was printed in the *Komsomolskaya Pravda* on September 9, 1942. It is a rich and revealing document and tells much not only of the morality but of many other social attitudes of Russia's fighting youth.

Let the letter speak for itself.

"Dear Comrade Editor:

"We are enclosing a letter which the army physician, Comrade V., has recently received from his wife, Lida Aristahovna Livanova. We have read this letter collectively. It has made on us a heavy and painful impression. With the permission of Comrade V. we wrote his wife a reply which we are also enclosing. We are asking you to print it.

"Of course there are not many girls in our midst like Lida. We know that our girls are pure and honorable. Yet here is a situation which we feel should not go unchallenged."

Here followed Lida's letter to her husband Valentin:

"GREETINGS, MY ESTEEMED VALENTIN:

"I am hastening to send you my warmest greeting and I beg your sincerest pardon for my long silence. I shall begin with a description of our present life in Saratov, particularly my own life.

"At the moment things are different from what they were formerly. The war has made our life more severe. Yet despite this severity our young hearts have not been drained of emotion, especially of those delicate sentiments which we call intimate, or, to speak more concretely—erotic.

"How much power and courage, enthusiasm and creativeness our remarkable times and our wonderful nature have stirred in our young hearts! They have emphasized the necessity of an idyllic setting for our lives. Yet, though the war has upset it all and our life flows along its new course, our mind is intoxicated with the beauty of nature, our hearts brim over with joyful sensations. That is why, Valentin, I think you won't be too severe and too cruel and you won't condemn me for falling in love with someone else.

"How accidental it all was! How involuntarily it all happened. Even now I can hardly remember how it came about. But a fact is a fact. I carelessly flirted with my own heart, and uncontrollable as it was, it ceased to obey my reason. I cherished the tenderest sentiments for you. But time has shaken them out of me. Am I guilty because they proved so unstable? Perhaps I am, who knows? But after the appearance of this new man, his image has gradually crowded yours out of my mind.

"Don't be hard with me, please. I only hope that my departure will cause you no special pain. On the contrary, the sun of bachelorhood will again rise over you. I trust that fate will bring you a comrade after your own heart who will not overburden your life. So be happy. I wish you the best of health and the greatest of success.

"LIDA."

This was followed by a letter to Valentin's wife by his fellow physicians.

"LIDA!

"We who write this letter have never seen you, but we know well your former life-mate Valentin. Living together at the front has brought us very close to each other, and we have become excellent friends. Often enough in moments of leisure we remember the past and talk to each other about our friends. Valentin always spoke highly of you and whenever he was finished with an exciting and tiring task he would say, 'When we've smashed the enemy, I shall return to my darling girl.'

"The war has taught us to value life. We have become more mature, more independent, more serious. We are fulfilling our honorable duty—participating in the annihilation of the most evil foe of mankind—fascism. Great is the help we are receiving from our brothers, our sisters, our friends who work with such sacrifices in the rear. . . . Their letters invigorate us, their parcels cheer us. We love our people. We know they are proud of us.

"There is no war without sacrifices. We may find ourselves among those who have suffered. We know that if we perish we shall not be forgotten; if we return home wounded and mutilated our friends will not turn against us. Rare is the creature that can turn against the man who has been at the front and abandon him.

"Recently our friend Valya, after prolonged and fruitless waiting, received what in our company has become known as your 'famous' letter.

"You have not behaved at all like a Soviet human being. Your letter breathes the spirit of petty emotional egoism. Apparently you are interested only in things that transpire under your own shadow and that concern only you.

"A woman whose husband fights at the front and who really loves him would never act as you have done. Now, knowing your new husband we are in no position to cherish sympathy and respect for him. By shattering the family life of a soldier at the front he has committed a deed which is none too honorable. But we accuse him less than we do you. After all, by writing him letters filled with tender sentiment you have skillfully been deceiving your husband. Only a short time ago you sent him your photograph with a lyrical inscription. Why did you do it?

"You even have the impudence to write Valentin that 'the sun of bachelorhood will again rise over you.' Excuse us from such imagery. We are happy to know that there are few like you in our land and that our people despise them.

"In all likelihood your behavior during Valentin's absence was not especially honorable. Most likely your flaming heart, which brims over with 'erotic emotion,' will search for a new husband every summer. In the long run your fate does not especially interest us. We are disturbed only by the fate of our friend Valentin, to whom you have brought misfortune.

"Is he hurt? Of course. It hurts to know that there are in our midst people like you who cast a shadow over the honor of our women when they so courageously bear up under all trials of war.

"Probably Valentin regrets he married a woman like you. Before us he feels humiliated that his wife turned out to be what she is.

"This letter is no brilliant composition. We have so little time to write.

And we are not even endeavoring to pack our lines with quotations from *Anna Karenina*. But we are astounded and aroused by your conduct and your 'literary' letter to Valentin.

"We do not suppose that you will feel offended by this letter, for your soul is crowded with 'other sentiments.' Still we hope that someday you will realize the complete impudence and baseness of your action.

"[*Signed by*] Group of front-line soldiers:

"MILITARY PHYSICIAN THIRD RANK FRID
"MILITARY PHYSICIAN THIRD RANK YEVTEYEV
"MILITARY PHYSICIAN THIRD RANK KRIVOSHEIN
"MILITARY FELDSHER VOZNAYA."

CHAPTER 23: ROMANCE

DURING my stay in Kuibyshev I met a man in the street I had first known in Moscow as a high-school student. He had been in the Army and was recovering from a serious wound in the small of his back. He invited me to go with him across the Volga for a sun bath. The doctors prescribed it for his back. Daily, if the sun was out, he ferried across the river and lay for hours in the sand or on the grass. He knew a particularly good beach, he said, a few miles above the city, and we could lie there and bake ourselves in the sun and talk.

I went along, and hardly had we stretched out on the white and warm sand when, with a show of emotion, he asked if I had seen the production of *Anna Karenina* in the Moscow Art Theater. I told him I had not yet been in Moscow. The theater was at that time playing in Saratov, and he urged me to make a trip to that Volga city and see the play.

"The audience," he said, "will be a revelation to you—particularly the young people. They are a new generation here, and if you've been away from Russia six years you don't know them. There is no better way of knowing them than watching them at a performance of *Anna Karenina*."

We were alone—not another soul was on this white beach. As if buoyed by the silence and the beauty of the scene my companion talked on expansively.

"You don't know my ten-year-old daughter Ninochka," he said.

"No," I replied.

"Every time I went to see the play—I've seen it four times—I thought of her and of the happiness she will have when she grows up. She won't be like her father and mother."

I laughed. It seemed strange to hear a hardened Bolshevik speak in such a confessional and retrospective mood.

"It's nothing to laugh at," he said with impatience. "Maybe it's because I've been in the war and have seen plenty of death and am myself sick that I feel as I do. But——" He sat up, threw a few pebbles into the water, and went on: "We were such ignorant fools, the youth of my day. We laughed at Anna Karenina—called her *bourzhuika—beloruchka* (the white-gloved one)—a woman of no character, no decency, no importance. Neither her love nor her troubles nor her glories meant anything to us. . . . Perhaps you remember how we talked in those days of Tolstoy's women, of Turgenev's women, of Pushkin's Tatyana."

I remembered it only too well and told my companion of a visit to a high school in Stalingrad, where, during a discussion of Pushkin's *Eugene Onegin,* I heard only violent denunciation of the heroine, and when one girl rose in Tatyana's defense she was denounced by the class as ancient and corrupt in her thinking.

"That's the way we were," said my companion, and laughed. "But when I saw *Anna Karenina,*" he resumed, "I realized how absurd my generation was and how healthy is the present youth of our land. They don't laugh at Tolstoy's heroine—heavens, no! They weep, now sadly, now joyfully. They know that had she lived in our epoch she might have been an engineer or she might have been at the front line carrying wounded soldiers off the battlefield. . . . They feel sorry for her but they adore her—because—because—she is so feminine, so Russian, and so romantic——"

He paused and shut his eyes, then continued:

"After seeing the play once I took my wife to see it. Would you believe it? Watching our young people made us relive our own youth. . . . We held each other firmly, warmly by the hand. My wife wept, and I, too, felt a dimness in my eyes, and it was so wonderful to be transported if only for a few brief moments into the world of romance—which my generation had so wantonly mocked and repudiated. . . . Yes, we were innocent little fools." And after a moment he added, "Our Ninochka will be different. She is already reading Pushkin."

Some weeks later I was in Moscow, and one foreign correspondent, in denouncing Alexander Werth's *Moscow Diary,* a memorable and sensitive account of the Russian people in the early months of the war, said to me: "He doesn't have to talk of Pushkin on every page—Christ, Pushkin isn't Russia."

Of course not. Yet without Pushkin Russia would not have been what she is now. Thirty million copies of his works have been sold between 1917 and 1940. Hardly a home one enters, whether in city or village, but if there are books on the table or on the shelf something by Pushkin is more than certain to be among them.

"Have you read much of Pushkin?" I asked Zoya Vladimirova, the seventeen-year-old girl from Tula who had spent six months in carrying wounded soldiers from the battlefield and who is a celebrated heroine in her province. Instead of answering she started to recite *Eugene. Onegin*. "Do you know it all by heart?" I asked.

"Almost," she answered, and recited some more of the famous romance.

As in the good old-fashioned days before the coming of the Soviet, school girls and boys copy this poem over and over, now as an exercise, now for pleasure. They know by heart all the more romantic parts.

"Pushkin makes them feel happy and romantic," said a schoolteacher. "That's why our young people read him so much."

Let a factory club engage an elocutionist to give an evening of Pushkin, and the hall will be crowded to the doors—indeed will be overcrowded. Pushkin stirs the imagination and the emotions of youth and makes it appreciate, as perhaps no other generation in Russia has appreciated, the meaning and beauty of romance. No wonder my old friend lying on the white, hot sand of the Volga felt so happy that his ten-year-old Ninochka was already reading Pushkin. . . .

"Ah, girls," exclaims Yelena Kononenko, one of Russia's most brilliant publicists and short-story writers, "you don't begin to appreciate the eagerness with which the boys at the front read your letters. They swallow not only every word but every syllable, even as a man tormented by thirst swallows water out of a cool spring. Every drop of water stirs a mighty surge of fresh strength and fresh life. When the heart is warmed the greatcoat, though frozen to the earth, is also warm. When the emotions are refreshed, the parching heat is easy to bear and fatigue is washed away as by a spring shower. . . ."

In the rebellious, nihilistic twenties such glowing words in praise of romance were as unthinkable as the restoration of the Czar.

"You say love and war?" the author rhapsodizes further. "The Soviet soldier—hero, guardsman, eagle—a man with a melancholy heart? No, my dears, this is neither sentimentality nor weakness. This is life, girls! Powerful as granite in battle, the Soviet soldier is not made of stone. He is a living human being. He rejoices at the sight of a snowflower blossoming beside his dugout. He weeps when he lowers the body of a fellow soldier into the grave and feels no shame for his tears. . . . Your faded little portrait or the cambric handkerchief he has carried with him through the fire and smoke of war he presses to his lips. . . . He is a living human being and that is wonderful!"

A first edition of a million copies was printed of Kononenko's *Your Little Photograph,* from which I have taken the above quotations.

I have not seen a production of *Anna Karenina* in the Art Theater, but I have seen the opera *Eugene Onegin* and the dramatic version of Turgenev's highly romantic novel *A Nest of Gentlefolk.* Ages of thought and blood divide the Russian girls of today from Pushkin's Tatyana or Turgenev's Liza. There is much in both that is alien to them and, in the light of their own social philosophy, pathetic and even ridiculous. Yet they never laugh. Great is their respect and reverence for these two Russian girls of yesterday whose love is shattered through no fault of their own and who in their defeat and heartbreak remain true to their better selves.

"Of course I wept," said a Russian girl who is the assistant Komsomol secretary of a Moscow factory, "when I saw *A Nest of Gentlefolk.*"

"But," I said, "you are a Komsomolka, and Turgenev's heroine retires for life to a nunnery."

"I can understand why she did it," explained the Komsomolka. "I was very much in love with a man. We met in the mountains and climbed Elbrus together. I was engaged to him. We were going to be married, but his mother talked him out of it. She said I was no girl for him because I'd be going to meetings and doing social work while he needed a domestic woman. He was a weak man. That was why he allowed himself to be influenced by his mother. Nevertheless, he broke my heart. I cannot tell you how alone and miserable I felt. Had I lived in Liza's time I, too, would have gone to a nunnery, and if Liza had lived in my time she would have found solace and forgetfulness in work and work and still more work, just as I did."

What a far cry are such words from the lament of the heroine in Panteleimon Romanov's story "Without Flowers." Writing to a girl friend, Romanov's heroine says:

"There is no longer any love among us. There is only sex relationship. Those who seek in love something more than physiology are viewed with ridicule as mere half-wits."

Now boys are no longer ashamed to write poems to girls and to gather flowers for them. . . . They do it readily, rapturously—and the war, instead of abating, has only intensified the emotional responses of Russia's youth.

Viewed in retrospect nothing was more natural than the rebellion against romance in the early years of the Revolution. After overthrowing the old government, shattering the foundation of the old society and the old ways of life, the Soviet regime could offer youth at most passions, hopes, aspirations, blueprints, hardly anything more. Confusion reigned in the highest circles in the country. There were all manner of plans and schemes for running the nation, but there was no agreement among leaders on any one plan or scheme. They were in perpetual discord. With the Revolution halted on the Russian border, a new orientation was imperative. The Stalin and Trotsky factions were pulling in opposite directions, and the times seemed palsied with inaction.

The coming of the Plans put an end to confusion and uncertainty. At galloping speed the Soviets proceeded to industrialize the country and to collectivize the land.

The harsh intolerances of those years pressed with special vigor against the finer things in life. "We don't want Rachmaninov," the director of the rubber workers' club in Moscow said to me. "We don't want gypsy music. We can get along without Tchaikovsky. We want only the songs and the music that will help the Five-Year Plan."

Then came the Second Plan. Complete and sweeping was the reversal of this attitude, particularly in its later period. The composers and writers not only of Russia but of the world were literally pushed into the open, almost all of them. The solid foundation of industry and agriculture led to a solid foundation in education, in literary and artistic appreciation. An end had come, sudden and violent, to the apostles of proletarian art, with their fulminations against Pushkin, Tchaikovsky, gypsy music, and their tumultuous insistence on the creation of a new art in which, to quote one Russian cynic, "not only the sight but the sound of the hammer and sickle shall drown out all other sights and all other sounds."

History was introduced in schools as a subject of serious study. Pushkin was elevated almost to sainthood. Tchaikovsky became the glory of Russian music. Tolstoy was lifted to the highest and widest esteem. In literary and artistic circles there were quarrels and battles. There were public confessions of error and public condemnations of sin. The censorship continued rigid and demanding but only with living artists and writers. It left untouched the creations of men already dead.

For the first time since the Revolution, Russian youth knew normalcy and stability in duty, in thought, in social elevation, in artistic appreciation. However much this normalcy was colored by political consideration, and however lacking in finality, it laid an end to rebellion—though not at once to the grumbling against school discipline and family unity, against other institutions and loyalties which previous generations of youth had sought to talk or laugh or spit out of existence. The new youth was receiving a multitude of emotional and intellectual stimulations which former generations had known only partially and chaotically.

While on a collective farm in the summer of 1942 I was shown by the director of the high school the new textbooks that were in use. Particularly was I impressed by the textbook on literature for the higher grades. In this fat volume not one line is accorded to the once-brash apostles of so-called proletarian literature. Not one line! The Russian school child of today does not know of the existence of these once-loud persecutors of truth and integrity in art. Of the five hundred and eighty pages of literary selections, four hundred and seventy-nine are allotted to prerevolutionary writers and Gorki, who had done his most distinguished work in prerevolutionary times. Only ninety-one pages are accorded to Soviet writers. Only those of talent somewhat comparable to the literary masters of prerevolutionary times are included, and they are Mikhail Sholokhov and Alexey Tolstoy. The poet Mayakovsky is also included, as are short excerpts from the writings of four men of non-Russian nationality. Reading a book like this, nurtured on the very best writings in the Russian language, the Russian youth of today can only be outraged by the behavior of the student in Romanov's "Without Flowers," as much so as was the girl who was the victim of his physiological coarseness.

CHAPTER 24: LOVE LETTERS

RUSSIANS have never been ashamed or afraid of their emotions. Freely they speak of them, freely they write of them, now more than ever. The war with its forced separations, its threat of death, has intensified the ardor of Russian men for the women they love and of Russian women for the men they love. The letters they write to one another are replete with this ardor.

Nor do they conceal these letters from friends. Soldiers at the front do not mind passing them to friends, from man to man, or having them read collectively, so that all will know about the love of Tamara or Katya or Zina for her Shura, her Boris, her Pavel. There is great intimacy among the men at the front, and they gladly share with one another not only the parcels they receive but the letters to or from their best girl friends.

Some of these letters appear frequently in the *Komsomolskaya Pravda*. They portray the romance and fighting spirit of Russian youth and much more—their hopes and ambitions and the kind of human beings they are or want to make of themselves. Tender and enlightening is the letter from I. Petrov, a soldier on the western front. It was published in the *Komsomolskaya Pravda* (May 10, 1942).

"I feel so agitated, for it's a long time since I've written you. I only hope the emotion I feel will flit past all fronts, all military roads, all guerrilla paths, and reach you as a fervent token of our trust and love for each other.

"The war has changed our life, has flung us apart, but it hasn't really separated us. We are Komsomols—our conscience will never torment us for spending our time in futility, not even in the days of our prewar youth. Remember how exciting our life was, how full of work and dreams and adventure and how nothing we ever did satisfied us? Always we reached out for something bigger than we had, something more stormy, more all-embracing.

"When Valentine Grizodubova, Marina Paskova, Polina Osipenko achieved their heroic flight in the plane *Rodina* [over the Siberian forests] you envied them, you wanted to be in their place. You dreamed of the plane as of something fabulous yet real. We could always find something real in our dreams, couldn't we?

"I remember our last meeting in Petrozavodsk after we graduated from school. We were waiting for assignments. We were very pensive. We knew that we were going in opposite parts of the country, you to Pudozh, I to Sortavala. But we were not gloomy, indeed our last meeting was by far our happiest. Falling into our old habit, we discussed the new film *The Great Waltz,* and—why conceal the truth?—you imagined yourself as Karla Doner and I as Johann Strauss. A great life presupposes great dreams.

"Outside the window gleamed Lake Onega. Somewhere on the opposite side was sheltered the primitive unknown Pudozh. On the western coast of Ladoga lay Sortavala. You were about to start for the east, I for the west. And we parted—as young romantic Johann Strauss and Karla Doner. Who'd have imagined it was to be our last meeting before the outbreak of the mighty events of our times?

"Hard were the days on the new job. In the faraway and forsaken Pudozh you felt dispirited. You dreamed of a gigantic task and struck nothing more than a menial industrial job. But you didn't allow yourself to become despondent. With your customary sense of humor you wrote me 'London–Pudozh–Paris.' We laughed together. I understood you, yet I was worried and wondered if you'd become sullen and go to pieces.

"Days passed. I grew to love Lake Ladoga and Sortavala. You grew to love Lake Onega and Pudozh, and each day earnest and fervent letters went from Lake Ladoga to Lake Onega and back. Every evening before retiring I sat down at my table and wrote you a brief letter. And you knew what I had done during the day, what I had planned to do the next day. Each morning I received a neatly folded letter from you, and I, too, knew what you did the day before and what you'd do today.

"Distance didn't interfere with our friendship, our love. As before, we dreamed together, worked together. The icy waters of Savain-Yoki and the aroma of the Sortavala forests had become a part of my home. When I left for the Army I was lonesome for the primeval forests of the Karelo-Finnish Republic and for the cream-capped waves of Lake Ladoga. You saw me off to the Army and you spoke only a few words. You said you believed in me, that I'd prove myself worthy of everything. I have repeated your words again and again, and every time I have done so I discovered something new and fresh in them. When the first German bombs whistled over my head and the earth groaned and quivered I knew I was face to face with a stern ordeal. Perhaps you were still asleep and didn't know what had happened. But I saw the blood of our men, the flame of our burning homes. My brief letter was already on its way to you. I don't remember what I wrote but I do remember that the letter was an oath!

"I got no reply. Those were horrible times. The German fascist army was deluging the earth with lead and fire and was pressing forward. Fighting rearguard action, we were retreating step by step. We abandoned our native town where you were spending your vacation. That's how I lost you.

"It's a long time since I've written you, and it seems I've forgotten how to set my thoughts down on paper. But I think of you constantly and have taught myself to believe that our warm, exciting correspondence has never actually ceased.

"Meeting one day a girl at the front I imagined it was you and that you are precisely like her—quiet, brave, modest, and fearless. I felt as though I were beside you, and the thought gave me courage and strength.

"Another time, seeing in the paper the photograph of a girl who achieved the highest record in a certain industrial establishment, I imagined that you, together with all the others in our land, were forging armaments for the war. Carefully I put away the photograph of the unknown girl. I accustomed myself to the thought that you are that girl, and that, too, adds to my courage and strength.

"When I first heard the story of Zoya Kosmodemyanskaya I saw in her your image. Zoya died but she conquered. Her death, like the fiery heart of Danko,[1] has lighted up the way to victory for hundreds and thousands of her friends and comrades, boys and girls. Like all others, I flamed with a passion to avenge her brutal death. This, too, gave me strength and confidence.

"The seductive beauty of the sea reveals itself in time of storm. The nobility of a soldier reveals itself in time of war. Every Soviet person is now a soldier, and I know you, too, wherever destiny may have taken you, have found your place in the front line. That's what I think, that's what I want to think.

[1]Gorki's short story in *Old Woman Izergil*.

"I know that wherever you are and whatever you may be doing you are not satisfied with your achievement. You want to do more and more. If you are in the rear, you work day and night, forgetting yourself and wanting to go to the front. If you are at the front or in a guerrilla brigade, you want to face the enemy, fight him, destroy him. That's the way our girls have been reared. The blood of Zoya Kosmodemyanskaya, Liza Chaikina, Tanya Petrov, and hundreds of others is crying for revenge.

"Let the wrinkles cut themselves into our young faces—we shall know them as a mark of valor. Despite the ordeal we are facing, our eyes shall preserve the light and fire of our love. I know I shall feel no shame in looking once more into your dark and trustful eyes. Equally am I confident that when our eyes meet your long and beautiful lashes will not drop. You are as much of a soldier as I am.

"I feel so stirred. It is such a long time since I've written you. I meant to say something loving and beautiful but I don't think I have succeeded. I trust you will understand me. Remember, dearest, we have always understood each other in half syllables.

"I. PETROV."

The letters from Gavrusha—an assumed name—commander of a guerrilla detachment, to his wife Natasha and her letters to him were given to me by Natasha herself. I have known Natasha for many years. She is a highly cultivated, literary young woman with a fluent command of several foreign languages, including English. She married Gavrusha seven years ago, and it has been one of the happiest marriages I have known. Gavrusha was impatient with college and never graduated. He took up advertising— a profession, incidentally, which before the war was commanding some of the finest artistic talents of Russia, even though there is no private business in the country. Tiring of advertising, Gavrusha took up camera reporting. He was hard at work on an assignment when the war broke out. Like every young Russian in good health he had served in the Army for two years and was a reserve cavalry officer. An expert skier and horseman, he always kept himself in excellent physical condition. He also kept his army boots ready for service. He never wore them in civilian life but kept them on a shelf, freshly greased, ready for use in an emergency. Mobilized as soon as the war broke out, he fought at the front, then joined a guerrilla detachment and soon became its leader.

For a long time Natasha had not heard from him. She did not know whether he was alive or dead. Then came a letter, and here it is:

"MY DEAR, DEAR NATASHA:

"This is the happiest day in my life—our Soviet plane has come and will carry this letter to you. I am now writing it but find it hard to believe that

I am, or that I see the plane, that the letter will be in your hands, that you'll be reading it, discussing it. . . . I have spent many long, sleepless winter nights thinking of you. In my mind I saw the terrors of your life, but I was unable to help you. But now I mustn't think of privations, of burdens, of terror. This is a day of joy, of rapture. I have been waiting for this day, this minute, for months—for a minute when I could write and feel certain the letter would reach you. Yet now that it has arrived—I seem unable to write anything but nonsense.

"My dear, my only Natasha! Where are you? Where are you? Is there anything wrong with you? Where are all of you? Perhaps some of my dearest friends are no longer alive or are sick, wounded. Nothing is impossible in these times. Day and night I keep asking these questions. . . . Answer them, answer them.

"You don't begin to realize what joy and luck it is to be once more on our own Soviet soil, to be freed and far away from the fascist jackals. Nor can you imagine the animal fury with which our people and I with them are annihilating and chasing the enemy from our native land. I am so full of hate for Germans that I kill them not only in battle but in unarmed tilts when I leap upon them unawares. I do not take them prisoner. I wouldn't know what to do with them if I did. Guerrillas are guerrillas— they are like beasts in the woods. They hunt and are hunted—and have no use, no thought, no place for prisoners. With a swift and single thrust of the knife or bayonet I put an end to them. I do it with no little satisfaction because of the pain and degradation they have been heaping on our people. Yet do you remember that before the war I could not even kill a chicken? Strange, isn't it? But for people like them I have only a heart of stone. No, they aren't people. They are cowards, jackals, barbarians. I have no words and not enough paper to tell you all I feel for them. Besides, it is getting dark and I'm writing by the light of a burning fagot.

"My dear Natasha—the day will come when we shall again be together and we shall live happily as we did before the war. Believe me, the day is near and soon we shall once more be in each other's arms.

"It is six or seven months since we parted. You'll never know how much I've lived through, how many times my life hung on a thread. Yet I always go forward. I haven't yet been defeated—there isn't a single scratch on my body. I only had two teeth knocked out—that's all the loss I've suffered in the guerrilla war.

"I have a lot to say to you, but the burning fagot is mercilessly fading. I hope to see you soon. Good-by, my dear one, all my dear ones.

"How I should love to spend one minute with you. I kiss you, my dear darling wife.

"Your husband, guerrilla named GAVRUSHA."

Natasha's reply:

"Many times I've said to myself that if I'd only heard from you the one word 'Alive' I'd be supremely, divinely happy. And now here you are, blessing me with a letter. I didn't even dream of it. Such great luck! How can I explain to you what this letter has awakened in me? It has stirred so much fresh strength and hope. It has given me wings, my dear boy. It's so wonderful to be alive! You know how I love our native land, how dearly I love it. Yet since I learned that you've remained behind the lines as a guerrilla, I've asked myself many times, 'How is he there?' I know how brave you are, how unafraid of anything. Still I spent many sleepless nights worrying and imagining things. I knew how great was the burden you'd taken on yourself and how great was the test to which you were subjecting yourself—a test which not all men can endure. That's why your letter is such an incomparable gift.

"Knowing that you are alive, that you are always thinking of me fills me with joy and pride. You and I, my love, are like two parts of one whole. Only war could part us. When it is over we shall be together and more firmly welded than ever."

Gavrusha's letter No. 2:

"April 5, 1942, 4 A.M.

"NATASHA DEAREST:

"It is early morning and the peasants are celebrating Easter. But I am wandering the fields on skis! Imagine that! Yet it's a fact, unbelievable as, it may seem. Sometimes I have the impression that everything is topsy-turvy now, not only in the world of man but in nature. It's getting to be hard to know the beginning from the end—winter from spring.

"In my white camouflage robe I look like a living ghost. The moment the Fritzes lay eyes on this robe they get the jitters. They know who I am—a guerrilla—and the least movement of my body brings out a hurricane of fire. I lie down and they grow silent. So deadly is the silence that you imagine you hear the beating not only of your own heart but of the heart of every German soldier. Thus minutes pass—minutes that seem like hours. Then comes another machine-gun volley. I crawl unobserved into a ditch. The machine gun ceases firing and there is a hush all around me. I lie in the ditch. That's how I spent Easter night.

"And how did you pass yours? Write and tell me everything, tell the truth. If any of my dear ones are no longer alive, don't conceal it from me. You've got to be brave in everything, unafraid of anything. I am accustomed to facing the truth and to taking the blows of fate without the flash of an eyelid.

"My dearest Natasha! I am longing for the day when your first letter will reach me. I am yearning for it as for the newborn son which you have promised me. . . ."

Gavrusha's letter No. 3:

"MY DEAREST, MOST BELOVED NATASHA:

"Your image is always and ever with me. How many times have I in my mind bidden you good-by—not only you, Moscow and all of my dearest and closest friends. But fate has been merciful. I am safe and sound, I still sing 'La Cucaracha'[2]—which means that I am in my right mind and am confident of myself and of victory.

"Sometimes a Moscow newspaper reaches us and I learn something about life in the capital, how normally it flows, with theaters and cinemas open and people attending them.

"Dearest Natasha! Please think of me often and write and tell me how you are. Hearing from you will make it easier for me to carry the burden of war. Think of me at least half as much as I think of you. Think of me on arising and on retiring. No, that's not enough. Think of me five more times a day. I think of you every time I want to close my eyes, which happens many times a day, because I have no definite hours of sleep and have to get it in winks and snatches. It's so long since I've seen a bed or pillow that they seem like something I've never really known, something which never existed."

Natasha's letter to Gavrusha:

"MY ONLY LOVE,

"It's more than six months since I've heard from you. A weary vigil has started again. When I get no letter from you I reread a dozen times the ones I have—there are only five—and I repeat by heart without looking at the paper the lines which have become so precious to me. I know them all to the last word, to the last syllable, and it isn't enough any more. . . . If only you knew how much your letters mean to me, how I live for them, how proud I am of you, and how difficult it is for me to live without you. I want you to know this, my dearest man, and I must have soon another letter from you telling me that you have received mine.

"In search of something tangible that would remind me of you, I went yesterday to our apartment, though I do not at present live there. I stood in your room, fumbled the pages of your book *Action of Cavalry in Battle*. I touched your cameras, your enlargers, your photographic paraphernalia. Yet I found nothing there—our home, our bright, cozy home, which was

[2]A Mexican song Russians learned from an American film which was very popular before the war.

so pleasant that people told us it smelled of happiness, means nothing any more. It is cold and empty and deserted. When you come back, my beloved, I don't want to live there.

"Then walking across the little valley I went to the subway station. Passing the home where you lived before we got married, I suddenly remembered the time when our love first began, the first weeks of our acquaintance, the evening when I first came to your house and you met me at the gates of our present home—at the iron staircase in the basement—do you remember? And you told me you were waiting for me as for 'a blue dream'? And I did wear a bright blue cloak!

"You led me up the dirty and winding stairway and you explained that the stairway was dirty because it was being repaired and of course there weren't repairs. The stairs were dirty because they had not been cleaned.

"Anyway, I wanted to recapture the flavor of those days. I walked into the hallway. It was pitch black there, and with difficulty I groped my way to the winding stairs. I climbed them and came to the well-known and elusive curve. Here you had stood. I remembered so vividly your wide-open eyes. They were so bright in the darkness. I came to the door. It was no use knocking—no one would come out. No one would open it.

"A neighbor stuck out her head and asked what I wanted. Awkwardly I mumbled some excuse and left.

"I am so happy that no wild flesh has grown over my heart, that my actions are so fresh and earnest and that I still not only live and feel but suffer and dream. You, too, are a dreamer, different from myself, more earthly but a dreamer nonetheless, my dear Gavrusha. That was why our life was so wonderful! Life can be beautiful, so beautiful.

"Is it for me, a civilian, to tell you this? However much I may hear or read of the terror of war, it can never be as vivid and real to me as it is to those who have been in the abyss and who have known the hour, the minute when the dividing line between life and death was so thin as to be hardly visible. Only such people, only men like you, Gavrusha, can have a true appreciation of life. Yet I cannot help speaking of it because it is, it can be so beautiful.

"Write, my joy, about everything—how you feel, look, live. Who is your best friend now and how are your military campaigns? You seem to be doing well, to judge from the news that now and then reaches me.

"Go forward, my dear, as ever without knowing defeat—forward to the bright and noble goal which lies ahead of all you men and women in the forests—emancipators of our native land. Let my love guard you against all evil. There is an old Russian saying that bullets spare those who are in love. Then you must be immune to harm because deep and real is your love.

"Someday you'll suddenly come to me. Do not frighten me. It is true people die of happiness. Yet I am not sure what will happen to me when I hear your voice.

"Your loving wife,
"NATASHA."

DON'T WEEP MARIANNA

Such was the headline which caught my eye in the *Komsomolskaya Pravda* of September 14, 1942. I read it and I pass it on to the reader word for word as one of the most moving stories I have read in the months I was in Russia.

"DEAR EDITOR:

"Please be good enough to print my letter in your paper. But first read the enclosed letter from Lieutenant Ivan Andreyevitch Artemenko, my commander and friend, to his sweetheart, Marianna Shleyeva. What a noble soul the man was. Read his letter, then I'll tell you his history.

"'GREETINGS, DARLING MARIANNA,

"'Today I have received your letter and am hastily sending a reply. Darling, if you only knew how happy I am. I always rejoice when I get a letter from you or when together with my soldiers I win a victory over the Fritzes and annihilate as many of them as are within the range of my sight.

"'Marianna! Whenever a letter from you reaches me, my hate of the enemy mounts—how long will it be before our separation ends? I believe that the hour is not far away when the enemy will be destroyed and we shall return to our native Don.

"'Marianna! The friend of whom I talked so much to you in the hospital has already returned—he is only about thirty steps away from me—sitting with the boys, showing his small white teeth and talking to them. I let him see your picture and he said, "In the picture she is very pretty and in life she must be even prettier." And I said to him, "Precisely." He wants very much to meet you.

"'Oh, if you only knew how eager I am to see you again and to thank you and all your friends for the splendid care in the hospital. It is true, is it not, darling, that it was the hospital which drew us close to one another? Often I shut my eyes and I see cots all around me and next to me Vasya Kretov—such a redheaded chap—and I see you reading a book to us wounded men or tucking the blanket around someone and whispering a few words to him.

"'Marianna, six months have passed since then, and to me it seems that it is no more than two. I see myself as I was then—a wounded man with you sitting beside me, nursing me, making me fall asleep. Nothing has remained of my wounds. I am quite well and it's no use your worrying about me. My soldiers are such splendid boys—and——'

"The lieutenant never finished his letter. As he was writing it on his leather case, the political commander Yereneyev stepped over and they started talking about the forthcoming battle. Then he showed me Marianna's photograph and said, 'If anything happens to me, deliver it to my family and tell them to love her even as they loved me.' But he gave me no address. Either he had no time or didn't think of it. A few minutes later he was in command of our company and we started to advance.

"He met Marianna in the hospital in December. He had been wounded, and she was studying to become a nurse. She has since graduated and is on duty somewhere in Moscow. Vanya told me much about her and of the tenderness and solicitude with which she looked after wounded soldiers. We went into battle. The lieutenant led us bravely, skillfully, as always, for he was an experienced officer. We were advancing. We gave the Germans plenty of hell. Suddenly I saw Vanya slip. He was wounded in the breast. I rushed over. He was still alive and lay under a bush, pressing with his hand Marianna's photograph which he must have just pulled out of his pocket. I am enclosing this photograph. He was saying something, but it was hard to understand him. I understood only a few words. 'Marianna—be assured. . . .' Then he said, 'Forward.' And then again, 'Tell . . .' but what he wanted me to tell and to whom I never learned. He passed out before he could finish his thought.

"So, dear editor, I beg of you to print it all in your paper—Vanya's letter and my story about him. I want so much for Marianna to know that her beloved friend died with her photograph in his hand and that he was thinking of her the last moments of his life.

"Remember, Marianna, love must be answered by love. Vanya loved you with all his soul. He lived with your letters. When they arrived he was happy, and . . .

"I beg your pardon for writing so incoherently. I am a bit uneasy. For a time I couldn't decide whether to send this letter. Then I decided to send it. Let our youth and especially our girls know how their letters cheer and lift the fighting spirits of our men and with what exalted emotion they go into battle and fight for the protection of those they love. . . .

"FEDIN."

"DEAR FRIEND FEDIN,

"I have read your letter. I have also read the unfinished letter of the man who was so close to my heart, my own Vanya. If only you knew how deep today is my hate of the German savages—I cannot express it in words.

"I know that Vanya gave his life for the fatherland, for freedom. Vanya is dead. For me his death is a great calamity. But let the enemy know that however great the sorrow, however heavy the grief it will not bend the backs of Soviet girls. They do not weep, they avenge the death of their loved ones. No Germans shall hear our laments or see our tears. I assure you, Comrade Fedin—Marianna does not weep—Marianna avenges. . . .

"I am now at work in the Army as an instructor in sanitation. By saving the lives of Soviet soldiers I shall revenge myself on Germans for the death of the man I loved. I only regret that I have not the opportunity to shoot them. Therefore I appeal to you, my dear Fedin, and to your fellow soldiers—shoot Germans. I beg you to send me the photograph which Vanya held in his dying hands. I want to have it—I want it always to be mine. Please do not refuse this request. Send me also the original of his unfinished letter. . . .

"MARIANNA NIKOLAYEVNA SHLEYEVA."

CHAPTER 25: FAMILY

ALEXEY FYODOROV was born in a factory district known as Powder Square, in the very shadow of a powder plant. The destiny of his family in czarist days, and even more since the coming of the Soviets, has been bound up with the factory. His father, now over seventy and retired on a pension, started as a worker there and spent forty-three years within its walls. All the children worked there. Alexey was still a boy when he had his first humble job there—brushing dirt off bricks. From this job he moved to another and to still another, and finally the factory sent him to study engineering in a mining academy.

To this day Powder Square is noted neither for its geographic grandeur nor its architectural beauty. The streets are gray, the cottages are ancient and small, some so low that they bloom dwarf-like out of the earth. Like other families in the district, the Fyodorovs out of their mutual earnings and savings "shook together" in prerevolutionary times a wooden cottage of their own. As homes go in certain new industrial districts in Russia, there was nothing pretentious about the cottage. It was small with only a few rooms. But there was a garden in the rear, a porch in front with the inevitable family bench. It was home, Alexey's home, the home of his two brothers, his two sisters, his aged father and mother.

With the coming of the Soviets the Fyodorov family, like so many others in Russia, was lifted out of Powder Square. The children worked but also studied. Leonid, the eldest, became an engineer and was elevated to an important position in the Putilov factory in Leningrad. Vladimir, younger than Alexey, also became an engineer. The older sister, Sofya, rose to an important administrative position in the powder plant. The younger, Nina, after a sojourn in Irkutsk, Siberia, returned on the outbreak of war to "the family nest" and now works in a manufacturing shop in the plant. Alexey himself chose the Army for a career. At the age of thirty-one he rose to the rank of colonel in aviation. He is married and has a family.

The war has fallen on the Fyodorovs with no less harsh a hand than on other families. The demands of the war have mercilessly ground down the family standard of living. Satisfactions which the old father and mother knew in the prewar days they must now deny themselves. All those of working age hardly know any rest—so busy are they with the tasks before them. Leonid, a reserve officer in the infantry, went to war. While leading his company in an attack at Tikhvin he was killed.

Alexey himself has been in the thick of the fighting since the outbreak of the war. He has seen much horror and blood. He has attended the funerals of many of his closest friends. He has been in many battles and has attained distinction in the Army. His mother and father are proud of him and write him frequent letters. A religious woman, the mother never fails to send him her blessings and never omits writing "May the Lord protect you, my little son." No doubt she prays for Alexey and lights a candle in his behalf before a favorite miracle-working ikon. "One may believe or be an infidel," says Alexey, "but a mother's blessing is always sacred."

A husband and a father, a man of eminence in the Army and in the country, a participant in the most stupendous conflict of our times, an eyewitness of the unending race between life and death, in field and forest, in the air and on the earth, Colonel Alexey Fyodorov has been asking himself many questions. He has pondered on war, on life and death, on many other subjects which, with the fierceness of a Russian blizzard, often enough assail the minds of thinking folk in Russia.

"Whenever," he writes in the highly authoritative *Red Star,* "I think of reasons for the self-sacrifice with which my fellow soldiers are fighting, there invariably rises before my eyes our family, and I know it is the thought that behind the back of each fighting man there is his family—a home, old folk, a mother and father, little children. The thought of family stirs the courage and the fury with which they fight—a fury which can conquer everything, including death."

To this doughty soldier, reared, educated, elevated to authority, risen to eminence in a Soviet society under the direct dispensation of the Communist party, of which he is himself no doubt a member, "the family is a hallowed institution." He rhapsodizes about the family with the fervor of a Protestant evangelist. "Family and fatherland," he writes, "are two words that lie closest to the life and heart of a Russian."

Nor is he alone in these days of bereavement in his words of appreciation and laudation of the family. In a letter to a girl in the rubber factory in Moscow with whom he started a correspondence, Lieutenant Vladimir Demyanovitch, a native of White Russia, writes:

"My native land is trampled and drenched in blood . . . my father, mother, brothers, sisters are there . . . I have had no letter from them in over a year. . . . I receive no letters from friends, from girls, because they have remained behind in White Russia. Some of them may still be alive, but others, no doubt, have lost their heads in the fight against the Hitler brigands. . . . My fellow soldiers are receiving letters from home. One of them has an old mother waiting for him, another has a father, a third has a wife. . . . But who can possibly wait for me? I have no wife, and whether my father and mother are alive I do not know."

I have met many Vladimir Demyanovitches, not only from White Russia but from the Ukraine and other parts which the Germans have overrun. . . . Their greatest concern and grief have been for their families. . . . On a collective farm I met two wounded soldiers who had come there to regain their health. Both were married and had children. Both had lost all knowledge of and all contact with their families. One afternoon we were together in the home of a peasant woman who had fled from German bombings in the Kalinin province. Graphically she described how she and the children had run to the riverbank and hidden there in the bushes to escape bombs. Many other mothers had gathered their children and run there to seek protection in the same bushes. But the bombs struck some of them, and the riverbank was "a terrible thing to behold." Both men were in tears, and one of them said:

"If this is what has happened to my wife and children—then what is the use of living? . . . I loved them so. . . . I loved them so. . . ."

In no other country in the world is the family as an institution so openly, and eloquently hallowed as in the Russia of today. In the press, in literature, on the propagandist's platform, it is accorded high tributes, the very highest. Now it is no less a pillar of society and of individual self-expression than in any other country. At a meeting which I attended of a group of political commissars in the Red Army Home of Moscow, Osipenko, Chief of Political Education in the Army in the Moscow district, asked if all of those who had gathered were married. One young man arose and said:

"I am not."

"How old are you?" asked Osipenko.

"Twenty-six."

"You are a little late, aren't you?"

The young man laughed, and so did the others.

"Nichevo," remarked Osipenko. "He will catch up—when the war is over—he will have a family."

To any Russian a young man of twenty-six without a family is deserving not of commendation but of commiseration! "The Pioneer," reads one of the commands of the Pioneer decalogue, "must be the pride of his family and his school!" In the present meaning of the word, Sovietism is as unthinkable without the institution of the family as without the institution of collectivized ownership and control of property! The family is accepted, revered, glorified.

Unlike religion and the Church the family has triumphed over all tests and storms to which it has been subjected since the rise of Soviets on November 7, 1917. How shattering at times were these tests is eloquently attested by Babel's powerful tale of the civil war, "The Letter." A young Cossack writes to his mother describing how his elder brother and fellow soldiers in the Red Army captured their father and executed him.

Deep and explosive were family conflicts all over the land. Son rose against father, daughter against mother, brother against brother. Political fury submerged all other emotions; social wrath overpowered conventional attachments and age-old loyalties.

Rich is the documentary and powerful is the fictional literature depicting the family cleavages of those times. No community, except possibly that of factory workers, was immune. The village was only slightly less riven with the conflict than the city. An underground society of young people, chiefly university students, took it upon themselves to assassinate the leaders of the Revolution, including highest officials of the Cheka in Leningrad. The son of one of these officials was a member of the society and he volunteered to take the life of his own father. He called on his father, talked to him, but at the last moment lost his courage and left without fulfilling his mission. Later he and a girl planted a bomb in a public building in Moscow and fled in the direction of Poland. In the forests of White Russia they were captured and shot. Exceptional as might be the instances of sons seeking to murder their own fathers, the fact of their occurrences only attests to the acuteness of the family feuds that raged in the land.

Yet unlike the Orthodox Church, which had been a mighty weapon in the hands of the Czar, the family as such, as an institution, was not under attack—not directly. None of the outstanding leaders had proclaimed it an outmoded institution with no claim to recognition by the new society. There were voices of denunciation, loud and fierce, which pointed to the family as the embodiment of the worst evils of capitalism and deserving annihilation. There were groups, some of youths, which in their disdain of the old society rebelled not only against its economic order but against its morality, its art, its social usages, and of course the family. But these rebels, whether Bolshevik or not, did not reflect the opinion of high Soviet authority. They spoke and acted in the light of their own fevered thoughts and emotions. Neither Lenin nor any other leader of note ever uttered a word in condemnation of the family as an institution, though not a leader but heaped abundant excoriation on the family as it had developed under the old society. The external compulsions with which it had become invested they all denounced and promised ruthlessly to exterminate.

With the end of the civil war the family, wherever it was broken or mangled, started, plant-like, to mend itself. The church was crashing, private enterprise despite the Nep was doomed, but the family was regaining its stability. Attacks on it continued, now mildly, now vehemently, but again without official sanction and receiving support neither in law nor in the utterance of the men holding power.

The promise of leaders to shake the family loose of legal and all other external compulsions was quickly and wholeheartedly fulfilled. Divorce became easy and could be obtained for the asking. No cause was necessary. Desire alone counted. There was no limit to the number of divorces a man or a woman might obtain. The cost was nominal, within the reach of the poorest pocketbook. The procedure was simple. It was easier to obtain a divorce than to buy a new pair of shoes—far easier. If the wife did not care to inform her husband of the legal separation she did not have to. He learned it through the reporter card which *Zags*—the registration bureau—mailed to him.

I first heard the divorce law expounded at great length by the prosecutor of Nizhnii Novgorod.

"Do you mean," I said, "that on the way to work in the morning a man or woman can stop at a registry office and obtain a divorce, and the union is at an end?"

"Precisely," was the proud reply. The compulsions, he explained, with which the former society had beset the individual, a socialist society could not tolerate.

In family life the behavior of the individual was a matter of his own choice or his own discretion.

Registration of marriage was not obligatory, though advisable for a purely administrative reason, so that the state would have more or less reliable records of marriages and divorces.

Women were favored with facilities to divest themselves of family obligations. Birth control was fostered, abortions were free and legal. Since there was much prejudice against the common forms of birth control and since, in addition, there was an inadequate supply of reliable equipment, abortions gained wide acceptance as a substitute.

The special disabilities which the old laws had imposed on women were now wiped out. Woman was the equal of man in all matters pertaining to individual self-expression and social accommodation. The husband boasted of no prerogatives which were withheld from the wife.

Under the impact of this new freedom, divorce, though chiefly in the cities, was rampant. Families broke up, started again, and once more went on the rocks. The Russian press now and then launched a tirade against men and women with a "frivolous" attitude toward marriage and the family. It called on young people and on people not so young to mend their ways of marital behavior. But for many the wine of the new dispensations was too exhilarating to heed these admonitions. Russian satirists found in the new condition of life a fertile field for dramatic exploitation, and plays in mockery of the existing family life drew capacity audiences.

Fantastic were the situations that now and then arose. One of my friends in Russia in those days was a young Hindu writer named Azis Azad. I often went to see him in his apartment. One evening, leaning out of the window, he pointed to an apartment across the courtyard and narrated an amazing tale about the family living there. The father, a man of fifty-five, had been a bank executive in czarist days, and because of his financial training was employed at a high salary in a government bank. He lived in a two-room apartment with his wife and his young son. Neighbors had always regarded them as an exceptionally happy family. One summer the father went on a vacation to the Caucasus. There, in a summer resort, he met a Georgian girl. He fell in love with her, courted her, and she fell in love with him. Unknown to his wife in Moscow, he went to a registration office, obtained a divorce, and married the girl. The clerk in the office sent the usual postal card to the wife in Moscow to inform her of the separation.

His vacation was coming to an end and he had to leave the Caucasus. His Georgian bride went to Moscow with him. He arrived faster than did the postal card informing his first wife of the divorce; so he told her what had happened.

A brave woman, the first wife resigned herself to the inevitable and moved to a room of her own in the same apartment house. The son was so infuriated with the father that he went to live with his mother. In the course of time the son became reconciled with his father and often visited him and his Georgian wife. Soon this attractive young woman intrigued the son. He fell in love with her and she with him, and one bright day they went off together. In the registration office the Georgian girl obtained a divorce from her husband and married the son.

Finding himself abandoned, the father went back to his first wife. He begged her to return to him and resume her former place in the home. Disdainfully she spurned the offer, and the father remained alone to rue the error of his ways.

Despite its fairy-tale flavor and its fairy-tale morality, this incident, however exceptional, is illustrative of absurd situations that might arise under a condition of uncontrolled marital association.

In those days endless volumes were written on Russian morality, Russian free love, Russian desecration of the family. Yet the family remained. Its roots were never shaken and were never in danger of being torn out. Despite easy divorce, the right to free and frequent abortions, the overwhelming mass of Russian humanity, in the village almost all of them, fell in love, married, and even when they did not record the union in the registration office, they stayed married. They raised children. They built a home in the best way they could. Stripped of the family compulsions that their fathers and grandfathers had known, they chose of their own accord to continue the ancestral habit and tradition of family life.

When I discussed the subject of the Russian family in those days with the late Havelock Ellis his comment was enlightening. He said: "The family is such an organic part of man's biology and psychology that nothing and nobody can destroy it."

Even those free-minded Bolsheviks—none, incidentally, in a position of high authority—who privately and sometimes publicly decried the family as a relic of a dark and bygone age were astonished at the oaklike strength it displayed.

The First Five-Year Plan again administered, indirectly, a fresh assault on the family. Consider the case of Pavlik Morozov, the boy after whom many children's institutions have been named and around whose life Eisenstein has built a motion picture. Pavlik was a peasant lad of eleven. His father had hidden grain which by law he should have sold to the state. Pavlik reported him to the Soviets. Children did such things in those days. Shortly afterward the father's brother killed Pavlik.

Here was a family tragedy of tremendous magnitude. Tragedies of a more or less similar nature were not infrequent in the days of the First Five-Year Plan. In retrospect, as I have already indicated, that period looms, among other things, as a continuation of the civil war and the completion of the task it had failed to achieve—namely, the permanent destruction of private enterprise not only in the city but in the village. This stage of the civil war was fought not by opposing armies—there was no White Army any more—but as a campaign violent and decisive—against holders of private property, or those who had been associated with it.

Hates and conflicts not only within the community but within the family which the civil war had kindled and which during the period of Nep had subsided, erupted afresh, in the village now as much as in the city. Sometimes more.

The newspapers all over the country flaunted announcements of children denouncing and disinheriting their parents for being kulaks, businessmen, priests, or members of some other group which the Soviets sought to extirpate. By breaking ties with their parents, often to the point of changing their names, these children could overcome the barriers to advancement in a career or to admission in a university. All over the country families broke up. Children left their homes. Fathers and mothers were abandoned to loneliness and to public contumely. . . . Not until the idea of a new constitution had begun to stir the country did an end come to these family feuds and ruptures.

I knew then an elderly man who had two brilliant sons, both university students. He had once been a successful salesman for a Moscow merchant; so his sons repudiated him publicly. For five years neither son came near the parents, and all efforts of the father and mother to win them back ended in failure. But when the battle for state-controlled industrialization and collectivization was won, when kulaks, nepmen, and all other immediate foes, however indirect, of the Plan were no more and the country began to talk constitution—the year was 1935—the two militant sons went back to their home, and the family celebrated a reunion.

There were many such reunions in Russia. The murk of abandonment and loneliness that had hung shroudlike over many a home gave way to the brightness of a new comradeship and a loyalty of which older folk had dreamed but had seldom hoped to attain.

Severe as were these fresh blows to the family, it remained unshaken. Thus, as an institution, the family survived all crises. Its economic foundation was gone. The former taboos were no more. There was no mention of adultery in the legal code. All children were legitimate, whether born in or out of wedlock. Neither fathers nor mothers any longer enjoyed the right to make matches unacceptable to their sons or daughters. Men and women remained the arbiters of their own sexual and moral behavior. Polygamy, of course, was forbidden and was punishable by "deprivation of liberty and compulsory labor up to one year" and a fine of up to 1,000 rubles. Birth control continued easy; so did abortions.

If a couple wished to go off on a love life of their own for a week or a month the law did not interfere. Nurseries and kindergartens by the thousands were leaping up all over the country. Mothers left their children in one or the other, depending on the age of the child, went to their offices or to the fields to work, and on returning called for the children and took them home. More than ever the state participated in the education and the physical and ideological molding of the child. Yet, contrary to forecasts and denunciations, the family shook off all assaults and all "liberations." Men and women found compensations in home life which transcended the seductive indulgences on the outside. There were transgressions, of course, not a few of them. But the appeal and the hold of the family on the individual, the comradeship, the love, the joy in children and in the common destiny it offered, provided unbreakable bonds of unity.

The removal of certain former compulsions aided the process of natural selection. Religion and church were on the decline. With the masses of the people both had lost authority. Their rulings were voided. The provision that a person christened in the Orthodox Church must marry only a co-religionist, or one who was willing to become one, no longer held sway. A Russian may now marry a Jew without any interference from family or Church. The same was true of the Mohammedan youth. On a trip in the Tatar Republic I was amazed at the freedom with which Tatar and Russian young people associated with one another and the frequency with which they intermarried. A mullah with whom I once stayed lamented the new blasphemies and irreverences, but admitted his helplessness to remedy them. The youth had ceased to believe in Allah, he said, and did as they pleased.

The proclamation and practice of equality of race and nationality further aided the process of natural selection. Russians, as a people, have never been haughtily or even conspicuously race conscious. Yet the legal and economic position of the multitude of racial minorities, particularly of Jews and of peoples in Asia, had kept them from amalgamating with one another or with Russians. This is all at an end. In the factories, in the new cities, in the educational institutions, particularly in the universities, in Europe and in Asia, there is constant intermingling of races and peoples. Young people meet on a plane of social equality. There are no exclusive schools, clubs, or fraternities. The result is intermarriage on a scale such as Russia has never before known.

The elimination of the middle and aristocratic classes and the economic and social prerogatives they enjoyed or erected for themselves has removed one more obstacle to free and natural selection. There are no exclusive groups or classes. There is no caste. At one time the communist stood out as a privileged person, but hardly any more now. He may enjoy greater authority, may have more ready access to persons of influence, but his special privilege scarcely extends any further. There was a time in the early and furious thirties when marriage to a communist was regarded in some families as a stroke of luck. Not any more. The non-party man has been lifted to a position of prestige which is no lower than that of the party man. Of course the engineer, the writer, the motion-picture director, or the man of rank in the Army, is much sought after by many a girl as a husband. They all command special standing in the community and also superior earning power. But they constitute no caste because there is nothing hereditary about their career or their position. They cannot pass on to offspring their earning power, their prestige, or any of their attributes or acquisitions. Nor are there any exclusive neighborhoods, exclusive social sets, or exclusive families who hold themselves aloof or above others. With all the power and prestige that Stalin commands, his children are never seen or heard of in the limelight. They set neither the pace nor the mode for anything—dress, custom, manner, sports, or any social diversion or usage. They are swallowed in a pall of obscurity, out of which only the unfolding of special gifts or special achievements, in no way related to their father's position, can lift them.

"Would you like to marry Stalin's son?" I asked a factory girl in Moscow.

"If I fell in love with him—yes." That was all she said.

For the present, Russia is liberated from any and all manifestations of social superiority, social snobbery, social exclusiveness. The son of a janitor may and does marry the daughter of a factory director. The daughter of a university professor will not disdain the hand of the son of a coal miner. It is all essentially a matter of mutual attractiveness and nothing more.

The economic independence of a woman is still a further liberation of marriage from external compulsion. Unlike the girls in Ostrovsky's dramas, whose families are always scheming to marry them to older men of means and position, the girl of today does not have to marry anybody of whom she disapproves or whom she does not love. She is not necessarily dependent on a husband for her economic security. She can always earn her own living. Careers and jobs from the highest to the lowest are open to her all over the land on a plane of equality with men, including equal pay for equal work.

Not that there are no self-seeking girls, jealous girls, scheming girls who will stop at no guile to inveigle into marriage the men who can offer them the most advantages and whose earning power will make it unnecessary for them to work. There are such girls, not a few of them, though during the war, unless they work, regardless of whom they have married, they lose their food cards and do not eat. Sovietism has not drained out of women their feminine wiles, though it has narrowed substantially the opportunity of exercising them. Yet no girl is ever under any compulsion either because of religion, race, pressure of family, public opinion, or some other external consideration to accept in marriage the man she does not want as a husband.

Yet in 1936, the very year when it was officially proclaimed that the enemy classes were no more, and when there was an air of growing tolerance in the country and the Constitution had become the most joyfully discussed subject, the Soviets imposed a set of unexpected external compulsions on the family. Not because the family was in danger of disintegration or collapse. Not at all. In a striking editorial of May 1936 *Pravda* proudly proclaimed:

"Fatherhood and motherhood have long ago become a virtue in the land of the Soviets. This is noticeable on first sight even in the external scene. Go for a walk on a rest day in the streets and parks of Moscow or any Soviet city and you'll see plenty of young people with their rosy-cheeked babies in their arms."

Nor had the new liberties and the right of woman to decide for herself whether or when she wanted any children seriously threatened the nation's birth rate. *Pravda* denied it. Again it proudly declared: "The birth rate is continually rising, the death rate is continually falling." It further added: "Marriage and divorce are of course personal affairs." Marriage and the family, by the admission of this editorial, had become stabilized and deep-rooted in the new society. Then why the external compulsions and the abrogation of newly won liberties? Many and varied were the explanations.

"In a Soviet society," said *Pravda* again editorially, "the playboy who marries five times a year cannot command public esteem, nor can the girl who flits with the ease of a butterfly from one marriage to another." In other words, there were abuses of existing liberties. True, the overwhelming mass of people had made a competent and decorous adjustment to them. But too many were the exceptions. Women were taking too dangerous advantage of the right of free abortions. They were injuring themselves. They were injuring society, and they were setting a baneful example to the younger generation. The experiment of laissez faire in sex was in the official opinion a failure.

"So-called 'free love,'" said *Pravda* editorially, "and disorder in sexual life are completely bourgeois and have nothing in common with socialist principles nor with the ethics and rules of conduct of a Soviet citizen. Such is the teaching of socialism, such is the eloquent confirmation of life itself."

A Christian clergyman could be no more direct and firm in his condemnation of promiscuity and licentiousness.

Hence the new law of June 26, 1936. So drastic a departure was it from established and formerly sanctioned, almost sanctified, usage that it shocked the outside liberal world and multitudes of Russians. Abortions were banned completely for the healthy woman. A physician violating the law was liable to from one to two years' imprisonment. The woman who sought the operation was subject to public censure; if she committed the offense a second time, she paid a fine of 300 rubles. If anyone, presumably the man with whom she had cohabited, coerced her into the operation he was brought to trial and faced a sentence of from one to two years' imprisonment. Thus all parties to the offense were penalized.

Birth control remained legal, but the literature on the subject suddenly disappeared from newsstands and bookshops. Physicians were not forbidden to offer information on the subject to patients, and women with weak hearts and other serious ailments were advised to seek such information.

Divorce was tightened. The postal-card system of notifying a divorced husband or wife of the separation was outlawed and had been outlawed even before the passage of the new law. The registry clerk was empowered to question the applicants and to act less as a dispenser of divorce and more as an arbitrator of differences and difficulties. Both parties to the contest were required to appear at the hearing. Both were questioned, both were encouraged to compose their quarrels and continue their marital life for their own good, for the good of the children and the new society.

The fee was boosted to 50 rubles for the first divorce, 150 rubles for the second, 300 rubles for the third. No fee for a fourth divorce is mentioned. Presumably there was to be no fourth divorce.

The method of alimony payment for children was stiffened. In the village it is paid at the source—deducted from the man's earnings, whether or not he and the woman work or live in the same kolkhoz. If there is one child in the family, the man pays one fourth of his earnings; if there are two children, the alimony amounts to one third; if there are three or more children, he pays one half of his earnings. Failure to make payments involves a jail sentence up to two years. The jail sentence is only a part of the penalty. Spirited public denunciation and the resulting contempt or coolness of friends and associates in the factory or office invariably precede and accompany court sentence.

It must be noted that under the Soviet system the custody of the child may be awarded to the husband, in which event the wife pays alimony in the same amount and under the same conditions as the husband.

Despite official vigilance not only by the government but by the Party, the Komsomol, the trade unions, and other organizations of influence and power, abortions have not been as completely banished as the law sternly demands. For a high fee some Russian physician will gamble on not being discovered. But if only because there is comparatively little private practice of medicine in Russia, the number of physicians who will risk defiance of the law is limited.

With the coming of the war and the heavy drain on the health of workingwomen, the interpretation of the law has been liberalized and abortions are more freely permitted, though never for other than reasons of health. No official or physician with whom I have discussed the subject but stanchly declared that there is not the least likelihood of an abrogation or a liberalization of the law at the end of the war, if only because of the heavy casualties the country has suffered in the fighting. On the contrary, there can be only a tightening of regulations.

In this writer's judgment the new marriage law, despite all official proclamations and explanations, must be linked to the rising clouds of war in Europe and Asia at the time the law was passed. War had haunted the Soviets since the day they came into power. With Japanese occupation of Manchuria and Hitler's ascent to supremacy in Germany and Mussolini's invasion of Abyssinia, the specter of war loomed to the Russians more and more menacing. Nazi Germany was subsidizing large families and was in other spectacular ways encouraging new births. The Anti-Comintern Pact of Japan, Italy, and Germany had already been formed. To the Russians it meant only one thing—an alliance of fascist nations for war against them.

In *Mein Kampf* Hitler committed himself to the seizure of Soviet Ukraine. In a speech in Nuremberg at a Nazi party Congress he orated about the Urals and Siberia. Other capitalist nations, so Russians then held—America, for example, and England and France—might be only too pleased if the fascist powers launched war on Russia, might even aid or join the crusade against the country that had "destroyed capitalism." Russians saw themselves ringed around by powerful capitalist nations and in danger of being obliged to wage war simultaneously in Asia and Europe. The war, they reasoned, might come soon or might be postponed for some years. But they had to be prepared not only with guns, planes, and tanks but with man power—with population. Hence the impetus for the new law, the restrictions, despite an already high birth rate, on interference with pregnancy, and the generous subsidies to large families.

These subsidies are a distinctive feature of the new marriage law. A woman who gives birth to a seventh child is allowed a sum of 2,000 rubles ($400 according to the official rate of exchange) for a period of five years. For every additional child, including the tenth, she receives an equal sum. When an eleventh is born the amount is raised to 5,000 rubles for the first year and to 3,000 rubles for each of the following four years. Additional children command similar subsidies.

Not only is motherhood subsidized, it is lauded and glorified in the press, in posters, in the theater, in the cinema. Motherhood and fatherhood have become as much a part of Russian ideology as patriotism.

"The woman who has no children deserves our pity," says *Pravda* editorially, "for she misses the full joy of life."

Formerly university students who were married while in college might, on their graduation, be sent to work to places far apart from one another. Not any more, except in rare and unavoidable instances. The same is true of diplomats. Invariably the wife accompanies the husband when he goes to a new place. Nothing is permitted which may contribute to the weakening or disintegration of any family.

And so like romance and like morality, with which it is so closely interwoven, the family has regained all the strength it ever had in the old days, and it has tapped fresh reservoirs of support, such as financial aid, and prestige which it never had known in pre-Soviet times.

The war has lifted the family to a new eminence and a new appreciation. Were anyone now to speak of it as a relic of a bygone age and fit for oblivion and annihilation, he would be deemed a maniac or an outcast. Now there are no more hallowed words in the Russian language than *semya* (family) and *rodina* (fatherland). Life and happiness, the greatest man can know, emanate from these as inevitably as light and warmth emanate from the sun. Without the family and the fatherland there are only emptiness and futility, even as without the sun there is only darkness and death. The fatherland makes the family secure, the family makes the fatherland invincible—such is the view and the attitude now.

Soldiers especially, married and unmarried, feel the importance of family life. On reconquering a village they are happy to go into a home, sit at a table and eat the soup that a mother or a grandmother serves them. It brings back memories of home somewhere far behind the lines. The most impressive feature of Russian newsreels showing soldiers entering a reconquered village is the enthusiasm with which they embrace not the girls but the mothers and grandmothers. A mother who for some reason happens to be in the services at the front is always more idolized than the girl sniper or the nurse or some other young woman in the service.

Here, for example, is Maria Ivanovna Karpenko, a mother from the village of Belev. She has no husband and only one son, Sasha, eighteen years of age. When the Germans occupied Belev the son wanted to join the guerrillas, but the mother pleaded with him and persuaded him to remain at home. One night he failed to return. The mother felt lonely and desolate. By dawn Sasha returned smelling of gasoline. He had spent the night pulling stoppers out of the tanks in German trucks and letting out the gasoline. Soon the Germans came and inquired of Maria Ivanovna where her son was. She swore he was not at home and detained the Germans long enough at the door to enable Sasha to leap out of the back window.

Shortly afterward the son was in the Army, and the mother followed him to the front. She became a cook in the regiment in which Sasha served. A kindly, determined woman, she became the friend of soldiers in the regiment, of the privates as well as the officers. Her son was a sniper and lived in a dugout with other snipers. Maria Ivanovna often visited the dugout. Bombs and bullets no longer held any terrors for her. They had become a part of her daily life and of the surrounding scene.

The dugout was warm and quiet, and the snipers, including Sasha, gathered around the *matushka*—little mother—and told her of their day's successes and listened to her words of praise and counsel. She stayed late, and no matter how inclement the weather—rain or blizzard—she went back to her own quarters to cook breakfast for "her boys."

The next night she was back again, Sasha's mother—everybody's mother—hearing tales of fresh adventures, offering more words of cheer and counsel to the snipers. Her very presence imparted a flavor of home and family to the faraway and solitary dugout. Nor is she the only Maria Ivanovna at the front.

CHAPTER 26: YOUTH AND CULTURE

WHEN THE Russo-German war broke out, Zoya Vladimirova was only a little more than sixteen years old. She was a high-school student, a poet, and an actress. In the theater of her native city of Tula she had played the part of Seryozha, the son of Anna Karenina in the play based on Tolstoy's novel.

To look at her, so small and pretty, with soft brown hair, firmly rounded mouth, and milk-white teeth, one would never imagine she had the courage and strength to go to the battle front and in the thick of the fighting to carry wounded men to safe shelters in the rear. Yet she did this for six months. Through hails of bullets she carried out one hundred and sixteen wounded soldiers!

"Were you never afraid?" I asked.

She shook her head, laughed, and said:

"I didn't have time to be afraid."

She laughed again, as though amused that anyone should put such a question to her.

Readily she told me the story of her experiences at the front. On June 22, 1941, she was in a village outside Tula, visiting an aunt. They had a late breakfast, and when the meal was over she gathered the dishes and started to wash them. The radio was on and she was listening to the program. Suddenly came the announcement of war.

"The dishes fell out of my hands," she recalled, "and the light went out of my eyes. When I recovered I said to my aunt, 'I'm going to the front.'"

The aunt was dismayed, but without losing any time Zoya packed her belongings and returned to Tula. Instead of going home she tried to enlist. Neither the Komsomol nor the Army would accept her because of her youth and size.

When the Germans were approaching Tula the authorities wanted to evacuate her together with other women, children, and the old people. Zoya refused to go. She would remain home and fight, and nobody could stop her. For weeks she persisted in her effort to enlist and one day she was told:

"Be ready at twelve o'clock."

She ran home and packed, saying nothing to her mother for fear there would be a scene. When she was ready she went to the door and, turning to her mother, said:

"Mama darling, I am going to the front."

Quickly she ran out of the house and never looked back. She did not want to meet her mother's pleading eyes or hear her pleading words.

She reported at the military depot and was assigned to the volunteer regiment just formed in Tula. The soldiers teased her, saying: "Little one, you'd better go back to Mama," or "Little one, you'll run the moment you hear the first bullet." Her only reply was, "We'll see."

She walked alone to staff headquarters, and all the time she kept wondering whether the fighting would scare her. Suddenly she heard a noise and looked up. Overhead was a huge black Messerschmitt, and swiftly she dived into the snow. She heard the rattle of machine guns, but they did not frighten her, and she then knew that bullets would hold no terror for her.

At headquarters she was outfitted with a woolen vest, a sheepskin jacket, felt boots, a greatcoat, and a winter cap. She was given arms—a gun, ammunition, hand grenades. In her new outfit she looked larger than she ever had in her life, and she was pleased—soldiers would no longer tease her for being so little. But she *was* little. The greatcoat was sizes too large for her. At first as she walked she got tangled in its flowing fullness, and soldiers teased her more than ever. But she marched on to the fighting position, walking immediately behind the lieutenant.

Informed by a scout that a German tank had become stalled near by, the lieutenant ordered the machine destroyed. She saw a Russian soldier leap on top of the tank, bayonet a German, and blow up the machine. . . . Suddenly bullets flew over their heads. The Russians ducked into the snow, and so of course did Zoya, and again she was elated that she felt no fear.

Luckily none in the company was wounded. When the firing ceased they resumed marching. On the way to Kaluga they encountered a stranded German soldier. Dirty and frozen, he stumbled out of a snow heap and surrendered. Asked where the German Army was, he said: "Tula, Kaluga, *tuptup*"—meaning "on the run." The Russians were happy that the Germans were at last fleeing. They laughed and amused themselves and kept repeating the word "tuptup."

It was so cold that they put American petroleum jelly on their faces to prevent frostbite. Finally they reached the Oka River. The Germans saw them and opened fire, but again luck was with them and no one was hurt. Later, under cover of darkness, they crossed the river and settled down in an empty factory building, bedding on the bare, cold cement floor.

The Germans were still holding Kaluga, and the Russians were waiting for the order to attack. At noon the order came. The attack started, and Zoya, by command of the lieutenant, remained in the factory to wait until wounded were brought back. She heard the crack of guns, the boom of cannon, and couldn't remain inside, so she ran outdoors. Under a tree in the snow she saw a sergeant, a giant of a man, lying prostrate. She started to lift him, but he said:

"Little fourpounder, how can you lift a giant like me?"

"I'll show you," she said, and carried him inside, where she dressed his wound.

"Little fourpounder," the sergeant said gratefully, "you're wonderful!"

Soon other wounded arrived. She hurried from one to another, dressing wounds and cheering the men. Now they called her "brave little sister," "kind little sister." One man, badly wounded, said brokenly, "Zoyechka, don't let me die." She laughed and stroked his head and shoulders and said, "You'll outlive me." But he did die, the poor man!

Six wounded men arrived at once. Their commander had been struck down and they had tried to save him, but the Germans fired and all fell.

"Where is he?" asked Zoya.

They told her where the commander was lying, and she ran out to save him. Being small, she had the advantage of appearing only a small speck as she crawled to the wounded man in the woods. But he was dead.

The Germans saw her and opened a volley of machine-gun fire. Branches crashed, trees burned, and she snuggled for protection under the dead body. Strange, too, for she had always been so afraid of the dead that she never had dared to pass a cemetery alone in the night. Now she had no fear of anything or anybody.

On returning she felt gloomy, but proceeded again to administer to the wounded men. Then a shell burst into the room, the ceiling caved in, debris in heaps was over everything. She was stunned for an instant, and blood dripped from her jaw. A tooth had been knocked out. Otherwise she was unharmed. Quickly she started pulling wounded soldiers from under the debris. Two, with whom only a few minutes earlier she had talked, were dead.

She moved them to another room. It was so cold that she built little fires on the cement floor. But the wounded kept complaining of the cold. She was warm because she had been working. So she pulled off her own sheepskin jacket and woolen vest and wrapped them around the men. Others were still freezing. She glanced at the dead bodies—here were greatcoats, sheepskin jackets, woolen vests. Without a moment's hesitation she stooped down and removed the clothing from them. The wounded men saw her and protested, "Don't, little sister, they are our brothers."

She paid no heed to the protests. She had to do something to keep the live men from dying of the cold.

When the Germans were driven out of Kaluga the place was a shambles. Despite their broken and burned homes the people were glad. They wept and talked and told horrible tales.

"One woman fell on my shoulders," said Zoya, "and wept and told how the Germans had burned her home and her seventeen-year-old daughter. I was so full of hate for Germans that I whipped out my gun and went around emptying bullets into their dead bodies, wishing they were alive so that I could kill them. . . . Oh, I'll never forget Kaluga."

She paused, solemn and contemplative, as if seeking to hold back her wrath. Then, with a start, she resumed:

"I must tell you about Lyova Volkov."

"Who was he?" I asked.

"One of my very best friends—no, not my sweetheart—I was interested in another boy, and he was interested in another girl. But we were in high school together. He was only a little older than I. We had always known each other. We went on long hikes, had long talks, quarreled endlessly—but only in fun. . . . He was one of the best marksmen in the Tula Komsomol. He killed twenty-five Germans before he was struck down with a bullet in the abdomen. We ran up to carry him off the field, but he wouldn't go. He was lying almost on the crest of a hill. He didn't think he was seriously wounded. He wanted to lead his company over the hill. He compelled himself to lift his head, but his body wouldn't rise. 'Onward for the fatherland,' he cried. A bullet struck his head and he fell over. I'll never forget Lyova—such a wonderful boy." For once tears glistened in Zoya's eyes. . . .

I met Zoya in Kuibyshev. The Komsomol had recalled her from the front and sent her to this Volga city to continue her education in summer school. I saw her several times, and we talked at length on the subject of which she never tired talking and to which I never tired listening—the Russian youth, her generation, the generation of Zoya Kosmodemyanskaya and

Shura Chekalin and other outstanding war heroes of the country. Unlike any generation of high-school boys and girls Russia has known either under the czars or the Soviets, it was making history not only for itself but for Russia. An eager, fighting, crusading generation, it loved to live and, despite the low material standard of living, it found a multitude of ways of making itself happy. When the war came it clamored to go to the front. Thousands went in one capacity or another. Thousands more joined the guerrilla brigades. Those in the rear are doing yeoman work in the field of production.

In the autumn of 1942 I was visiting Tula and made inquiries about Zoya Vladimirova. She was no longer in Kuibyshev, I was informed. She had finished her summer-school courses and been sent to a rural district in the Tula province to supervise the gathering of the harvest and the planting of winter crops.

Zoya Vladimirova is representative of young Russia, not only in her fearless participation in the war, but in the persistent pursuit of knowledge which, whenever possible, Russians continue in the midst of armed conflict.

There can be no doubt that the stabilization which the Plans achieved—in industry, in agriculture, in family life, in morals, in other social relations, above all in education—had much to do with the molding of the spirit and the mentality of this generation of Russia's youth. To them culture was no longer an abstraction or a generalization. No longer were there any conflicts among leaders as to its meaning and purpose. The institutions that disseminated it were as multitudinous as they were varied. A survey of these institutions offers a further insight into Russian life and Russian character.

In 1913 there were 859 newspapers in Russia, of which 775 were in the Russian language. Their combined circulation was 2,700,000. In 1938 there were 8,500 newspapers, or almost ten times as many as in the old days. One fourth of them were printed in seventy non-Russian languages. The combined circulation of all these newspapers was 37,500,000. It would have been several times higher had it not been for the acute paper shortage.

Under the Plans schools leaped up with mushroomlike swiftness. In 1914–15 there were 1,953 secondary schools with 42,803 teachers and 635,591 pupils in Russia. On September 15, 1939, there were 15,810 secondary schools with 377,337 teachers and 10,834,612 pupils.

Many of the new schools were hastily built. Their equipment was inadequate. Not always were teachers properly trained. Often enough they were too immature for their duties. But in education, as in everything else in the country, the aim and the law were *mass procedure*. Adjustment, adequacy, and improvement would inevitably follow. Such was the belief.

The important thing was to draw the nation's millions into the new culture and the new schools.

In 1914 there were ninety-one institutions of higher learning in Russia. Of these thirty-five were in St. Petersburg, twenty in Moscow. In all there were 112,000 students in these institutions. Now there is hardly a city without one or more colleges or universities. North and south, east and west, the Russian as well as the non-Russian peoples have hundreds of universities. Particularly rapid has been the growth of scientific schools—in engineering, in medicine, in agriculture. It is these schools that have developed much of the talent and the competence that helped make the Red Army the fighting organization that it is. Stupendous are the problems of such an army—transport, supplies, military equipment—and only men of highest training could cope with them as adequately as the Russians have.

In 1939 there were 111,000 clubhouses in Russia which ministered to the social and recreational needs of the population, especially the youth. There were also in that year 86,266 public libraries with 166 million books.

Book publishing has always been one of Russia's large enterprises, under the czars as well as under the Soviets. In 1913 a total of 26,200 books and pamphlets were published, with a combined output of 86.7 million copies. In 1938 the number of titles grew to 40,000, the number of copies to 692.7 millions. This number includes the many political and technical pamphlets that are issued in editions of six and seven figures.

Illuminating are the figures on creative writing, chiefly prose, published in Russia in the years 1917–40. Let us first glance at Russian literature, including only the best-known authors, all of whom, except Gorki and Mayakovsky, lived and wrote in the nineteenth century, long before there was even a dream of the Soviet Revolution, though the literary and intellectual atmosphere teemed with revolutionary ideas.

It is interesting to observe that the so-called "proletarian literature," which was at one time violently fostered by RAPP—society of proletarian writers—receives no recognition in the surveys that Russians are now making of the authors and books that have been published under the Soviets.

AUTHORS	COPIES IN THOUSANDS	NUMBER OF LANGUAGES IN WHICH PUBLISHED
Herzen, A. I.	1,312	2
Gogol, N. V.	7,731	33
Gorki, A. M.	39,876	65
Griboyedov, A. S.	771	3
Lermontov, M. Y.	5,383	42

AUTHORS	COPIES IN THOUSANDS	NUMBER OF LANGUAGES IN WHICH PUBLISHED
Mayakovsky, V. V.	7,150	31
Nekrasov, N. A.	8,250	27
Pushkin, A. S.	29,840	72
Saltykov-Shchedrin, M. E.	6,755	26
Tolstoy, L. N.	20,916	57
Turgenev, I. S.	9,906	39
Chekhov, A. P.	15,326	56
Shevchenko, T. G.	4,817	33

Foreign literature has always been popular with the reading public of Russia but never so much as since the coming of the Soviets, if only because there has never been so much literacy in the country or such a large reading public. Not to burden the reader with too many figures, I am choosing only leading foreign novelists, dramatists, and poets, and again only the classics of the more modern times. In this table in adjacent columns I am giving figures for one period before and one since the coming of the Soviets.

AUTHORS	IN THOUSANDS OF COPIES	
	1894–1916	1917–40
Byron	178	488
Balzac	106	1,743
Heine	124	1,070
Goethe	246	482
Hugo	466	2,886
Dickens	850	2,086
Zola	939	2,109
Maupassant	1,517	3,234
Rolland	24	2,036
Cervantes	126	567
Stendhal	25	774
France, Anatole	523	1,788
Shakespeare	611	1,209
Schiller	455	525

From 1917 to 1942 leading American writers were translated into the Russian language, and the number of copies of their works published (which in Russia means sold) were as follows:

Sherwood Anderson	45,000
Pearl Buck	140,000

Erskine Caldwell	73,000
Paul de Kruif	900,000
Harriet Beecher Stowe	
(*Uncle Tom's Cabin*)	135,000
Upton Sinclair	2,633,000
Fenimore Cooper	284,000
Bret Harte	220,000
Longfellow's *Hiawatha*	123,835
Ernest Hemingway	70,000
John Steinbeck (*Grapes of Wrath*)	325,000
Richard Wright	75,000
Eugene O'Neill	25,000
Stephen Leacock	230,000
O. Henry	1,144,310
Theodore Dreiser	263,000
Sinclair Lewis	194,800
Mark Twain	2,004,850
John Dos Passos	97,850
Jack London	6,428,000 (figures incomplete)

The theater has played no small part in the education and edification of Russians and has contributed its share to the mental and emotional stabilization of the youth. It has been the outstanding source of public entertainment and social festiveness. With churchgoing out of vogue except among small groups of worshipers, attendance at the theater has become more and more a fashion, almost a rite. The absence of comfortable homes has no doubt contributed to the popularity of the theater. When a new city rises up in the wilderness, whether in the mountains of the Urals or in the steppes and forests of Siberia, a theater also rises soon after new factories have been put up, sometimes as soon.

In January 1941 there were 850 theaters in Russia—all of them permanent and professional. Territorially, though not linguistically, they were divided as follows:

Russian Republic	469
Ukrainian Republic	119
White Russian Republic	16
Azerbaijan Republic	31
Georgian Republic	48
Armenian Republic	27
Turkmenian Republic	14
Uzbek Republic	45

Tadzhik Republic	23
Kazak Republic	40
Kirghiz Republic	18

Of these theaters 173 are devoted to productions exclusively for children and young people and 276 are in the countryside. The theatrical studios cannot keep up with the demand for actors in the new theaters. In prewar days special committees were continually journeying to Moscow, making the rounds of theaters, dramatic studios, Komsomol, Party and trade-union quarters in search of actors, singers, and dancers for newly built theaters. The government supports the cause of the theater energetically and so do the Party, the Komsomol, and especially the trade-unions.

Before the war noted architects worked on special designs for theaters with the aim of making them decorative and appealing to the eye and the imagination. It can hardly be said that in the theaters already built they have achieved anything particularly noteworthy. There are, of course, exceptions—such as the new theaters in Novosibirsk and Rostov, the Tchaikovsky Concert Hall in Moscow. These are landmarks in the architecture of public buildings.

Exposed to a multitude of new stabilizing influences—in the home, in schools from kindergarten to the highest courses in the universities, in the press, in literature, in the theater, in music—Russian youth, especially since the time of the adoption of the Constitution, has found a new world in which to cultivate its ideas, its imagination, its habit of living, its social attitudes, its tastes, its romantic proclivities, without the danger of encountering the uncertainties and conflicts which so often harassed preceding generations of Soviet youth. With all its faults, particularly its inadequate information about foreign countries and foreign peoples, Russia's new youth, in this writer's judgment, is the most highly cultivated, the best mannered, the most normal, and most imaginative that the Soviets have yet known.

The nationwide sweep of what the Russians call *samodeyatelnost*—amateur art—has aided in the stabilization and refinement of this youth. Particularly is this true of music, drama, and dancing. The clubhouses of the country teem with samodeyatelnost; so do schools, including the kindergartens. Some of the liveliest and most spirited folk dancing I have ever seen in Russia was in kindergartens. I know of no children in the world who love dancing more than Russian children or who more readily learn new steps however complicated and strenuous.

There is not a collective farm or factory, however remote its geographical location, but has its amateur performers. Some of them are raw and coarse. There are not enough trained leaders to direct their work. A new and popular profession has sprung into existence and has enjoyed wide growth during the first three years of the Third Five-Year Plan—namely, instructors in samodeyatelnost. Yet there have never been enough of them. The movement is encouraged by all the organizations of the country, and large sums of money have been appropriated by them for the purpose.

The distinguishing feature of this amateur effort is its emphasis on native folk art—in music and the dance especially. This is particularly true of choirs, of which there are more than Russia has ever before known. Factories, trade-unions, the Army, the Navy, the Komsomol, the Pioneers, the schools—all cultivate choirs. Choral concerts are the most popular form of entertainment. Yet every one of these choirs features folk music, and every audience I have watched during a concert of such music responds with enthusiasm. Some of the songs by the most popular composers in Russia might as well have been ancient folk tunes: for example, Zakharov's sensational love song, "And Who Knows." So might the recently and equally sensational war song, "She Is So Little, She Is So Lovely, Our Darling Little Girl." An orchestra or a choir can always count on a loud ovation after the rendition of either of these melodies. They are lively and tuneful and humorous, and soldiers at the front or in the camps roar with laughter when they hear them. So do other audiences.

Invariably these choirs have not only singers but dancers. Invariably, also, they feature not only folk music but folk dancing, and not until one attends a choral concert in Russia does one begin to appreciate the liveliness and humor of ancient Russian dances and melodies. Foreigners who have gathered the impression that Russian music is an endless dirge would swiftly change their opinion if they attended a concert of any present-day Russian choir.

During one of the saddest moments in the summer of 1942, when Germans were surrounding Rostov, Maurice Lovel, a British journalist, and I attended a concert given by the choir of the Central Trade Unions of Moscow. Despite the sad mood aroused by the war news the audience laughed joyously. A sailor dance performed by a boy and a girl elicited as clamorous an outburst of mirth as I had heard in any theater. So did a number of ancient folk melodies rendered by a barytone soloist. Lovel and I couldn't help remarking how out of tune was the gaiety of the audience with the gloom of the people outside the theater.

Wherever I went in Russia I found struggle, sorrow, undernourishment, and resistance to the foe amid an unbroken preoccupation with culture as the Russians understand the word.

PART FIVE: RUSSIAN WOMEN

CHAPTER 27: THE NEW ROLE

DURING MY STAY in Moscow I visited a woman who within a week had received the saddest news that can come to a woman. One of her sons, an aviator, had been killed in battle, another in the infantry had died from wounds.

Yet, though grief-stricken, she did not break down. She showed me photographs of the two sons she will never again see. She kissed the photographs, tried hard not to weep, and said:

"Last summer my husband died; now my two sons are gone. But I am a Russian woman—I can stand it."

During the summer of 1942 I was at a railroad station near a village from which newly mobilized men were leaving for the war. The mothers, wives, sisters, and children came to see them off. Loud and heartbreaking were the wails. I can still hear the piercing screams of wives and mothers as the train departed. Home-loving as Russian peasants are—not given much to traveling far from their native village, not knowing, therefore, lengthy separations from the family—a journey, any journey to a distant place, usually stirs older women to tears. But this was no ordinary journey. Men were going to the front. They might never come back. This was a calamity beyond compare. Loud, therefore, almost earth-shaking, was the grief of young and old.

After they had had their cry, explosive as it was, these women went back to their homes. Some of them carried babies in their arms. With a new gush of love they kissed and hugged the babies and spoke tenderly to them, as though in them lay the strength and the solace they now needed. On reaching their homes they dried their tears, washed their faces, and

plunged into work! They were so busy working in the fields that they had
no time for tears. I stayed several days in this village—called on some of
the women. "If only the war would end, if only it would end," was their
universal wish. Except for the *babushki*—grandmothers—hardly any of
them cried any more. Hardy and deep-feeling they went about their duties,
the heaviest they had ever known, with unshrinking fortitude. They were
alone and they knew it. The summer's work was ahead—the weeding, the
cultivating, the hoeing, the reaping, the threshing—and they would neither
falter nor fail in the performance of the strenuous tasks. Many of them
knew that they would never again see their husbands; they said so and
grieved over their lot. But they toiled unendingly from sunrise to dusk.
They looked after their families too. Daytime, usually under the supervision of a trained nurse, their children romped in the meadows, chased
butterflies, gathered daisies, cornflowers, sorrel leaves, wild onions; they ran
races and shouted with anger or with triumph as children do when they
are outdoors and at play. But on returning from the fields the mothers
washed them, cooked for them, scolded them, and put them to bed.
Neither weariness nor sorrow interfered with the maintenance of the
home and all that the word spells to the Russian people in these sad yet
romantic times. These women remained as unbowed by the burdens and
the agony of war as the ash and the fir that towered so high over their
shrunken huts. . . .

How many widows are there in Russia? No official records are available,
but one can venture an estimate. Russians marry young. Many girls marry
at seventeen; many men loathe the thought of remaining unwed after
twenty. Neither the Revolution nor the machine age has stunted or
thwarted the urge for early mating. A man or woman of twenty-five who
is still single rouses concern even in Bolshevik circles and sometimes stirs
the good-humored mockery of close friends.

The officer corps in the Russian Army, from sergeants to generals and
marshals, is almost entirely made up of married men. Soldiers from the age
of twenty are more than likely to have wives. Since about four million men
in the Russian Army have died it is more than logical to conclude that
they have left behind no less than three million widows. I have seldom
visited a Russian home in which friends gathered but there were widows
among them. This is especially true of Moscow. Heavy were the casualties
of the volunteers who participated in routing the Germans from the outskirts of the capital.

Yet these widows are Russian women. They can stand it! They do stand
it with a dignity that rouses in the understanding foreigner more than
admiration!

How they have worked, these sturdy and deep-feeling Russian women!

There are no idlers among them. They do the lightest and the heaviest
tasks of the nation. They have replaced men in industry by the millions. By
still more millions they have been doing the work of men on the farm.
Wherever one travels in Russia one sees them not only "turning" but "running" the wheels of the nation. At least thirty million of them are doing the
menial and the creative work of the land. Thirty million! Without them
Russian transport would be in the doldrums, Russian industry would be a
holocaust, Russian agriculture would be a nightmare. With them at the
factory bench, the flow of ammunition to the front, to all the fronts, rises
higher and higher. With them at the plow, in the pasture, in the cow barn,
the flow of food to the fighting men—meat and butter, cheese and vegetables—never ceases to reach ever-rising levels. They are *the* power and *the*
glory of the ever-seething Russian rear. Because of their energy, their skill,
their universality, their adaptability, the Army can with safety summon
all but the most imperatively needed and disabled men to the fighting
ranks. Monumental is the work and the dedication of the Russian women.

Not that there are no schemers and, as the Russians say, "vipers" among
them. There are. Russian women can be mean and jealous, cruel and
vitriolic. As I was passing a hotel in Moscow, I saw several uniformed
men and several militiamen carry a stately, well-dressed young woman
into an ambulance. I caught a glimpse of the young woman's face. It was
all red and burned, and the eyes looked as if they were pressed deep into
the sockets. What had happened? No one knew the details, but the loquacious doorman had his explanation. He said, "A jealous woman poured
acid over her face—that's what she did—the beast." Throwing acid into
the face of a competitor for a man's attentions seemingly has not passed
out of fashion in Russia.

Russian women have their failings. I know the secretary of a noted
foreign correspondent who would be a champion shrew in any country,
more than a shrew. No one I have ever known or read of can nag more
vehemently or more viciously. Some of the "floozies" who hover with
the persistence of moths around the foreign colony in Moscow could give
lessons in gold digging to any of their benighted sisters in America or England.

Some Russian girls, even in these awesome times, would rather look at
themselves in the mirror than go to work in a factory or office. Now and
then the press, especially the *Komsomolskaya Pravda,* will lift into the
limelight for a verbal trouncing a woman or a girl who is out of tune with
the times.

One of the subjects of their crusade is Masha B——, in the town of
Askhabad, Central Asia. Masha is over twenty, a typist by profession.
Masha works—she has to or forfeit her food card. Other girls in Askhabad

gather in the evening in some office or schoolroom and sew warm mittens
for the soldiers. But not Masha. She has heard that working under an
electric light dents the face prematurely with wrinkles—and she dreads
the thought of wrinkles. On rest day crowds of girls go to the railroad
yards and unload wood. But not Masha. Her hands are soft and tender, and
handling wood would make them coarse and hard; Masha trembles at the
thought of such hands. Hardly a girl in Askhabad, especially in the offices,
fails to work overtime. But not Masha. Overtime, she says, induces fatigue,
and fatigue is bad for the complexion. Masha—according to the *Komsomolskaya Pravda,* which of course prints her name in full—thinks and worries
about and loves only herself. She is so preoccupied with feminine indulgences that she has no time for fighting Germans or saving the fatherland
from conquest and desecration. "Masha is absorbed only in her beauty. But
what is there beautiful in the pathetic life of a drone?" Bitter and mirthless
is the portrait of Masha B——.

Not that the *Komsomolskaya Pravda* is against feminine decorativeness.
For years it has been eloquently preaching the gospel of personal attractiveness. Now more important items have crowded such subjects from
its pages, but it has not risen in revolt against this gospel.

Moscow has many girl militiamen. Some of them—in the Lenin Library,
for example—might have come from the chorus of an operetta. They know
their duty only too well, these trim and pretty militiamen, and woe to the
man or woman who seeks to slip by without showing a passport or identification card. Yet these girls neither disdain nor neglect the use of lipstick.
I have seen girl chauffeurs driving trucks of ammunition to the front,
and when they looked out of the hood over the driver's seat their lipstick
made one think they were on their way to a dance.

The use of cosmetics is in abundant evidence—even in villages. Nor
have the beauty shops been closed. The crowds of women who patronize
them testify to a wish to keep up appearances regardless of the burdens
and the sorrows of the times.

Go into the headquarters of the Central Committee of the Komsomol
in Moscow, and even the girls who have just come on missions from the
guerrilla detachments in the primeval forests of Bryansk, or in the equally
primeval swamps of Kalinin, surprise you with their comeliness. Their first
thought on reaching the capital, it seems, is to make themselves attractive. Olga Mishakova, the blonde, handsome young secretary of the
Komsomol, could step into any fashionable gathering in America or England and feel at home. With all the burdens that rest on her shoulders,
she finds time to look after her purely feminine self. Neither the *Komsomolskaya Pravda* nor any other paper has ever found fault with such procedure.

Once I stepped into the room of Alec Werth, the British correspondent. He was entertaining the young editor of the *Komsomolskaya Pravda*. No foreign correspondent in Moscow was more stylishly or more neatly dressed. The propaganda in the last years of the Plans, on the eve of the war, for style and decorativeness has borne rich fruits, as a visit to any theater so visually testifies.

But a Masha B——, who will be only a "wallflower" even in these volcanic times, stirs the wrath and the execration of leaders and of others too. Such girls, or shrews like the secretary I have mentioned, or the "floozies" that sprint in and out of Moscow hotels, live in a world all their own and have nothing in common with the millions of women who work and sacrifice for the success of the Russian armies.

The self-assertiveness and the spirit of independence of Russian women, which have attained their highest fulfillment in the present war, are not new in Russian history. Manifestations of them in ancient times rise almost with the frequency of rainbows after showers. In folk tales and ballads women knights rise up in defense of their people and slay their enemies even as does the man Ilya Murometz, the greatest of all Russian folk heroes.

There is hardly a rural district in Russia that hasn't its historic folk heroes and folk heroines. Legendary or not, they emphasize the acceptance by the common folk of the countryside of the concept of the physical heroism of women and their devotion to the people.

"Think of our Yaroslavna," exclaimed a Russian teacher of literature, "and what she represented even in her dim day." Wife of Prince Igor, Yaroslavna was endowed with certain virtues which even hardheaded Bolsheviks like this schoolteacher regard as worthy of remembrance and emulation. In the Soviet textbooks on history and literature Yaroslavna is accorded no little attention and no little appreciation. A romantic and patriotic woman, she not only loves her husband with a deep devotion but grieves for the lot of his army and of the people. When Igor is captured and imprisoned by the Polovtsy, his fiercest enemies, she sends her tears "on the waves of the Dnieper 'to wash' the bloody wound on his [Igor's] mighty body." She prays to the Dnieper, to the wind, to the sun for a miracle to liberate her husband. She weeps for the sorrows of his men and of the people at home. Love of her husband, devotion to family, concern for the welfare of the people—all of which are now extolled as basic virtues in any human being but particularly by a citizen of a socialist society—mark the life and the character of this eminent historic woman.

Early in the nineteenth century, after the Decembrist Revolution, the wives of the rebel leaders, women of highest social origin and highest social attainment, followed their husbands to their faraway Siberian exile.

In doing so they not only shocked Nicholas the First, Emperor of Russia, but their own families, some of whom were among the emperor's closest friends and associates. They established a sentiment and a tradition of independence which to this day are regarded as epochal as well as heroic.

In the city of Chita, Siberia, I visited the museum in which has been gathered—in photographs, in household belongings, in an array of other exhibits—something of the history of the life of these Decembrist men and women in Siberia. It was one of the most stirring exhibits I had ever seen, and Kuznetsov, an old revolutionary in charge of the museum, told endless stories of these remarkable men and women and of the great contribution they made to Siberian civilization.

Nekrasov's long poem, *Russian Women,* which is a tribute to the martyrdom and the heroism of the Decembrist women, rouses the emotions of the Soviet youth even as it did of the youth in pre-Soviet times. The action of these high-born women in breaking with the court and their families, even though motivated not by social but by personal longings, had always aroused the admiration and the fervor of the pre-Soviet intelligentsia and had deepened their appreciation of the intelligence, the courage, the inexhaustible potentialities of women.

The most remarkable aspect of the Russian movement for the complete emancipation of women is the fact that its most eloquent advocates have been men. Throughout the nineteenth century, during which Russia emerged as one of the great literary nations of the world, Russian publicists, literary critics, novelists, and others raised high their voices in praise of woman's intellectual and social attainments and in the advocacy of their right to a position of equality with men in all the affairs of the world.

In this one respect alone Russian "feminism," if it may be so called, differs from the feminism of Western countries. In Russia it was men who did so much of the creative writing and the social thinking in the past century, who championed the concept and the doctrine of woman's absolute equality with man. They not only had faith in the talents and the capacities of women to acquit themselves with competence and glory of the tasks which had been the exclusive monopoly of men, they longed and cried for the day when women will grasp or be granted the chance to demonstrate the powers and the glories of their unused and frustrated energies.

In his didactic novel *What Is to Be Done?* the great Chernyshevsky, who spent the most fruitful years of his life in Siberian exile, says: "The history of civilization would have moved ten times faster if the true, devoted, and powerful mind with which nature has endowed women . . . had not been denied and frustrated but had been permitted natural action."

In Goncharov's celebrated novel *Oblomov,* depicting the indolence, futility, and hopelessness of the Russian gentry, the only character of promise, energy, and initiative is the heroine. Dobrolubov, who died at the age of twenty-five and who to this day is read as an apostle of a new social age for Russia, in writing of Goncharov's heroine, says: "In her we perceive that there is a suggestion of a new Russian life; from her we expect words that will scatter and burn up *oblomovism*. Personal family cannot satisfy her . . . a tranquil and happy life frightens her . . . like a muddy abyss that threatens to suck you in and swallow you."

In Russia women did not have to fight *against* men for their emancipation from inequalities before law and society. The czarist regime discriminated against them only a little more harshly than against the masculine part of the population. For a long time it kept them out of the higher institutions of learning; but the national minorities, the peasantry, the factory workers, with relatively few exceptions, were for a long time likewise unwelcome to these institutions. The gentry, the rising middle class, the clergy, because of their superior economic status alone, found themselves the chief beneficiaries of middle and higher education in old Russia.

The revolutionary parties in Russia, however violent their disagreement with one another as to the society that was to succeed czarism, championed complete equality of the sexes. To the leaders and followers of these parties the so-called inferiority of women in any form, including intellectual, did not exist. Women, they said, were the equals and the comrades of men in revolutionary battle; in the enjoyment of the fruits of the battle, whatever these might be, they would also be equals and comrades.

That is why throughout the nineteenth century and afterwards, until the day of the overthrow of the Czar, there hardly existed in Russia a "feminist movement" in the sense in which it rose to action in the West and in America. There was no need for it, because men were championing women's equality with themselves.

Remarkable and singular as was this condition, which has no parallel in any Western land, it arose in part out of the absence from Russian history of certain forces which had molded and colored the attitude of men toward women and of the notions that women themselves cherished of their destiny in the world.

Chivalry, for example, never reached Russia. It halted on the Russo-Polish frontier. It was principally and exclusively a Western movement. It struck roots in Western life, in Western history. Therefore the gulf that it had established between men and women never could exist in Russia.

Puritanism likewise never permeated Russian thought or Russian social life. There were, there still are, puritanical sects in Russia. But their influence is local and has never attained national or mass magnitude.

The Russian Orthodox Church was free from the influence of puritan-

ism. The ribald tales about priests that the churchgoing peasantry told with
zest in pre-Soviet days testify to none too high a folk appreciation of the
morals of the Russian clergy. The discrimination and inferiorities which
puritanism had leveled against women never had become a part of the
religious or social thought of Russia. The intelligentsia at least were never
impregnated with them. When Western progressive and revolutionary
ideas, including the equality of sexes, reached Russia, they found more than
fertile grounds for propagation. The mind of the Russian intellectual was
comparatively free from the hostilities which school, church, social usage,
and home life had been implanting in the youth of Western countries. It
absorbed the new ideas with the passion of an impoverished man exploiting a newly found treasure on his own lands.

Greatest of all, in this writer's judgment, was the folk influence on the
Russian intelligentsia. The word *narodnichestvo*—meaning populism, love
of the people, or rather of the peasantry, faith in their gifts and virtues, and
dedication to the cause of lifting them out of the frustration with which
they had been artificially saddled—exercised a magic sway over the intelligentsia throughout the nineteenth century. Westerners and Slavophiles
were impressed in different ways by the potentialities of the forlorn and
neglected muzhik. All of this led to studies, observations, cogitations on the
village and on the peasants which colored the social thought and creative
writings of the times.

Village lore might abound in sayings like the one that "a hen is no bird
and a woman is no human being"; but in everyday actuality, because they
were workers and because the peasant, too, had never known the doctrines
and usages of chivalry and puritanism, women remained the social equals
of men. They lived together in the same hut, in the same room. They
worked together in the fields and in the gardens. They ate together at the
same table, at the same time, and slept beside each other on the floor, on
the oven, in the bed, in the haymow, in the meadow. They walked to
church together, worshiped side by side and walked home together. When
unmarried, boys and girls freely visited one another in the homes, in the
streets, in the market places. On Sundays, on holidays, of which there was
an overabundance in old Russia, they danced and sang and jested and
laughed together. Women were the equals of men and freely they asserted
this equality. The boy a girl did not like she did not mind snubbing. If
he was too persistent, she might be provoked into swearing at him or striking him on the face. She could not be held legally accountable for such
action nor did she invoke on herself social censure.

Parents often enough compelled their daughter to marry a man she did
not love. But only a little less persistently did they also compel the son to
marry the girl he did not love. Only in the event that the girl was known

to have lost her chastity, or, still worse, given birth to a child out of wedlock, did she find herself a burden and a disgrace to her family and usually
unwanted as a wife by the young men of the village.

With her chastity inviolate a girl hardly faced any discriminations from
men in the years before her marriage. After marriage, as a bearer of
children, as a worker in the fields and in the barns, she maintained in the
home and in the village a status of equality of which men despoiled her
only if she was a person of frail health or of weak character. But the man
of weak character and frail health might likewise find himself dominated
by his wife.

Thus throughout the centuries the fact that the peasant woman was always a worker and a producer kept her status in the community and in the
home almost always on the level of equality with men—a condition which
the women of the Russian merchant and trading classes, for example, seldom knew. The Tatar invasion came and went, the *domostroy* proclaimed
in the time of Ivan the Fourth and imposing heavy strictures on the liberties of women came and passed; czarism came and vanished. But in the
village men and women continued to live and work and worship and associate in many other ways on a level of equality. Such equality, however
incomplete, appealed to the imagination of the Russian intelligentsia.
That is one reason why women were accorded a place of equality with men
in the various populist movements.

But whatever the historic cause, the fact is that in Russia the emancipation of women was inextricably interwoven with the emancipation of the
people. Unlike women in Western countries, in Russia they did not have
to fight *men* for the recognition of the idea and the principle of equality.
Bolshevism, with its doctrine of economic determinism, its appraisal of
men and women as co-producers in the factory, in the field, in the office, in
the laboratories, proclaimed from the very start the absolute equality of the
sexes and the right and the duty of women to make the most of this
equality.

Women flung themselves into the professions, into industry, into the
Party, into the Soviets. Severe and intense as the Kremlin dictatorship has
been it has never abated in the vigor with which it has sought to draw
women in ever-increasing numbers into careers. The Plans never could
have been as ambitious as they were without the contribution women
made to their fulfillment, in industry, in education, in farming, in the
Party, in Soviet administration. Without the earnings of women, Russian
families in those fierce and tumultuous days could not have fed and clothed
themselves as well, according to Russian standards, as they did. In the home
and in the family women have become even more of a pillar than they
had ever been. Propaganda for the idea that women must have careers and

that, in the words of Dobrolubov, "a tranquil and happy life threatens like a muddy abyss to suck you and swallow you," has been so all-pervasive that even girls in the kindergarten talk of the work they will do when they grow up.

Tolstoy's *War and Peace* is by far the most popular novel in Russia, especially since the outbreak of the war. School children read it as eagerly as adults. "I have been teaching literature," said a Russian woman, "for twenty-four years, and every year I reread *War and Peace*. When I was a girl I wanted to be like Natasha and ever since my daughter was born I have wanted her to be like Natasha as a girl." Powerful is the appeal of this novel to Russian men and women.

The warmth, the beauty, the capriciousness, the vivacity, and the romantic intensity of Tolstoy's heroine fascinate and stir the admiration even of high-school girls—that is, Natasha the girl does, but not Natasha the wife. They feel sad about or contemptuous of her overabsorption in domestic affairs after her marriage.

"She should have been like Vera Pavlovna," I heard a fifteen-year-old Russian miss remark. She was referring to the heroine of Chernyshevsky's *What Is to Be Done?* The idea of careers for women is as much a part of the mentality of Russian youth as collective control of industry. . . .

Women have as yet failed to attain the very highest positions in government and politics. Not since the beginning of the Soviet regime has a single woman been chosen to the all-powerful Politbureau, which actually rules the country. Only a few women have risen to the rank of commissary, which corresponds to membership in a cabinet. Yet outside of these two categories and the Army women have been conspicuous in all walks of life. In the Supreme National Soviet they hold 189 places. In the supreme Soviets of the various republics they number 1,436 deputies. In all the local Soviets—city, village, district, and provincial—there are 422,279 women. In the trade-unions they hold even more official positions than in the Soviets. In the Party they constitute 29 per cent of the membership, and they hold about one fifth of the executive positions in the government and in the Party.

High is their representation in all branches of industry and agriculture, and in the liberal professions. In 1939 over ten million women were employed in industry; among them were over 100,000 engineers and technicians. They made up more than one third of all the workers in industry. In agriculture they constituted before the war one half of the labor force; and tens of thousands of them have held executive positions such as agronomists, veterinarians, livestock experts, and chairmen of collective farms.

In 1939 four hundred women were station agents on the railroad; 1,400 were assistant station agents. The medical profession is crowded with women. In pre-Soviet days there were no more than 2,000 women physicians. Now of the more than 150,000 physicians in the country at least one half are women. In scientific institutes, in the law profession, in teaching, women are eminent and conspicuous.

The war has broadened and deepened their representation in industry, in agriculture, in the professions. In some factories outside of the textile industry, which has always been regarded the world over as almost a monopoly of women, they have easily replaced the men who were summoned to war and make up from two thirds to three fourths of the labor and executive personnel. In agriculture it is they who shoulder the chief, often the exclusive, burden of the work. Had it not been for the energy and the skill which women have shown in their work in the factory and on the farm Russia could never have mobilized the large and powerful army that she has.

In no other nation at war, Axis or Allied, do women hold so decisive a position as in Russia. They are anywhere and everywhere in the depths as well as on the peripheries of the far-reaching Russian land. No work is too difficult for them, no responsibility too irksome or too dangerous. At a gathering of physical-culture leaders in Moscow I asked a short and unimpressive-looking woman with flaxen hair and gray eyes what she was doing for the war.

"For eight years," she replied, "I have been teaching our border guards bayonet fighting."

The small group of foreign correspondents who heard her gasped with astonishment. She was from Leningrad, married, the mother of three children; and she appeared too little and too inconsequential physically to instruct picked troops as masculine and as desperate a mode of warfare as bayonet fighting. She did not mind doing it, she said. "There is no reason why a woman should not be teaching men the expert use of a bayonet," was her calm comment.

High above Novosibirsk, capital of Central Siberia, is the district of Narym. In czarist days it was chiefly a penal colony. In 1913 the cultivated land came to less than seven thousand acres. Narym did not grow enough grains and other foods for the natives and the exiles. Hunting, fishing, and gathering wild berries and nuts were the chief occupations of the inhabitants. In 1939 Narym became a leading agricultural district with a cultivated area of 373,855 acres. Liquidated kulaks and other settlers came there, and with sixty-horsepower caterpillar tractors they cleared forests and brought under the plow many lands. Now Narym grows a surplus of rye and winter wheat. The population also hunts the rich supply of wild game. Some of the most renowned hunters in Siberia live in Narym.

When the war broke out the hunters, who were among the best marksmen in the country, were needed as snipers in the Army. Nearly all of them went to the front. As in other places women found themselves in charge of the new large collective farms. Not enough of them had learned to operate the tractors, combines, and other modern implements. Nor were there many livestock experts and agronomists among them, and time was precious. Because of the short summers time is always precious on the Siberian farms. So in the village of Verkhny Yar, in the upper part of Narym, girls started a campaign to train themselves in the new tasks that faced them. Within a brief time 8,000 girls responded to the campaign and hastened to enlist in the special courses that were opened to teach them all they needed to know about combines, grains, livestock, and farm management.

Nor was Narym an exception. Similar campaigns swept other parts of the country. In 1941–42, out of the 370,426 tractor drivers that were newly trained, 173,794 were women, mostly girls. The others were chiefly boys below military age. Of the 80,577 combine operators, 42,969 were women. Tens of thousands of women became expert mechanics of farm machinery. And 1942 was a banner year in Russian agriculture.

In prewar days girls and young women in Narym had also been hunting, though not as extensively as men. Now they took up the profession with a fresh earnestness. Furs are an important item in national defense, and the government encouraged women to take up hunting. Of course shooting Siberian squirrel for furs is no easy task. One has to be a skilled marksman to hit the squirrel in the eyes so as not to injure the skin. Yet many girls and young women more than fulfilled the quota of 500 skins for three months. Tatyana Kayalova, a young mother, left her children in the nursery and gave all her time to hunting. So skilled and successful was she that she delivered to the government 300 skins a month, not one with the least mark of damage. Other young women in Narym equaled her record in squirrel shooting.

During the weeks that the Germans were advancing on Moscow and on Tula women by the tens of thousands were digging trenches, building barricades. Old and young, they worked with spades, with picks, with crowbars, with handcarts, with saws and axes. They worked day and night, in wind and blizzard, often under enemy fire. The hundreds of miles of trenches they dug and the endless lines of barricades they erected evoked the highest praise of generals and civilian leaders.

In the winter of 1942–43 Moscow and other large cities had to be heated by wood. Beginning with the middle of the summer, tens of thousands of women were mobilized for cutting wood. "Where is your Marusia?" I asked a woman executive in Moscow.

"Gone to cut wood," was her reply.

Hardly a housemaid remained in the capital. Not a young woman in acceptable physical condition who could be spared from other work but journeyed to the forests, lived there in a camp for two or three months, and wielded a saw and an ax to prepare wood for the schools, the hospitals, the homes of Moscow and other cities.

Hardly a task related to national defense but Russian women are performing with no less success than men.

Women are hunters and farmers, coal miners and steel workers, welders and locksmiths, masons and carpenters, mechanics and chauffeurs, painters and plumbers, rail-builders and lumbermen, sailors and aviators. They are not only in the rear but at the front. They are snipers and guerrillas, among the best in the country. They are army telegraphers and telephone operators. They operate anti-aircraft guns, machine guns, artillery batteries. There would be regiments, divisions, armies of them in the fighting forces if the government would accept them as fighters. They are wanting neither in energy nor in aptitude, neither in courage nor in self-sacrifice.

The drama of their work and their life we shall glimpse from the stories of Russian war girls I have known or of whom I have heard and read.

CHAPTER 28: **A MOTHER-IN-LAW**

HER NAME was Fedosya Ivanovna. She lived in a village on the Dnieper deep in the heart of the Ukraine. She was a haughty woman with a sharp tongue and a ruthless manner. The girl whom her son Misha was courting hated and feared her. So did other girls in the village. There was something strange and implacable in the way she ordered people around and sought to dominate them. Foreman of the village kolkhoz, she incessantly hurried and chided the people under her so they would work harder, achieve more. Naturally, they had no love for her.

But the kolkhoz administration appreciated her sense of discipline, her administrative ability, her tireless effort to lift the productivity of workers and boost the yield of the land. They awarded her a high premium—a new house of two large, well-lighted rooms, a kitchen, a vestibule. Fedosya Ivanovna was more haughty than ever. She boasted of her attainments, of the high recognition they had won her. She set herself up as an example for others to emulate, especially in their love of work, in their devotion to the common good, in their loyalty to the kolkhoz.

The discipline she imposed on workers in the field she enforced in her home. At the threshold she placed a little rug, and woe to the person who entered without wiping his boots or his bare feet. So proud was she of her home that daily she rose before break of dawn, swept, dusted, and washed, until floors and walls and furniture gleamed with cleanliness. Keeping her home bright and spotless was the supreme pleasure of her life.

Because she was so pretentious and severe she hardly had a friend in the village, least of all in her daughter-in-law, the girl whom her son had been courting. Living in the same house with Fedosya Ivanovna was a burden and a trial to the young woman—so she assured Yurii Slyozkin, the Russian writer, to whom she told the story of Fedosya Ivanovna. She often complained to her husband, but he was tolerant of his mother. After all, she *was* his mother. She meant well, he kept saying to his wife, and couldn't be any different from what she was. He counseled his young wife to be patient with the old woman.

In May 1941 a child was born to Misha's wife. There was much rejoicing in the family, and Fedosya Ivanovna was proud of her grandchild.

Then came the war. Misha was mobilized. As is the way of peasant women when their husbands depart for the front, the young wife followed him along the village street weeping and wailing. But not Fedosya Ivanovna. She scolded the daughter-in-law for showing so much grief. Grasping the young woman by the shoulder, she led her home and scolded her some more for breaking her son's already sad heart with loud tears and loud words. Not a man in this pretty village on the Dnieper was as self-contained, as ruthless with himself and others in this moment of national catastrophe as was Fedosya Ivanovna.

For once her daughter-in-law lost all fear of her, all control of herself, and lashed out with all the violent words she could muster, unloading the gall and the mortification that had long been smoldering inside her. Unaccustomed to such denunciation, Fedosya Ivanovna felt deeply wounded, left the house, and didn't return until late in the night. After this, though the two women lived under the same roof, they hardly spoke to each other.

So many men from the village had to go to war that only four remained in the kolkhoz—old men. Together with the thirty-four women and the children they worked the land. Day and night they toiled, and their toil and sweat yielded one of the richest harvests they had known. They lived in the fields amid the shocks and stacks of wheat, amid the buzz and roar of engines and threshing machines.

One day in August a man on horseback galloped into the village. He was from the district Soviet and he brought the saddest tidings the village had ever heard. The Russian armies were retreating and the village would have to be evacuated. The grain that was threshed must be hauled to another district; the unthreshed sheaves must be set afire. Cattle must be driven deep inland. Not only shepherds but milkmaids must go along in carts and with pails, strainers, milk cans, so that cows could be milked as regularly and as well as when they were on the kolkhoz. Everyone in the village, the Soviet messenger further advised, must secrete as much of his personal possessions as he could and make himself ready to leave.

Again there were wails, and again Fedosya Ivanovna remained cool and self-possessed. More, she announced that she wouldn't be evacuated. No, she wouldn't. The village was her home. It was the only home she had known or cared to know. She had roots in it and couldn't and wouldn't pull them up. She wasn't afraid of Germans. What could they do to her when she was no longer young, when land, work on land, was the only thing she knew and loved? No, she wouldn't go away, wouldn't let her grandchild—blood of her blood, flesh of her flesh—be taken away. They belonged to the village on the Dnieper and nowhere else in the world. So Fedosya Ivanovna remained in her home and her daughter-in-law, wishing to avoid a sharp conflict, likewise remained with the child.

Late one afternoon the Germans came. There were five of them. They roared into the village on motorcycles. Soon they discovered Fedosya's home and knocked at the barred gateway. Fedosya herself came out, opened it, and found herself facing a man who pointed a gun at her. Calmly she told the man she was no soldier but a woman—only one woman—and since one bullet was enough to finish her there was no need pointing the gun at her. The man put it away.

The Germans entered the cottage. The daughter-in-law and the child went into an adjoining room. She heard the Germans talk and open bottles of liquor. They drank one bottle after another. The more they drank the more loud and the more hilarious they became. They asked Fedosya for food and she brought it to them. They opened more bottles of liquor, drank more toasts to one another, and ate food—Ukrainian food, which Fedosya had willingly given them.

Then the daughter-in-law heard the Germans ask for more food and Fedosya say she would cook Ukrainian borsch for them. She built a fire in the oven and started cooking the borsch. The daughter-in-law chafed with indignation but dared not say a word in protest. Then she heard the Germans ask Fedosya for vodka and heard her say she would find it for them.

The daughter-in-law boiled with rage. Her own mother-in-law, a full-blooded Ukrainian woman, the mother of her husband Misha, who had gone to war to keep Germans from devastating their beloved Ukraine, was according five German intruders the hospitality reserved for friends and guests. She suspected Fedosya Ivanovna of dark intentions, the darkest possible. She hated her more violently than ever.

When she heard her go up to the garret she followed, and when she saw her mother-in-law lift a bottle from a trunk she was beside herself with fury. She wouldn't permit any vodka to be given to the Germans— vodka that was left over from her own wedding gifts and which she was saving for the day when the child would cut its teeth so she could rinse its mouth with it and alleviate the pain. Seizing the old woman's hands with all her strength, she whispered madly, "Let go or I'll kill you." The old woman clung to the bottle, wouldn't let go of it. "Stop, you fool, they'll hear you!" But the young woman was too enraged to care. So the two women struggled for possession of the bottle and finally the daughter-in-law succeeded in pulling out the cork, tipping it so its contents started to pour. From the odor the young woman knew it was not vodka. It was kerosene.

Her breath stopped. She withdrew her hands from the bottle and let Fedosya Ivanovna have it. . . . She didn't know what to make of her mother-in-law, but she was certain the old woman was no friend of the enemy. She was even more certain of it when Fedosya Ivanovna said, "I made you stay. It was a mistake. You must go. Take your son in your arms and run, wrap him in the downy shawl so he'll keep warm, and run— crawl out on the roof here—the ladder is in the corner, climb down and run."

For the first time in her life the young woman had no wish to resist her mother-in-law. Indeed she felt drawn to her as to someone who was really close to her and for whom she cared. . . . But why had she persuaded, yes, stopped her from leaving the village with the others when they were evacuated? Fedosya admitted now it was a mistake, but why had she done it? The young woman couldn't answer the question and there was no time to talk about it, least of all to reproach Fedosya Ivanovna for her error. Instead the younger woman said, "And what of you, little mother?" Never before had she addressed her mother-in-law so tenderly. With her wonted coolness Fedosya Ivanovna motioned with her hand and shut a near-by window.

In the dark of night the younger woman, with her baby bound close to her breast, made her way out on the roof. Quietly she descended the ladder and made for the near-by woods. She walked all night trying to get into Soviet territory. At dawn she found herself, to her dismay, back on the outskirts of the village. She didn't know how she had gotten back. She was certain she had plunged into the woods and was on her way farther and farther from the dreaded enemy. But she was an inexperienced woodsman and she must have taken a turn which only brought her close to his lair.

Then, remembering the bottle of kerosene, her heart sank. The more she thought of it the more terror-struck she felt. With her coolness and audacity Fedosya Ivanovna was capable of anything—yes, anything—even of—— But no, impossible! She loved the cottage too much—that's why she wouldn't leave it. . . . Still——

Overcome with foreboding, the young woman rushed back to the village, and when she came to the old home she felt as if the earth were slipping from under her feet. There was no house any more. There was only a mass of smoldering ruins with the blackened chimney sticking defiantly into the air. Not far away, lying on the ground, were the five motorcycles on which the Germans had roared into the village. But Fedosya Ivanovna was nowhere around. Nor were the Germans—they had all burned to death.

CHAPTER 29: THE GIRL WITH THE COUGH

HER NAME was Nina Bogorozova. She was young and small, with bright eyes and a persistent cough. She wore a soldier's greatcoat and felt boots. That's all I know about her personal appearance. Vadim Koshevnikov, the Russian war correspondent, who wrote of her in an obscure literary journal, tells no more.

She was sitting in a hut somewhere at the front. The hut was crowded with soldiers all wrapped in greatcoats and fast asleep. Her cough gave her no rest, and to muffle it so it wouldn't disturb the sleeping men she kept pressing her hand against her mouth.

She was waiting for dusk to come and for the soldiers to waken. She had already told the intelligence officer, a young lieutenant, the purpose of her mission, which was to guide a scouting party to a point in the enemy's rear where they could gather priceless information for the Red Army. The lieutenant questioned her at length and, convinced of the importance of the information, agreed to grant her request.

At dusk the lieutenant came and aroused the soldiers. There was no light in the hut, and the men washed, dressed, and ate in the dark. Nina ate with them, slowly, small mouthfuls at a time, as though it pained her to chew the food and swallow. When the men finished eating they rose; she still had half of her food on the plate, but she rose too. Chevdakov, the noncommissioned officer in charge of the scouting party, urged her to finish the meal, but she said she had had enough and was ready to leave.

The cold was intense, and they bundled up heavily for the journey; Nina took particular pains to wrap a woolen shawl snugly around her throat.

There was a full moon in the sky, and Chevdakov swore at it. So did the other two men, Ignatov, a Russian, and Ramishvilli, a Georgian. A full moon might be beautiful, but not on a night when men were on a scouting expedition into the enemy's rear. Its very beauty might spell death. But Nina was unperturbed and walked along in utter silence.

She was so small, her greatcoat was so long, the snow was so deep, her cough was so persistent, that Ignatov felt sorry for her and offered to take her by the arm. Stiffly she spurned the offer.

"Why do you want to do that?" she asked.

Ignatov was lost for words—he didn't expect such a rebuff. Ramishvilli, talkative and ebullient, came swiftly to his aid.

"In the Caucasus," he said, "it is customary for a man to be chivalrous to a girl."

"This is the front," retorted Nina coldly. Many a Russian Army girl is like that—severe to the point of ruthlessness.

Chevdakov reminded his companions that they were on a scouting party and talking was neither permissible nor wise.

At midnight they passed the front line. They were in a dense forest in enemy territory. Nina took over the leadership. With hands tucked inside the sleeves of her greatcoat, she walked ahead, the men following, all of them waist deep in snow. The exertion drained Nina's strength. She breathed hard, her heart pounded, her cough grew worse, and this annoyed the vigilant Chevdakov.

Soon they reached the end of the path. Ahead lay a valley brilliant with moonlight. They were so close to the enemy position that walking was dangerous. The moon might betray them and they would be picked off by German snipers. They could only crawl across the valley. Falling flat on their stomachs, they started to crawl. The farther they crawled the more weary Nina grew, but not once did she pause for rest or complain of fatigue. It took them an hour and a half to creep across the valley, and Nina was on the verge of collapse.

Luckily her guidance was no longer necessary. The enemy position was in clear view, and so Chevdakov ordered her to remain where she was and rest. They would rejoin her as soon as they had done their scouting.

When the men returned they were beside themselves with rapture. Never had a trip into enemy territory proved so rewarding. The information they gathered would gladden the heart not only of the lieutenant but of the general. Full of fervor and gratitude to Nina, the ebullient Ramishvilli, remembering how small and frail and sickly she was, proposed they carry her back in their arms. But Ignatov, still feeling the sting of the rebuff, urged caution. Yet the girl haunted Ignatov. There was something strange and beautiful and heroic about her. Turning to Ramishvilli, he said,

"Do you think she is married?"

Chevdakov came up noiselessly behind them and ordered them to stop talking. The enemy was too near and the night so still. So they walked on in silence, all thinking of the wonderful Nina, of the gallant service she had rendered them, the Red Army, the fatherland. . . . They could love such a girl. They did love her.

Soon they were beside her. She felt rested and together they started back for "home." Trudging through the snow as bravely as the men, Nina led the way, for only she knew the forest and the road back.

They came to the village of Zhimlost. It was held by the Germans. Through the woods, some distance away, they saw a party of Russian prisoners cleaning snow off the road while German soldiers wrapped in Russian blankets, in Russian kerchiefs, stood guard over them. They halted, and the excitable Ramishvilli started to grope for a hand grenade. Observing the movement of Ramishvilli's hand, the disciplined Chevdakov said:

"No action without orders."

Then to the dismay of the men, especially of Ramishvilli, Nina broke in and said resolutely:

"Such orders will never be given."

The men glanced at her sharply, unbelieving. This little girl with the bright eyes, the irrepressible cough, standing up as if in defense of the German guards.

"What do you mean, never?" asked Chevdakov.

"Just what I said. Never. Do you want them to kill us?"

Chevdakov felt like laughing.

"These twelve frozen Fritzes? Ah no, we'll get them all—one swift attack and they are finished—they are dead."

Nina wouldn't hear of it.

"And I say," she insisted, "we'll do nothing of the kind."

Now Chevdakov was angry. "No more discussion," he said. And turning to the men he added, "Get ready for action."

To their amazement Nina drew up to her full height, whipped the gun out of her leather pouch, and holding it before them she said in a low, firm voice,

"I'll shoot."

Now even the chivalrous Ramishvilli was indignant. Yet Chevdakov wouldn't brook her interference.

"Drop that stuff," he said. But he didn't know Nina. Standing across the road so as to bar the way, she pointed her gun at her companions.

Beside himself with fury, Chevdakov leaned over and with a swift thrust of his arm knocked the gun from Nina's hand. The mere sight of the Russian prisoners cleaning the highway for German trucks on their way to the front with gasoline, tanks, ammunition, other instruments of death, cried for action and revenge. Chevdakov was in no mood to be stopped.

"Don't you see," he snarled, "how our people are suffering?"

Nina was unmoved. She was without a gun but she was not helpless.

"Don't dare," she said. "I'll scream."

Scream! So the Fritzes could hear her, turn on them, mow them down! That's the kind of girl she was—this Nina, this innocent-looking, malevolent Nina! But there was no mistaking the fierceness of her resolution.

Chevdakov lifted her gun out of the snow and turning to the men said disconsolately:

"This hysterical female. I guess we'll have to obey her. Otherwise—we stand a good chance of losing our precious information, perhaps our lives."

Ignatov, who only a few moments earlier glowed with tenderness and love for the girl, was bursting with rage.

"Louse!" he snorted.

The romantic Ramishvilli muttered in disgust:

"Bad girl, very bad."

Nina swallowed the insults without a word of reproof.

In gloomy, angry silence they resumed the journey. Now the men hated the girl. They wouldn't speak to her. When she slipped and fell they wouldn't offer her a helping hand. She was an enemy—their enemy, that's what she was.

On their arrival at staff headquarters Chevdakov told her to go to the hut and get some sleep. Nina did so, and the men went to make their report to the intelligence officer.

The young lieutenant was more than overjoyed with the report. Not in a long time had such momentous information come to him, and it was all a result of the proposal laid before him by a strange, sickly little girl who had come from the village of Zhimlost. He was happy he had trusted her, very happy. He asked about her.

Glumly Chevdakov said: "She gave us plenty of trouble." He told the lieutenant in detail what had happened. The longer he talked the redder in the face grew the lieutenant. It was obvious that he was smarting with indignation and exasperation. When Chevdakov finished he said severely:

"She was right, absolutely. You had no business jeopardizing the information for the sake of a fight with a dozen Fritzes."

Chevdakov was stunned. So were his men. The lieutenant was upholding the girl who had come close to forfeiting everything—not only the information but their lives and her own. The lieutenant went on:

"Haven't you noticed that she coughs all the time?"

Chevdakov nodded.

"Have you seen how lacerated her neck is?"

Chevdakov shook his head. How could he have seen it when she had a heavy scarf around her neck?

"Two days ago," said the lieutenant solemnly, "in that very village of Zhimlost, the Germans had her on the gallows, and guerrillas barely saved her life."

The lieutenant spoke no more. He dismissed Chevdakov and his men.

The three soldiers walked outdoors. For some moments they didn't speak. They didn't know what to say to one another. Then Ignatov proposed that they go to the girl and offer her an immediate apology. Ramishvilli heartily concurred in the proposal. But Chevdakov was deep in thought. Words never held much meaning for him. It was acts that counted.

"Well?" asked Ignatov. "Shall we go to see her?"

"Yes, let's go," burst out the irrepressible Ramishvilli.

Chevdakov shook his head.

"Don't you want to?" asked Ramishvilli with disappointment.

"Not now," said Chevdakov. Then he added soberly, "We've got to go first to Zhimlost and take a good look at the Fritzes."

There was silence, tense with reflection.

"Maybe," said Ignatov, "she'd rather have it that way."

"When we come back," said Chevdakov, "we'll shave and wash and put on clean collars and go to see her. But first we must take a real look at the Fritzes. Of course we'll lose a night's sleep. But that's nothing. We've done that before and we can do it again. To Zhimlost, boys."

The three men turned and marched off in the direction from which they had just come.

CHAPTER 30: CAPTAIN VERA KRYLOVA

DURING the German drive on Moscow the company in which Vera Krylova served found itself encircled by the German pincers. There was panic among the men. Vera, holding the rank of captain, was the highest officer left alive, and though she had never led in battle or commanded men, she leaped on the back of a horse and by her courage and strategy fought her way through the pincers and brought her men safely to the Russian main lines in the city of Serpukhov.

In the Red Army and among the youth of the country she is one of the romantic legends of the war. So when my telephone rang and I was informed I could meet her I hastened to the place of appointment.

"The girl with the pigtails," was the way Marjorie Shaw, a British journalist, had christened her, for in all the pictures we had seen of Vera Krylova in newspapers and magazines the feature that instantly caught the eye was her pigtails. A girl with pigtails and a redoubtable soldier—the essence of femininity and masculinity dramatically blended in this young and renowned girl.

"You've been quite a successful soldier," I said to her.

"Not so very successful," came the modest and calm reply.

I had heard many estimates of the number of "Fritzes" she had disposed of with her own hands. I mentioned some of these estimates and asked if they tallied with her own records.

"I have accounted for plenty of them," was her answer.

The calm with which she spoke emphasized the hardness that has come over Russians when they speak of German soldiers. Yet it seemed incredible that "the girl with the pigtails," who celebrated her twenty-first birthday on October 19, 1942, had taken the many German lives with which she had been officially credited. In her handsome captain's uniform, with her large blue eyes and luxuriant dark hair accentuating the smoothness and paleness of her face, there was nothing severe or formidable about her. To men in the Western world the expression "woman soldier" is likely to evoke the thought if not the image of an amazon. Yet there is scarcely a hint of the amazon in the stature or appearance of Vera Krylova, particularly when seen without the loose, bulky Russian greatcoat that always adds inches to the size of the wearer. As I saw her sitting before me without military cap, with the pigtails falling over her finely curved white neck, her face unscarred and pale with fatigue, I could hardly imagine her in hand-to-hand combat with a German officer and winning the battle of life and death.

She was above medium height and her back was broad. Otherwise, with her heavy dark brows, her vivacious manner, the extraordinarily thick and shiny black hair tumbling over her forehead and coming down in short braids over both cheeks, she seemed too lovely, too kindly, too feminine, too life-loving to care for the meanness, roughness, and cruelty of war.

"Do you dance?" I asked.

"I love dancing. I can dance all night," and she proceeded to ask questions about dancing in America, which to the Russians, because of American jazz, remains the classic land of social dancing as much as the classic land of the modern machine.

"What other recreations have you?" I asked.

"Theater, ballet, parties, concerts, books, and of course outdoor sports."

VERA KRYLOVA

She laughed the gay, rolling laugh of a person eager for play, fun, adventure.

Yet she was a celebrated soldier with two military decorations and four wound stripes on her tunic. One of the wound stripes was yellow, indicating a serious injury. I asked her about it. In the course of battle a piece of metal had penetrated her back and remained there. She came to Moscow for treatment. A well-known surgeon told her he would have to perform an operation. She asked how long it would take to recuperate.

"From three to four months," said the surgeon.

"No," she said. "I cannot be idle that long. I want to go back to the front."

She is postponing the operation until the end of the war—"That is," she added, "if enemy bullets keep on sparing me. If they don't——" She smiled, shrugged her shoulders, and did not finish the sentence.

Vera Petrovna Krylova never attended a military academy and never had any wish to make of herself a professional soldier. If she's alive when the war is over, she expects to take off her uniform, put away her holster, and return to peaceful civilian life. The daughter of a factory worker in Kuibyshev, she displayed from her earliest years a passion not for guns but for books and social work. At first she thought of medicine as a profession and became a *feldsher*—one whose knowledge of the art of healing is equivalent to that of a trained nurse in America or England.

But tending the sick quickly lost its appeal for her, and she subscribed to a Leningrad institution for a correspondence course in history and geography. On completing this course she became a schoolteacher in the Ukraine, teaching geography and history, the subjects she had studied by correspondence and which to this day remain her favorite intellectual pursuit.

She went to Siberia, taught history and geography there, and was promoted to the position of district supervisor of education. She enjoyed executive work and saw herself rising to a career of distinction in education. But as in the case of millions of others the war interrupted her peaceful life. As soon as she heard Molotov's speech announcing the war, she offered to enlist in the fighting services. Her application was refused—too many girls were volunteering. Impatient and indignant, she wrote to Andreyev, of the Central Committee in Moscow, protesting the refusal and appealing to him to intercede in her behalf with the military committee in Novosibirsk, where she was then living. The letter brought its reward, and Vera Petrovna was taken into the Army but only into the medical corps. From first-aid nurse and stretcher-bearer she rose rapidly to the position of medical inspector of the regiment with the rank of captain.

She brought to the Army an array of talents and experiences that fitted well not only into her medical work but even more into the profession of soldiering. She had always been a lover of sports..Her grandfather lived in the Caucasus, and while visiting him during vacations she had learned to ride horseback. In time she had become a skilled and daring horsewoman. In the rifle ranges in school and on the playgrounds she had acquired competence as a sharpshooter and had won the Voroshilov badge for marksmanship. She skied and skated and excelled in both sports. The winter preceding the war she had won second place in a ski tournament in Siberia; that was high testimony to her skill in that sport because Siberia, with its deep snows and long winters, is a land of magnificent skiers.

Then, too, she was endowed with a true sense of mechanics. Within a short time she acquired knowledge of machine guns, automatic rifles, pistols, and bayonets. She watched, studied, and trained in many branches of fighting.

As a stretcher-bearer and medical attendant she often worked only thirty yards away from the German line. Flares blazed all around her, bullets whistled, shells screamed; but Vera hugged the earth like a worm, crept up to a wounded man, dressed his injuries and, grasping him by the coat collar, dragged him to a crater or ravine and crept forward once more to the firing line to bring out another stricken soldier.

"Darling little sister," the men used to say to her, "thanks, thanks."

Their words poured energy into her and made her feel as though no task were too difficult or too dangerous.

Young and eager, strong and fearless, imbued with a sturdy zeal for work and sacrifice, the Siberian schoolteacher of history and geography carried hundreds of wounded men out of the firing line.

"My mother tells me that I deserved to die at least a hundred times—but," and she laughed, "bullets seemed afraid to touch me. I've been very lucky."

Then came the most momentous and most stirring experience in her fighting career. Only a person of enormous courage and brilliant military skill would have dared assume the responsibility that was suddenly thrust on her.

The German advance on Moscow was in full force. The company in which she served was retreating with its regiment, with many Russian regiments. The German long-armed pincers were pressing forward with astounding swiftness, and Vera's company found itself completely cut off from the main Russian armies. It was surrounded! The commander and the commissar were struggling to extricate the company from encirclement. With their caravan of supplies they were journeying eastward. For six days they marched and drove themselves through mud and swamp all in the effort to elude contact with German forces. It had rained hard and they all were wet, but they never ceased to press forward.

Vera was sitting in a rattling cart in the rear of the caravan. She had been wounded by a stray bullet and felt weak and miserable. Wrapped in a greatcoat and blanket and wet with rain, she sat with her eyes shut, her mind a blank, hoping she could doze off and get some relief from her misery. They were approaching a village and suddenly there was shooting. From a belfry shielded by a birch tree German soldiers were cracking away with automatic rifles. The commander of the company was killed. The commissar rushed over to Vera and ordered her to destroy all her supplies. They could no longer move along as a company, he explained, and would have to break up and save themselves as best they could in groups or individually. The supplies which they could not carry they would have to destroy so the Germans would not seize them. Shocked by the commissar's order, Vera asked if he meant she was to destroy her medical stores. The commissar never answered the question. A bullet hit him and he fell over before Vera's eyes.

The situation was desperate. The soldiers were leaderless.

Instantly Vera forgot her chills, her wound, her misery. Panic had already seized some of the men and they were running toward the woods. The Germans were hurling fire and death at them. Unless checked instantly, the panic would spread with flamelike swiftness and they would all be captured or killed. There was not a second to lose. The situation cried for instant, decisive action. Though in the course of her duties as nurse and medical inspector Vera had had occasion to use rifle, hand grenade, and pistol, she had never been an active soldier on the firing line. Nor had she ever commanded men in battle. Yet now she was the highest officer in the company; all the others were dead.

Without a second's hesitation she leaped on the back of a horse, whipped the gun out of the holster, and brandishing it before her cried out, "Follow me!" and galloped forward into battle.

Her act electrified the lost and undecisive men. They darted after her. Some of those who were taking to the woods turned back and joined in the advance. No one thought of her now as a girl medical inspector or a nurse. She was leading them against the attacking Germans.

Vera quickly surmised the strategy of the Germans. The village they were holding was sheltered on two sides by forests; it was their aim to outflank the Russians and prevent them from escaping into the woods!

Vera broke up her company into two units and directed each to assault with all fire power at their command the creeping pincers that were seeking to outflank them. Men whom she knew by their first names were falling at her side. If only she had time to dress their wounds! But now she was in command of the fighting. She had not a moment to spare, and she hardened herself against the cries and groans of the wounded. Her horse was shot under her. She leaped on another horse and continued to ride about and cheer her men and give orders.

So successful was the counterattack that the Germans fell back, though the automatic riflemen in the belfry continued to shoot. Only artillery could silence them, and Vera had no artillery.

Yet it was something to have driven off the Germans' main fighting force and to have turned them into flight. Vera knew only too well that the Germans would soon return with reinforcements and would launch a new offensive. She could save her company only by plunging into the forest. Turning over the command to a soldier named Petrunin and ordering him to keep the Germans off the highway at all costs, she galloped away to the caravan and hastened to organize it for departure into the woods.

As she rode around inspecting the carts, she observed several trench mortars and cannon. Quickly she had these guns dismounted. Meanwhile, other Russian soldiers, likewise encircled and in retreat, came up the road. They had artillery. Vera ordered them into action, and directed their fire on the belfry. The big guns boomed, and soon the belfry and the birch tree flew into the air. The Germans were puzzled. They had not expected Russian artillery in action. They fell by the scores, but they turned around and fought back.

Inflamed by the success of their artillery, Russian infantry and cavalry fired and slashed away, and the Germans once more rolled back. Their attempt to surround the Russians and cut them off from the forest failed.

Vera hastened to prepare for retirement to the forest and to move eastward in a roundabout way before the Germans who were closest to their main armies could return with a larger force and again engage them in battle. As she was rounding up the men she galloped into the village to recall Petrunin and the men under him whom she had sent there on a mission. Of course it was a careless act. She could have sent someone else on the errand. But she had never particularly cared for caution when her own safety was involved. This time she paid for her recklessness with the most unexpected and most hair-raising experience she had known in all the sixteen months of her soldiering.

As she was entering the village a German officer with five men ran out of a peasant house and surrounded her. Here she was on horseback with plans all made for further retreat eastward, with the caravan already moving into the forest—and she a German prisoner with five German soldiers pointing their guns at her! The officer seized her horse by the halter and the soldiers dragged her to the ground. The officer ordered her to go into the house out of which he and his men had come. Bluntly she refused to go. She knew that once inside the house all thought of escape or rescue might be futile. She was at the mercy of her captors, and she decided to fight it out with them in her own way.

They started to drag her toward the house. Still she resisted more and more fiercely. With the butt of a gun she was struck such a violent blow that three of her teeth fell out. Blood filled her mouth, and though she could hardly stand on her feet, she gathered all her energy and emptied the blood on the officer. She was beaten violently and lost consciousness.

But again luck came to her rescue. One of the Russians in the village, a soldier named Shulbanov, saw what was happening and with his automatic rifle he picked off the Germans. When he joined her she asked for a drink of water. The drink refreshed her, and she and Shulbanov rode away into the forest and joined the company. Though her mouth ached and her body smarted from the beating, she turned to her men and said, "Sing, boys." She, too, sang, and they all felt heartened and happy.

"It's amazing what a song can do to men after a battle," Vera Petrovna said. "It lifts their spirits, makes them gay and confident, and prepares them for the next fray with the enemy. I never fail to call on my men to sing as soon as a battle is over."

The first encounter with the enemy was now over. They were in the forest and for the moment out of danger. But they were still inside the vast German pincers. All surrounding villages were in the hands of Germans, and in many of them large German garrisons were stationed to combat guerrillas and to round up and liquidate such companies as she was commanding. Vera and her men could save themselves by following the compass eastward and avoiding contact and communication with villages. Yet this was not always possible. Now and then a village lay directly in their path. There was no way of avoiding it.

Presently they came to such a village. Vera knew Germans were holding it. But how many of them there were and what kind of armaments they had, she did not know. She sent two soldiers to scout around and bring back the needed information. The soldiers never came back. What was she to do now? Wait longer? Suppose the Germans had made them "talk" and learned how comparatively weak and small was her army? After all, they could always bring powerful reinforcements, but she was cut off from the main Russian army and could hope for no outside aid. If she couldn't avoid a fight it was best to force it on the enemy and take him by surprise before reinforcements could reach him.

Together with seven men she went to the village and ordered the company to follow slowly. On reaching the village, a dog barked and she shot him. On beholding a little light in a window in the third house from the end of the street, she made for it. She was certain that there were no

Germans in the house; no Germans would openly expose themselves to discovery and attack.

Entering the house, Vera saw an old woman. So glad was she to see Russian soldiers that she started to weep. Vera asked if she would boil the samovar for them. The woman gladly agreed but warned them there was a machine gun on the roof of her house! Vera did not believe it and went out to investigate. She was greeted by a volley of machine-gun fire. In the darkness the Germans could not take proper aim and neither she nor any of the others were hit.

Her column was in the village and answered the fire. Many of the Germans were asleep but the firing woke them. They ran outside with their weapons. The night was pitch dark, but the fighting went on, wild and scattered, Russians shooting, Germans shooting, Russians running, Germans running, weaving in and out of each other's lines. It was a dangerous affair and Vera leaped on her horse and started to bring order out of chaos. Again she set artillery into action. That never failed to impress the Germans. The big guns boomed and blasted away with shattering effectiveness. The fight lasted long, and when it was over Vera commanded the company to pick up the wounded and hurry into the forest. As soon as they were beyond the reach of the enemy she dismounted from her horse, and though she was weary and knew her men were weary she called for a song. The woods resounded with the sprightly tunes of Russian marching melodies.

Even while they were singing a part of her men came along with a haul of prisoners.

"We've brought you a gift, comrade captain," said one of the men.

One of the prisoners carried a large birch cross on his shoulders. He was a gravedigger and was evidently going about his task of burying an officer and putting a birch cross over his grave when the Russians stumbled on him and took him along.

Of the seven prisoners two were officers.

"What did you do with them?" I asked Vera Krylova.

"What could we do with them?" she answered. "We were encircled. Our food was low. We were on the march. We had no way of accommodating prisoners or delivering them to our headquarters. In a moment of crisis they might do us deadly harm. So—well—Fritzes are Fritzes."

They rested in the forest and started marching again. The men were in good spirits particularly because they had artillery. Besides, Vera never allowed doubts of eventual success in their quest of deliverance from German encirclement.

"Nichevo; doydyom, rebyata [Don't worry, boys, we'll get there]!" was her fighting message and her marching slogan.

The farther they journeyed the more stranded Russian soldiers they picked up. Their numbers swelled but their supplies were running low. If only there was a way of replenishing them!

Again they came to an unavoidable village. They soon learned that there were no Germans there or anywhere within two miles. The village looked as though a tornado had swept over it—so empty was it of life and substance. As they were investigating it a seven-year-old boy made his appearance and told them the story of the German revenge on the village. They robbed it of everything, and when people resisted they murdered them. Five Germans entered his home and started taking ikons and pictures off the walls. The boy's mother protested and they started abusing her. The boy crawled under a bench in the corner; that was how he saved himself.

This boy went with Vera all along the village and pointed out the bodies hanging from trees.

Vera ordered a military funeral for all the dead. in the village. Soldiers dug a common grave, and the bodies were laid together. They were covered with soldiers' raincoats.

"I delivered a speech and you should have seen the men—the hate and the determination in their faces. We fired a rifle into the air in honor of the dead—fired it three times in the direction of the Germans."

They resumed their march and soon came to a river. Here they had their longest and toughest battle. It lasted twenty-three hours. Again luck was on their side, for the German forces were small and some distance from their main lines. But they fought stubbornly and sought to prevent the Russians from crossing the river.

The Russians knew that unless they made the crossing they were lost. They also knew that once they were on the other side of the river they would be out of the German encirclement. So they fought with desperation until the Germans fled. Sitting on her horse, Vera finally plunged into the water and cried out, "Follow me!" Russian infantry and artillery immediately started after her. They crossed the river. In their flight the Germans abandoned a truckload of ammunition and armament which the Russians were only too happy to appropriate.

This was Vera's third and last battle with the Germans. The road was now clear. They were near Serpukhov. After two weeks of marching and battling, of mud and blood, they were free of encirclement and they sang again, this time with heightened exultation.

Some distance away from Serpukhov Vera halted her company. She would not lead it to staff headquarters in its disheveled and dilapidated condition, so she gave orders for the men to wash, shave, clean their boots,. and brush their clothes. The cooks were ordered to bring out the field kitchens and prepare as sumptuous a meal as their provisions permitted.

While the camp buzzed with talk and mirth and the field kitchens crackled and blazed with freshly made fires, Vera herself galloped away to Serpukhov to report to General Zakharkin at staff headquarters. The guard would not let her see the general. Angrily she shouted at him and he shouted back, and the loud altercation brought the general outdoors. At sight of the girl commander, he burst out laughing. While on her way to Serpukhov her pigtails had come apart, as they often did in battle, and her hair fell all over her face and shoulders. "I must have looked very funny," she said. Her greatcoat was torn and full of bloodstains, and she limped as she walked. But she did not lose her composure, and standing at attention she reported to the general her march through the wilderness, her battles with the Germans, her emergence from encirclement. At once the general issued orders to supply her column with fresh bread, cigarettes, and drink; and he gave her directions where to set up camp in Serpukhov.

She galloped back to her men and pushed through her cleanliness campaign. Her men would not march through the streets of Serpukhov unless they made themselves as smart and as dignified as an army on parade should be even in wartime. For two days the company scrubbed and brushed and polished and sewed. Then they formed into line and marched into the town—infantry, artillery, the supply caravan—and "it was such a joy to see them," she almost chanted, "after those two terrible, exciting, heroic weeks." They set up camp and for three days they rested and feasted. They had not only cigarettes and spirits but hot food, chocolates, and entertainment. She was proud and happy, and it was then that she was awarded the Order of the Red Banner.

But the triumph of this company did not mean the end of fighting for Vera. During the Russian offensive in the winter of 1941–42 she marched as captain of a guard division with the advancing column. On her arrival in Kondrovo she joined a battalion of ski troops and was with them when they piloted their way deep into the German rear to encircle a German unit at the railroad station. This was one of the most strenuous marches she had ever undertaken. It led through frozen swamps and rivers and newly blown snowbanks, but she kept up with the men to the end.

In one battle near the village of S—— she saw the commander's adjutant on horseback and beside him a riderless horse. She asked whose horse it was, and the adjutant said it was the commander's.

"Where is he?" she asked.

"Dead," was the reply.

"No," she protested, loath to believe that Colonel Brynin was no longer alive.

"On my word he is," said the adjutant. "He was just struck down and his body is in the field and it is impossible to recover it."

Instantly Vera mounted the riderless horse and galloped away to the battlefield. When she reached the shrubbery where the soldiers were holding the line she dismounted from the horse and asked where the colonel's body was.

"See that black spot over there in the battlefield?" replied a soldier. "That's it."

It was a dangerous trip to make. The battalion commander advised Vera not to attempt it, for the field was in clear view of German vision and German guns. Vera disregarded the advice. Dressed in a white gown, she crawled over the white snow and finally reached the stricken colonel. He was unconscious but still alive. Lying flat on her back, she dressed his wounds; then, taking him by the sheepskin collar, she crawled back.

Had she been on hand when he was first wounded he would have been easily saved, so Vera assured me. A bullet struck him in the arm and he fell. In the effort to rescue him the adjutant lifted him from the ground, "which he never never should have done," said Vera. "When enemy guns are near a wounded man must never be lifted, or he becomes a ready target." The Germans hit him again—this time in the leg. Once more the adjutant committed the unpardonable error of lifting him, and the Germans fired a third time and struck him in the head. The adjutant thought he was dead.

Vera got him safely to the Russian lines, put him into a conveyance, and hurried to headquarters. Chief Surgeon Kazantzev examined him, but it was too late. The commander died in Vera's arms. That very day he was awarded the Order of Lenin, but he did not live long enough to receive the award.

Winter passed. Summer came. Vera continued active duty. The last battle in which she participated before she came to Moscow was on September 18, 1942, on the western front. The Russians had withdrawn temporarily from a certain sector, but she remained to gather the wounded. She knew that the Germans would leave them to die or might even hasten their end. She crawled from one man to another, dressing his wounds, cheering him, promising evacuation from the battlefield. Several Germans came out, and Vera at once lay very still, watching every move of the Germans. She saw two men coming closer and closer. They bent over some Russian bodies and took out the watches. She had an automatic rifle and could have disposed of them, but that would have invited instant and stern retaliation. So she lay still, wondering if they would find her. The

Germans stepped along leisurely and seemed oblivious to interference with their movements or to possible danger.

The field was scattered not only with dead and wounded but with ammunition. Surveying the field, Vera discovered near by several hand grenades and anti-tank bottles of kerosene. Cautiously she raked them together and waited. Soon she heard behind her the roar of Russian tanks. Her heart leaped with joy. But German tanks rushed out to meet them. One was moving closer and closer to where she lay. She seized a bottle of kerosene and waited. When the German tank was within striking distance she flung her bottle violently and the tank caught fire. She lay down again—holding a hand grenade ready for use on Germans or herself, depending on the situation. But the Russians pushed on and she rose and joined the advancing column.

"Does your mother know everything that's happened to you?" I asked.

"No, not everything—only what she reads in the press."

"Isn't she worried?"

"Of course. She keeps telling me the bullets are not always going to spare me. But you see—my father is at the front, my brother is at the front, my sister is at the front. We are a fighting family."

"Are you going back to the front?"

"Of course. I am postponing the operation on my back until after the war is over. I've got to fight."

"And what'll you do when the war ends?"

"Become a schoolteacher again."

"And get married?"

"Yes, of course!" She laughed gaily. "A husband, children, a career—that's my idea of a woman's place in the world."

PART SIX: RUSSIAN CHILDREN

CHAPTER 31: LITTLE PATRIOTS

EXCEPTING INVALIDS there are few non-combatants in Russia. The man at the plow feels he is as much a part of the nation's fighting army as the soldier. The girl who cuts wood for the Moscow factories feels as much a warrior as Ludmila Pavlichenko, whose superb marksmanship has ended the lives of over three hundred German "supermen."

This universal consciousness of war and of the need to fight it to the bitterest end has communicated itself to children as well as to adults. I never visited a school or went for a walk in Moscow but I was made aware of the extraordinary part children were taking in national defense.

Soon after my arrival in the capital I came on a group of children in a side street who were leaning over a huge sandbox. They talked so excitedly that I stopped to watch them. "What are you doing?" I asked a lively, blue-eyed little girl of no more than six. "Putting out bombs," was her reply. "But I see no bombs," I said. Lifting a little round stick with rings of birch bark she answered, "Here it is," and with a scoopful of sand she proceeded to demonstrate how to put out a bomb.

Presently other children in the street gathered. Older folk also stopped. There was much spirited talk and discussion as there always is when a Russian crowd assembles. From this talk I learned of the momentous part children played in extinguishing incendiary bombs and preventing fires in Moscow in the days when the Germans were making nightly raids on the city. Very young children were of course barred from the work, though often they managed to slip out of air-raid shelters and join the older children. But those of fourteen and older organized themselves into brigades and stood guard on roofs, in courtyards, in the streets. Neither sirens nor

317

explosions bothered them. They showed not the least manifestation of fear. Bombs by the thousands rained from the skies—all manner and sizes of bombs. But, soldierlike, they remained at their posts. They ran from one incendiary bomb to another and hastily put them out. They got so they could get hold of a bomb by the "tail" and dump it into a barrel of water.

Endless were the stories I heard at that street gathering of the courage of children. Some of them were blown to pieces. Some lost an arm, a leg. But these accidents did not stop others from going on with the task. If Moscow shows comparatively little damage from fires, it is in no small measure due to the extraordinary vigilance and heroism of school children who night after night remained on guard and put out thousands of incendiary bombs as soon as they fell.

But fighting fires is only one of the multitude of ways in which Russian children actively participate in national defense. Recently I visited one of the leading schools in Moscow. The principal showed me into a room that was lined with sewing machines and other accessories of a tailor shop. "In this room," she said, "our children sew for the Army."

They sew underwear, shirts, kits. They mend blankets, socks, uniforms for the wounded. They do this not only in Moscow but all over the country. In the city of Kuibyshev school children mended for only one hospital 450 pairs of socks, 120 blankets, and 275 shirts. School children also gather gifts for soldiers—from safety pins to safety razors and books. In every school in Russia the pupils have conducted a number of energetic campaigns for such gifts.

Food is a serious problem. There is plenty of land but it has to be worked; so in the summer of 1942 millions of school children, instead of going to camp, went to farms. They pulled weeds and hoed. They gathered for the Army mushrooms, berries, and sorrel leaves, from which Russians make soup. They collected thousands of tons of medicinal herbs. In one northern district they raised 45,000 rabbits for the soldiers, and in another, in Siberia, they caught thousands of tons of fish. They planted gardens and grew cabbage, cucumbers, potatoes, and onions for the Army. In city and village they constantly collect junk for armament factories.

They do more—in a variety of ways they help in the care of the wounded. They visit hospitals, wash floors, run errands, write and read letters for disabled men. They read stories to them, recite poems. They play, sing, dance, and seek in other ways to divert and entertain them. They visit the wives of these men or of men at the front, help them in the care of babies, in housework, or stand in line for them at the shops. They do all this readily, gladly, not only in the capital but all over the land.

But the most astonishing thing they do is military espionage of one kind or another for the guerrillas and the regular troops. The Army, of course, bars children from regular service, but they come anyway—boys and girls not only of high-school but of grammar-school age—with priceless information, which the officers are only too glad to get. The guerrilla brigades take children as active members. Often enough the children join together with their fathers, mothers, older brothers, sisters. In fact, they make exceptionally useful and brilliant guerrillas.

On one occasion guerrilla scouts observed the movement of German tanks and troops over a main highway. There were so many enemy tanks and troops that the guerrillas did not dare engage them in battle. But they wanted to delay their advance, and the only thing they could do was to destroy a long wooden bridge over a river several miles away. The highway was so closely patrolled by German soldiers that it was impossible for an adult to make the journey to the river. Then two boys volunteered. With fishing poles over their shoulders and dressed in old, loose-fitting clothes, they started out supposedly to catch fish. In their pockets they carried cans of kerosene. On reaching the bridge they poured kerosene over the beams and pillars and set them afire. By the time the Germans reached it the bridge was impassable.

The Germans, as is well known, make constant use of parachute troops. Sometimes they land them in large detachments, sometimes in small groups, or a person at a time. It is important for the Red Army to discover these troops as soon as they descend to the ground and to annihilate them before they have time to carry out their destructive missions. Children have proven among the most vigilant searchers for enemy parachutists and enemy scouts. Outside large cities no stranger, especially if he speaks Russian brokenly, is safe from the suspicion and scrutiny of children.

In a village in a western province a squad of girls were weeding a field when they observed an unknown man pass by. They spoke to him and he came over and tried to make friends. In the course of conversation he made inquiries about a certain village. At once the girls grew suspicious. But they pretended to be friendly and talked and jested with him. Two of them arose and said they were going after water. They went instead to the village and reported the stranger to army headquarters. Soldiers hurried to the field and arrested the man. He proved to be what the Russians call an enemy "diversionist."

In another village a stranger wearing the uniform of a Russian soldier approached a boy named Vasia and said he had lost his company and was trying hard to rejoin it. He asked the boy if the company which had been stationed in the near-by village had already left. Instantly Vasia grew suspicious and said the company had left but indicated the opposite direction from the one the company had taken. To Vasia's amazement the stranger threw up a flare. Without a moment's thought Vasia started to run. The stranger fired and wounded him in the hand. But Vasia did not stop running. Out of the clouds, in response to the flares, appeared a German transport plane, and parachutists started to tumble out of it. Vasia ran faster and faster. On reaching the village, he told the peasants what had happened. The peasants, together with soldiers, went out to the fields and rounded up all parachutists.

Russia is a country of over one hundred nationalities, and the remarkable thing about the children is that their war-mindedness and militancy are not confined to Slavs, who make up about four fifths of the population.

Here is Aslan Sumbatov, a Circassian living in an *aul* (settlement) in the North Caucasus. He is not yet thirteen, but he is a patriotic little citizen; and when the Germans came to the peaceful Caucasian auls and devastated and burned them, Aslan's Circassian blood boiled with wrath. He vowed vengeance and went scouting for the Red Army.

One day, as he was lying in a grove of trees searching the sky with his eyes, he saw a plane pass over him. His ears were keen, and from the sound of the engine he knew it was not a Russian plane. He watched it circle around and around and then he saw four little black bundles tumble out into the open. The bundles kept growing larger and larger and were falling closer and closer to the earth. Soon more bundles tumbled out of the plane. Aslan counted eighteen in all.

He knew they were German parachutists, but before reporting them to the Red Army he wanted to learn all he could about them. He crawled closer and closer and counted them again. There *were* eighteen of them and each carried a short, stumpy black rifle. Aslan watched them until he saw them settle in a ravine. Then he started to run. Fighting his way through brush and briars which cut his face and legs, he ran without let-up, looking back now and then to make sure he was not being pursued. At last he spied a group of Russian soldiers armed with automatic rifles and belts of hand grenades. Aslan made straight for them. A young lieutenant named Beloborodov was in charge of the column. Breathless with excitement, Aslan informed him of the parachutists, giving all details as to how many of them there were, their armament, and the place where they had hidden.

To the lieutenant it was clear that the mission of the parachutists was to blow up Russian communication lines so that reinforcements could not reach the men in the mountains. The lieutenant asked Aslan if he knew a path that would lead to the rear of the ravine in which the Germans had settled. Aslan said there was not a path in the mountains that he did not know. Walking in step with the lieutenant at the head of the column, this thirteen-year-old Circassian boy led the soldiers to the height that overlooked the ravine.

The Russians lay down and listened. They could hear the Germans speak.

"We'll take them alive," said Beloborodov, "except those who resist."

With shouts of "Hurrah, hurrah," as is their custom when they launch an offensive, the Russians leaped on the Germans and demanded their surrender. Some of the Germans started to shoot. Six of them were instantly killed. The others dropped their weapons and surrendered. When the battle was over little Aslan went back to scouting.

Legion is the number of children who have worked for national defense as bravely and as audaciously as did Aslan. Nor is the number small of those who, under the spur of uncontrollable emotion, have insulted Germans and have suffered a sinister fate. A vast literature has grown up on the subject of children in "the people's war," a literature that is rich in adventure and humor, in triumph and tragedy. On reading this literature or observing the patriotic zeal of Russian children one unwittingly asks himself what there is in Russian civilization or in the Russian scene that makes children such eager and willing little soldiers who literally leap at burdens and risks which the world associates not only with the years but with the responsibilities and prerogatives of maturity.

Most certainly the material standard of living has not been a determining influence, though children have invariably received the best and the most of what the country has had to offer. Children behave as they do in spite of periodic and drastic shortages of fats, fruits, sweets, and other purely material satisfactions and pleasures which they may be craving.

The fact is that they are surrounded by a multitude of conditions and stimulations that make them responsive to their outside world and to any national crisis or catastrophe.

Russian children spend most of their waking time outside of the home—in school, parks, camps, playgrounds, Pioneer Homes, in other institutions that cater exclusively or overwhelmingly to their needs and their desires. Sports and games are endlessly fostered. One of the commands in the decalogue of the Pioneers, the national children's organization, reads: "The eye of the Pioneer is sharp, his muscles are like iron, his nerves like steel." Children are continually enjoined to keep healthy and harden their bodies through exercise, outdoor play, outdoor work, outdoor living. Walking, running, skating, gardening, dancing, all manner of war games, with all the suspense and excitement they afford, are some of the more common and more widespread modes of physical culture into which they have been drawn.

Rich in variety and appeal is the recreation that is at their disposal. Enlightening in this respect is a visit to a school or a Pioneer Home. Mechanics, dramatics, painting, elocution, choral singing, folk dancing—

hardly an activity or adventure that might interest or excite a child but is fostered in as large a measure as the capacity and the talent of the child allow. Neither instructors nor guides are lacking.

Since the adoption of the Constitution there is no longer any discrimination against children who happen to have been born in the wrong family or of the wrong class. Now the son or daughter of a priest, a former kulak, or a banker is as welcome to children's institutions as is the son or daughter of a factory worker. Such homogeneity has made possible a more unified direction of the life and the thoughts of the nation's children. Now they all belong. They all profess the same fealties. They are all made to feel that the opportunities and diversions they enjoy are the richest and the most advanced in the world and that they are the luckiest children in the world.

Of course the children whose fathers or mothers were caught in the purge of 1936–37 regard themselves as the most unfortunate in the world.

Children love group life, and in no other country is it so cheerfully and diligently fostered. Individual pastimes and work, except when the task requires it, are discouraged and condemned. The child who loves to be by himself or who for some reason puts himself above his milieu, whether in school, in the street, or in the apartment house, is viewed as something of a pariah. The "lone-wolf" type of mentality is assiduously discouraged, while collective effort, communal association, dependence on the group for individual fulfillment and individual welfare are assiduously inculcated. Thinking in terms of the community, feeling a part of it and indebted to it for all the good it has offered and for the promise of still greater good in the days to come, the child is attuned to its needs, and swiftly responds to its call for service.

By far the most powerful influence in the life of Russian children is the political education they receive. It starts in the earliest years and with age grows more diverse, more all-embracing, more concrete, more intense. This education transcends political ideology, or rather expands it to include everything that Sovietism represents. In words, symbols, formulas, exercises, experiences within their comprehension, children are made aware of the campaigns for steel, for machine-building, for tractors, for combines, for electrification, for schools, for books, for summer camps. They may be living in their own world of fancy and make-believe. They may be listening to fairy tales which have long ago been legalized and to exciting stories of legendary heroes in Russian history. They may be imagining themselves in the place of these heroes, in combat with the same enemies of the Russian people and achieving no less spectacular victories over them. But they are also drawn into the world of everyday reality and all that it implies in time of feverish peace and flaming war.

Because of this political education they mature early. Wholeheartedly they accept the Soviet order as the highest and holiest ever known to man and deserving not only their support but their sacrifices. Of course they have no basis of comparison. Nor do they seek it. After all, they are children, and the word of those whom they trust and who minister to their everyday needs and cultivate their tastes, their ideas, their spirit of adventure is to them truth and law.

"The Pioneer," reads the second command in the Pioneer's decalogue, "fervently loves his fatherland and hates its enemies." The child necessarily hears much of the fatherland. With it more than with father and mother, more than with family, with anything in the world, are bound up all the child's personal fortunes, all the good of today and of tomorrow. Without it there is no life, there is nothing.

Many-sided, ever-present, highly dramatized are the patriotic appeals to children. Flags and bunting, ceremony and song, drilling and marching, saluting and standing at attention are as much a part of the appeals as words and slogans, sermons and exhortations. And the children respond not only with verbal declaration but, as already indicated, with risks and deeds, with their very lives!

Because of the war and the air raids the schools in Moscow were closed in the winter of 1941–42. After a year's vacation they reopened a month later than usual. The children were needed in the harvest fields and did not return home earlier. They were glad to go to school; Moscow was in a festive mood on the day it opened.

But with decisive battles raging in the west, in the north, in the south, school could not be the joyful experience it had been in prewar days. For one thing, it carried burdens and responsibilities which in prewar times devolved on the shoulders of older people.

Here is school number 255 in the heart of the city. It is a new school pushed back from the street, so passers-by hardly observe it. The classrooms are large, spacious, airy. For a year it has been used by the Army. Now it is once more a school. For days before it opened squads of pupils, led by teachers and the principal, were scrubbing floors, washing walls, wiping windows, dusting portraits, hanging maps, shifting furniture from room to room and from hallway to hallway. The teachers were not merely directing and supervising the work but were on their knees beside their pupils wielding scrubbing brushes with no less diligence.

Gone were the janitors, the porters, the other caretakers. They were needed elsewhere—to dig coal, assemble rifles, make shells, work the land. Yet on the day school started not a desk was missing, not a portrait was misplaced. All the classrooms were as neat and serviceable as they had been when professional caretakers looked after them.

Pupils and teachers continued to be their own janitors and attendants. As soon as classes were out they donned their old clothes, got down on their knees, scrubbed and washed, and did not leave the building until the classroom had been made ready for next morning. Nobody growled. Nobody complained. Pupils knew there was nobody but themselves and the teachers to do these chores, and they divided themselves into brigades and did them.

The Moscow winters are severe, and to keep the school warm the pupils and teachers now had to look after their own fuel. With Donbas coal in German hands, with the coal in the Moscow district needed desperately by factories, the school could be heated only by wood or peat. The city brought the wood and the peat on barges, on freight trains, on trolley busses. But the children and the teachers had to haul the fuel to the school-yard. They did it on trucks when any were available, on handcarts when there were no trucks.

Despite the new labors and the multitude of social duties the children must perform—visit hospitals for wounded men, mend soldiers' clothes and help their wives or widows—the school curriculum has not been cut or altered. Yet children have not complained. After a year's vacation they were only too glad to be back in the classroom and to rediscover the world of books.

They knew the meaning of the war only too intimately. Hardly one in school 255 but had a father or a brother in the Army. Not a few were already orphans. Others had had no news for months from fathers or brothers and were slowly reconciling themselves to the worst. Never, said the teachers, had the children been so mature and so sensitive. Their nerves were on edge, and teachers had to be cautious in the language they used in reprimanding them. Severe words made them weep, the boys as well as the girls.

At the time I visited the school they were in the midst of a campaign to gather gifts for the Army. On November 7, the day of the anniversary of the Soviet regime, a committee of students and teachers were to journey to the front and deliver the gifts in person to the soldiers. The principal showed me into the room in which the gifts were kept and allowed me to examine them. Invariably they were neatly packed in boxes or wrapped in paper. I was amazed at the variety of useful things which, despite shortages, the pupils managed to obtain—safety razors, blades, pencils, writing paper, books, chocolates, soap, handkerchiefs, scissors, penknives—nothing luxurious, everything distinctively utilitarian.

The senders put letters into the boxes and packages. One girl named Velmira, aged nine, wrote the following letter: "Dear soldier, I have no father. You are my father now. Dear soldier, avenge the death of my papa, my mama, my little sister Sofochka."

There are many Velmiras in Russia, and their presence explains much about the fighting morale of the people and the fighting spirit of the children. We shall learn more of the nature and meaning of this fighting spirit from some spectacular stories I have gathered about Russia's children in wartime.

CHAPTER 32: VANYA ANDRYONOV

VANYA ANDRYONOV lives in the village of Novomikhailovskoye in the Moscow province. One cold day Vanya was lying on top of the brick oven beside his two little sisters, Nadya and Zina, when he heard the distant boom of cannon. Turning to his mother, he said:

"Mamma, Germans are coming."

Shortly afterward the Germans occupied the village. Vanya buried his new overcoat and helped his mother bury several sacks of potatoes and cereals. Deliberately he went around in ragged and loose-fitting clothes with deep pockets so the Germans would not be tempted to disrobe him and take away his clothes. Under cover of darkness Vanya kept going to the hole where the food was buried and filling his pockets with potatoes for his mother and for the guerrillas that were hidden in the woods. He saw the Germans shoot the chickens in the village and slaughter the cows and the pigs, and he so hated them that he could not bear looking at them and kept out of their sight.

One day he saw German soldiers bore holes in a barn on the edge of the village. Then he saw them cart machine guns inside the barn and set them into these holes. He knew they were preparing for something.

That something was the Russian counteroffensive. Vanya waited impatiently for the Red Armies to come. One day, lying on the top of the oven to keep warm, he saw through the window far away several men on skis and in white hoods. He knew they were Russian soldiers, and his heart sank. They were running straight into the machine-gun nest he had seen the Germans set up in the barn. Jumping off the oven, he ran into the dark vestibule and shouted to the approaching skiers, "Don't come this way!" But his voice never reached them, for they kept moving forward. Machine-gun fire mowed them down. Vanya saw them dead on the snow.

Later more men were coming on skis, many more, scores and scores of them. They were following the footsteps of the advance guard that lay dead in the snow. If they kept on they would be mowed down. Vanya decided to warn them. Without saying a word to anyone in the house he started to make his way in the direction from which the skiers were coming. The snow was deep, and, like so many children in Russia, he had learned in school the art of creeping through machine-gun fire. The deep snow helped to hide his movements. He crept and crept; then he rolled down a hillside. When he knew he was out of sight of the Germans he started to run again. As soon as he was within hearing distance of the Russians he started to shout, "Don't come this way!" The Russian officer stopped and questioned him at length. But one soldier was suspicious of him and said severely, "Maybe you are lying?" Stung and hurt, Vanya replied, "How can you say that when all the time I've been carrying potatoes in my pocket to the guerrillas in the forest?" Thereupon the officer said, "Very well, I believe you; now you show us the way to the rear of that machine-gun nest." Joyfully Vanya replied, "I will." He led the soldiers through a ravine which was out of sight of the Germans and brought them unseen to the rear of the barn. So swift was the Russian attack that the Germans had no time to resist.

Frightened of the shooting, the people, including Vanya's family, scampered into cellars. When they heard that their own men had retaken the village, they came out and crowded round the victorious soldiers. The officer placed his hand on Vanya's back and said:

"Vanya showed us the way here. He has helped us destroy the enemy. He is a heroic lad."

Vanya smiled, and his two little sisters gazed at him with proud and incredulous eyes.

The War Council of Russia, of which Stalin is chairman, bestowed on Vanya the Order of the Red Star.

CHAPTER 33: ALEXEY ANDREYITCH

ONE EVENING I was discussing the war activities of Russian children with a Moscow schoolteacher.

"Have you heard of Alexey Andreyitch?" she asked.

The name sounded so adult and so formal that I did not associate it with children, and so I said:

"No, who is he?"

"If you were a Mark Twain," she said, "you could write a *Huckleberry Finn* about him."

She scanned her bookshelf and pulling out a little yellow paper-bound book by Lev Kassil she gave it to me and said:

"Read it and remember that for purely military reasons the story of Alexey Andreyitch cannot be completely told. Only a part of it is recorded in these pages, and I do wish some Mark Twain would read it and write another *Huckleberry Finn* about him."

As soon as I came home I read the breath-taking story of Alexey Andreyitch. It seemed so incredible that I made a special trip to the teacher and told her what I thought about it. She was rather surprised and said: "Anyway, Americans ought to know the story because 'Alexey Andreyitch' is rapidly becoming one of the great folk heroes of the present war. Hundreds of volumes will be written about him and boys like him in the years to come." So I am following the advice of the teacher and am giving the reader the full story of Alexey Andreyitch as recorded by the Russian author Lev Kassil.

So mysterious a person was he for a long time that the Russian commander of a certain sector of the western front had heard much of him but did not know who he really was. At this sector only a river divided the Russian armies from the Germans and naturally the Russian commander was eager to know the strength and the disposition of the German troops on the opposite bank.

One day while one of the Red soldiers was scouting around the woods he ran into a barefoot Russian boy who stopped to talk to him. Out of his pocket the boy brought seven little white stones, five dark ones, three chips of wood, and a rope with four knots tied at one end. In a whisper the boy explained that the white stones represented trench mortars; the dark ones represented tanks; the chips of wood machine guns; and the knots in the rope, field batteries. This meant that on the opposite bank of the river the Germans had seven trench mortars, five tanks, three machine guns, and four field batteries. The soldier asked the boy where he had come from and the boy answered, "Alexey Andreyitch sent me over." He would not say who Alexey Andreyitch was nor where he was keeping himself.

The next day the boy came again and took out of his pocket more stones, white and black, and chips, and a rope tied with more knots. Astonished with the amount of information the boy had gathered, the soldier again asked him about Alexey Andreyitch. The boy only said, "This is wartime— too much talk is dangerous—and besides, Alexey Andreyitch ordered me to keep mum."

Thus day after day this barefoot boy kept coming into the forest with fresh information about the Germans, and invariably he explained that

Alexey Andreyitch had sent him over. The Russian commander concluded that Alexey Andreyitch was a grown man skilled and experienced in military espionage.

One evening as the commander was entering his tent he was informed that a strange boy of thirteen or fourteen wanted to see him. He ordered the boy shown into the tent. Barefoot and in short trousers the boy came in.

"Allow me to introduce myself," he said. "I am Alexey Andreyitch."

The commander was breathless with surprise. He never had imagined that the mysterious and mythical Alexey Andreyitch was only a boy. On questioning him he learned that he was captain of a brigade of eight boys who had been operating a ferry between the two opposite banks of the river that was between the German and Russian lines. The ferry was only a raft and was named "the tomb of the fascists." On this trip, explained Alexey Andreyitch, they had brought over from the German rear three wounded Russian soldiers. They were so heavy that the boys could not carry them, so he came for help.

He led the commander and the stretcher-bearers to a place in the woods where the wounded men were carefully hidden. The stretcher-bearers carried them at once to the tent hospital. But the commander was so intrigued by this remarkable boy that he interrogated him for some time on his "crew," his "ferry," his crossings between the two front lines. Calmly Alexey Andreyitch explained that they had managed to escape trouble because of the hill at the bend of the river which when passed shielded them from the view of the Germans. Getting to this hill was difficult and risky, but they had done it for some time and the Germans had failed to stop them.

The next day Alexey Andreyitch came again with his "bookkeeper" Kolka—the boy who had first come to the Russians with an exhibit of stones and wooden chips and knotted rope—and another boy, the "assistant bookkeeper," named Seryozha. Alexey Andreyitch showed the Russian commander a drawing he had made of the fresh disposition of the German troops. The commander asked how much equipment the Germans had, and Alexey Andreyitch called on Kolka to empty his pockets and count the white stones, the dark stones, the wooden chips, and the knots in the rope. "What about armored tanks?" asked the commander. Alexey Andreyitch called on the "assistant bookkeeper" to take out his shells. Seryozha showed thirteen shells. Alexey Andreyitch explained that it was dangerous to have military information centered on one person. If he fell into German hands the information they had so painstakingly gathered might be lost, so he divided it around among several members of the crew.

That evening Alexey Andreyitch also turned over to the Russian commander eighty German rifles. Asked how he got them, the boy explained that he always watched for German celebrations. Usually the Germans got drunk and if sentinels were not around or were also drunk he and his brigade wormed their way inside the German camp and helped themselves to rifles and other weapons they could carry.

Once they had to dump a load of rifles into the river. They were in danger of being discovered and did not want German rifles on the raft.

"We have a cannon too," said Alexey Andreyitch, and explained that the Germans had gotten stuck in a swamp with a large cannon. All day they had tried to lift it out but had failed. When darkness came they left— afraid to remain in the wooded swamp lest guerrillas attack them. But the cannon was there, and if enough men went along it might be gotten out and ferried across on "the tomb of the fascists." The commander sent seven men after the cannon. They pulled it out of the swamp and loaded it on the raft. It took them nearly all night to do it, and as they were going across the river the Germans saw them and opened fire; but it was too late—the raft had swung over the bend to the other side of the hill.

The boys and soldiers were all wet and muddy. The commander fed the boys and put them to bed in his own tent. When they awoke the commander said to Alexey Andreyitch, "What can I do to reward you for all you have done for the Red Army and the fatherland?"

For once Alexey Andreyitch found himself bereft of speech. He remained silent. The commander unbuckled his holster, took out his personal revolver and gave it to the boy. Alexey Andreyitch looked at it, as did the other boys, with hungry and covetous eyes. This was a gift of gifts—especially as it came from the commander's personal holster.

But Alexey Andreyitch would not take it. It was too dangerous, he said. "If I fall into *their* hands and *they* find it on me they'll know that I am a real spy."

So Alexey Andreyitch and his crew went back to the raft and ferried to the other side of the river.

What has since happened to Alexey Andreyitch and his crew I have not heard.

CHAPTER 34: SONG OF THE NIGHTINGALE

IMMEDIATELY after the dinner hour German troops swung into the village. In reality there was no village any more. Petras Tsvirka, who wrote this story in *Izvestia,* assures us that not a house was left standing—nothing but heaps of ashes and rubble and blackened brush and burned trees. Not a soul was around—not even a dog. All life seemed emptied out of it as completely as though no one had ever lived there.

Tired and sweaty, the Germans halted for a rest. The lieutenant took out his binoculars and searched the countryside in all directions. He was seeking to locate the village for which he was bound.

Suddenly there broke into the air the song of a nightingale. On and on the nightingale sang with increasing zest and melodiousness. The lieutenant and the soldiers turned their eyes in the direction of the near-by grove, but they saw no bird. Instead they beheld a bareheaded, barefoot boy sitting beside a ditch. He was whittling a stick and whistling.

The Germans came toward him and he looked up, frightened. Hastily he thrust his knife into his pocket and pressed the stick inside his clothing. The lieutenant ordered him to come over, and when he did the lieutenant saw him rolling something in his mouth.

"Let me see what it is," demanded the lieutenant.

The boy took out of his mouth a birch whistle, small and soaked in saliva. It was with this whistle that he had been imitating the nightingale. The lieutenant and the soldiers gathered around and scrutinized the whistle with deep curiosity. For once the lieutenant's stern expression gave way to a smile of satisfaction and he said: "Splendid, splendid, my boy." He asked the boy to whistle again. The boy, only thirteen years of age, was happy to oblige the German officer. "I can imitate a cuckoo too," he said, and forthwith produced a perfect imitation of that bird. The soldiers were amused, but the officer, bent on an important military errand, asked the boy whether he knew the road to the village of Surmontas.

"Yes," said the boy, "I used to go fishing in the millpond there with my uncle."

The officer took out a shiny cigarette lighter, held it before the boy, and promised to give it to him if he would lead them to the village of Surmontas. "But if you fool me," threatened the lieutenant, "I'll twist your head off your shoulders."

With the German field kitchen in the van and the boy marching beside the lieutenant, the company started. As they walked, the officer asked more questions. Pointing to some trees, he said:

"Are there *partisans* [guerrillas] in the birch woods?"

With an innocent mien the boy asked:

"You mean mushrooms?" Without waiting for an answer he enumerated the kinds of mushrooms that grew in the birch woods.

The officer asked no more questions.

Out of seeming playfulness the boy started to whistle again, thirty-two times like a nightingale, twice like a cuckoo. The Germans marching along liked the joviality of their youthful companion and said nothing. But deep in the woods the lurking guerrillas knew the meaning of the bird songs. They knew that thirty-two Germans and two machine guns were on the road.

As the party wound along the road and entered the woods, the boy, fleet as a startled rabbit, darted away; and from the concealing birches came a hail of bullets. Not one of the Germans remained alive.

CHAPTER 35: **TOLSTOY'S OLD HOME**

IT WAS AFTERNOON when we drove by the pond at the gateway of the Tolstoy estate in Yasnaya Polyana. We followed a road between tall trees and, emerging into a grassy clearing, drove up to the rear of a building and stopped. The building was once a wing of the Tolstoy residence. Now it is known as the literary museum. In it are kept exhibits depicting Tolstoy's bountiful literary creativeness.

As we got out of the car a Russian sergeant hailed us. He was standing beside the museum, which his detachment, then on an excursion, was visiting. He was tall, heavy-jawed, dark-eyed, and, like all Russians in these tense times, full of talk. Tolstoy, he began with a loud earnestness, as though addressing a mass meeting, was the pride and glory of Russia; and his old home, everything in it, was a sanctuary for which he and men like him would fight to the end.

"Yes," he rushed on torrentially, "we shall fight for our simplicity, our truth."

I couldn't help thinking how elated and touched Tolstoy would have been by the words of this earnest, stalwart soldier. To Tolstoy there were no more noble words in any language than *simplicity* and *truth*. In his Sevastopol stories he wrote:

"The hero of my novel, whom I love with all the power of my soul, whom I have striven to portray in all his grandeur, and who always was, is and shall be grandiose, is truth."

"Truth, simplicity, sincerity" were the trinity on which, as on an imperishable rock, Tolstoy built his life and his art. With it, as with a mighty

333

weapon, he battled for personal fulfillment and for the fulfillment of the brotherhood of man. In art this weapon had won him great triumph; in life, despite ceaseless battle, it had yielded unassuaged despair. Yet with the fealty of a prophet to his god he clung to it all his life; and when, at the age of eighty-two, he fled from his home in the dark of night, it was only to shake off the encumbrances of family and civilization and to be free to wield it anew in a more humble setting and with an undivided heart.

To Alexander Werth, who came with me to Yasnaya Polyana, and to myself, the words of the soldier speaking of "our truth," "our simplicity" as ideals for which he was willing to give his life, sounded like a message or a greeting out of the very earth of Yasnaya Polyana. Of course the soldier's truth was not Tolstoy's truth, though his *prostota*—simplicity—could not have been different. *Prostota* is one of those rich, meaningful Russian folk words which are invested with an array of virtues which their English equivalents do not possess. To this day it is as much a part of the speech and the spirit of the Russian people as it was in Tolstoy's time. No wonder that in this moment of national calamity and individual heart-searching it is constantly on the lips and in the souls of Russian men and women.

The Tolstoy estate is now a national museum, under the direct control of the Academy of Sciences and under the immediate supervision of Sofya Andreyevna Tolstoya, the novelist's granddaughter. It embraces an area of 866 acres, of which 750 are woods, the remainder orchards, lawns, gardens, fields. The land is deeded to the museum in perpetuity, and not an object on it, whether building, tree, chair, table, or bookstand, but must be preserved in its original condition precisely as it was in the days when Tolstoy lived there.

Recently the fifty-year-old oak brooding over the writer's grave dried up. It was dug out by the roots and hauled away, and in its place a new oak of the same age was planted. During the winter of 1939–40, the severest Russia had known in one hundred years, one third of the 5,000 trees in the orchard of which Tolstoy had written with so much joy and fervor were winterkilled.

I saw row after row of dry, brown, and lifeless trees. Some of them had been planted by Tolstoy's grandfather, some the great writer had set out with his own hands. All those that are dead will be replaced. Horticulturists from the Academy of Sciences will supervise the task, and in every instance the new trees will be precise duplicates of those they replace.

At the doorway of the Tolstoy home is "The Poor Man's Tree"—so called because peasants from near-by villages, pilgrims from faraway places, gathered on the green-painted benches that partly encircle the trunk and waited for the novelist to receive them and to listen to their requests, their

"THE POOR MAN'S TREE" BY THE RESTORED TOLSTOY HOME

Photo by Sovfoto

TOLSTOY'S BEDROOM, LEFT IN FLAMES BY THE RETREATING
GERMANS

THE DOMED ROOM IN WHICH TOLSTOY WROTE
WAR AND PEACE

petitions, their complaints. Many a humble and distinguished visitor, including Maxim Gorki, spent hours in conversation with Tolstoy, standing under this tree or sitting on the benches.

The Poor Man's Tree is a gigantic elm with four stalwart limbs uplifted as if in prayer and one stretching horizontally across the road as if in perpetual welcome to the visitor. Significant and dramatic episodes in Tolstoy's life are associated with this elm. But it is old, bent, ailing. Yet even in the midst of war distinguished scientists are continually journeying to Yasnaya Polyana and applying new remedies to the sickly trunk, the withering limbs. Like everything else on the Tolstoy estate it is invested with memory and sanctity, and science does all it can to prolong its declining years.

As long as there is a Russia, Yasnaya Polyana and everything in it and on it will remain as it was in the days of the novelist's life. Nothing else can so fittingly express and symbolize the simplicity and the truth which Tolstoy always preached and emulated.

Sofya Andreyevna, granddaughter of the novelist, came to meet us. She is a handsome woman of forty-two with the lofty forehead, the upturned nose, the deep-set eyes, the heavy brows that evoke the image of Tolstoy. Though she apologized for her English she spoke it fluently. She also speaks French, but her German, she confesses, she has almost forgotten. English, she told us, was the favorite foreign language in the Tolstoy family. All read it, spoke it. During the novelist's life the mail was always bringing newspapers and magazines from England and America, and about one fourth of the books in the Tolstoy library were in English. England and America were always closer to the Tolstoys than any other foreign lands, and during the novelist's last years distinguished visitors from both countries kept coming to Yasnaya Polyana and were always welcome.

Sofya Andreyevna showed us through the old homestead. All the more eager were we to see it because the Germans had occupied Yasnaya Polyana from October 29 to December 14, 1941. As we went through the wing in which Tolstoy lived we observed few outward marks of spoliation. But since the departure of the Germans the house has been completely restored; the heaps of refuse and wreckage have been removed; the walls have been repainted; the floors, windows, and doors have been repaired. But many old furnishings are gone. They were burned or stolen by the Germans and can never be replaced. Only those who, like Sofya Andreyevna, knew every detail in the old house in the days before the German occupation can tell how much is missing.

We climbed the wooden stairway to the second floor and entered the Tolstoy living room, modest in size and furniture but brilliant with the light that poured in from the main windows. In the middle is the long dining table with chairs all around, with the gleaming Tula samovar at one end, as if ready for the charcoal and the water that it had so often held. Leaning against the front wall and set in heavy, ornately carved frames are two ancient mirrors dating back to the eighteenth century. In the right-hand corner opposite the door is the mahogany table, made by serfs, around which evenings after dinner the Tolstoy family gathered for recreation. While Tolstoy's wife sewed or embroidered, Tolstoy himself or some member of the family read stories and articles not only in Russian but in French, German, or English.

A lover of chess, Tolstoy played the game with skill, and the low and ornate chess table which the Russian author Sergeyenko had sent him as a gift remains precisely where it had always been—midway at the rear wall.

At the entrance to the dining room are two pianos, one in the corner by the window, the other across the doorway against the wall. Both are shiny with black polish and bear the trade-mark of Beker—an old Russian firm. Taneyev, Wanda Landowska, Goldenweiser, and many another celebrity had journeyed to Yasnaya Polyana to play for the novelist on these pianos. A few steps away from the one set against the wall is the high-backed chair with the cushioned footstool on which Tolstoy sat during concerts. He himself often played the piano; so did other members of the family. They all loved music and cultivated its knowledge and appreciation.

On the walls are old framed pictures, also frames without pictures. The Germans stole the pictures or burned them. This room suffered least from devastation during the six weeks of German occupation. The Germans locked it and put on the door a printed sign which read, "Confiscated by the German High Command." It was the intention of the Reichswehr to ship everything of special value in the Tolstoy home to Germany. They therefore flung into this room and put under lock and key many precious museum objects. They did not stay long enough to carry out their intention. "But German officers," said Sofya Andreyevna, "kept sneaking in and stealing valuable museum pieces which we shall probably never get. Fortunately for us some of the things you see in this dining room and in this house were hidden in the basement, and the Germans were so afraid we'd put mines there they never dared go down."

We walked through Tolstoy's study, library, bedroom, his wife's reception room and bedroom. In the library is the stand of the dictaphone which Thomas Edison had sent Tolstoy as a gift in 1907. The dictaphone itself is in storage somewhere for the duration of the war. In the bedroom is the bowl and pitcher which Tolstoy had inherited from his father and which his father, while a prisoner of the French in the War of 1812, had carried with him wherever he went.

In the bedroom, leaning against the wall in black and sturdy solemnity, is the couch on which Tolstoy was born, on which all his children were born. It is over one hundred years old, and Tolstoy loved it and wrote of it with a nostalgic warmth in *Childhood, Adolescence, Youth, Family Happiness, A Russian Proprietor, Anna Karenina,* and *War and Peace*. It is one of the most famous pieces of furniture in the world's literature.

Now there is a large rip in the black leather cover, and Sofya Andreyevna told us that it was of recent origin. A German physician had decided to appropriate the couch. Two watchmen on the estate, Fokanov and Filatov, both peasants, were equally determined to prevent the expropriation. There began a tug of war, the German pulling the couch one way, the two Russians the other. Marina Shchegoleva, one of the scientific attendants in the museum, appealed to the German physician to be reasonable. But he was deaf to all appeals. In despair she turned to several German officers who happened to pass by and pleaded with them to help her rescue the precious couch from seizure. They mumbled something to the frantic physician, and he, abashed and angry, abandoned the struggle for its possession. Fortunately the rip in the leather covering is the only damage that resulted from the scuffle.

Tolstoy's study is almost bare of pictures, and Sofya Andreyevna pointed out the places where hung the ones of three Americans Tolstoy particularly admired and of the English author he singled out for special eulogy. The three Americans were Henry George, Ernest Crosby, and William Lloyd Garrison. Garrison's inscription on his photograph, "Liberty for all, for each, and forever," particularly appealed to Tolstoy. The English author was Dickens. Tolstoy used to say, Sofya Andreyevna explained, that if all the literature of the world were sifted until only one author remained he would be Dickens; if all of Dickens were sifted until only one of his books remained, it would be *David Copperfield;* if all of *David Copperfield* were sifted until only one chapter remained, it would be the one describing the storm. Praise of an author couldn't be higher. But then there is something about Dickens that has always struck a deep intellectual and emotional response in the Russians. To this day he remains the most universally beloved and most avidly read English novelist.

In the guestroom of Tolstoy's wife, whose name was also Sofya Andreyevna, is a famous little table. It is so small, so light, so obscure that had not the granddaughter spoken of it we might have passed it by with no more than a glance. This is the table on which Tolstoy's wife rewrote in longhand the full manuscript of *War and Peace* seven times. Tolstoy was a tireless, one might say an eternal and hopeless, reviser of his writing, and in those days, before the rise of typists and typewriters and before his daughters grew up, Tolstoy's indefatigable copyist was his wife. Observing the table, we marvelled how it could possibly have served as a writing desk. But it did—one of the most famous writing desks in the history of literature. On the top are two silver candlesticks and the pens, the blotter that Tolstoy's wife used, also the clump of dark horsehair on which she cleaned her pens.

I sat down at the table, touched it with my hand. It wobbled as though in protest against the intrusion on its silence and solitude. It was only because it had been removed to the basement that it was saved from destruction by the German occupants.

German officers had turned the bedroom of Tolstoy's wife into a clubroom. At the entrance they had put up a sign marked "casino." Here they drank French and German wines, played cards, amused themselves. Not a trace of their presence has remained. The room has been restored to the coziness in which the novelist's wife had always kept it. Pictures are back on the walls, but only those the attendants had managed to remove before the arrival of the Germans and in the early days of their stay. The array of feminine and housewifely accessories which the elder Sofya Andreyevna prized highly are in their old places again. In its diverse detail and authentic simplicity the room bears the imprint of the orderly and energetic woman the novelist's wife was. It is one of the liveliest rooms in the old homestead, bright though never ornate with embroideries and decorations.

We descended the wooden stairway, and Sofya Andreyevna led us to the room in which Tolstoy wrote *Anna Karenina*. It is a small room furnished with astonishing simplicity. Adjoining it is a guestroom, also small and furnished with a plain iron bed, a small square table, a stand with a bowl and pitcher, and a few pictures on the wall.

It was in this room that Tolstoy received William Jennings Bryan in Yasnaya Polyana.

"Grandfather," said Sofya Andreyevna, touching with her hand a narrow beam over the door, "pointed it out to Bryan and told him of the time he had thought of hanging himself on it."

Vividly I recalled that years ago, when I was a hired man on a farm in Central New York, the man I worked for, a stanch, irrepressible Bryanite, read aloud to the family the story Bryan had written in his *Commoner* on this episode in Tolstoy's life.

We walked through several other rooms and finally reached the one no visitor to Yasnaya Polyana with the least appreciation of Tolstoy's writings ever fails to see—the room in which Tolstoy spent four or five years writing *War and Peace*. It is only a medium-sized room with large windows and a vaulted ceiling. Tolstoy's grandfather had used it as a meat cellar and the rings from which the carcasses of slaughtered animals were hung were still there—fastened into the walls. The very shadows in the room seemed alive with the tenderness and brightness, the beauty and heroism of the men and the women—Russian to the core of their hearts—with whom Tolstoy had peopled that unforgettable, monumental tale.

Yet even this room the Germans did not spare from desecration. They made of it a barrack for soldiers. There was no need of quartering anyone here. On the hill beside the Tolstoy schoolhouse was a spacious, timber-built house which had been the dormitory of pupils from faraway villages. Though it was adequately equipped for rest and sleep, the Germans hardly used that building. Instead they hauled into the vaulted studio of the great writer—a shrine of shrines to the Russian people and to booklovers all over the world—hay, planks, and bedsteads from the Yasnaya Polyana hospital and turned it over to German soldiers. After their hasty departure the room, in the words of an old servant, "resembled the foulest den I've ever seen."

The moment the Germans arrived in Yasnaya Polyana they started to indulge in their favorite sport upon occupying a Russian village—chasing after chickens and shooting them. Squawking piercingly, the chickens darted under fences, into fields, into piles of firewood, behind trees and bushes. Pistol in hand, the Germans dashed after them. The grounds of the Tolstoy estate turned into a clamorous hunting ground with the squawk of chickens, the crack of pistols, and the screams of protesting servants fusing into a cacophony of noises that had never before disturbed the pastoral tranquillity of the place. Violently flapping their wings, stricken chickens floundered around in their death struggle, spattering lawns, footpaths, and flower beds with their blood. The hunt did not cease until all the chickens were dead.

Such a hunt always marks the beginning of German depredations in an occupied community. Food, whether in barns, cellars, storehouses, or on the hoof, is the first object of their attention. There were twelve cows on the estate. The servants had managed to lead away and hide three. The others the Germans slaughtered. Cattle, sheep, pigs that the Germans seized in the village they drove into the yard of the Tolstoy home. This yard became the slaughterhouse, and carcasses of dead animals hung from trees and posts at the very doorway of the man who in the later years of his life had forsworn the killing of living things and would not touch a mouthful of meat. There were other places more suitable for a slaughterhouse, but the Germans insisted on having it at the very door of the Tolstoy residence.

There were twelve horses on the estate, but only one was saved from theft. Grain, hay, oats, potatoes, cabbage, everything on which the Germans could lay their hands they seized. The Russians on the estate lived chiefly on potatoes, which they had managed to hide.

"Where did you keep yours?" I asked Marya Petrovna, the janitress.

"Under the mattress in my bed," was her reply. "I was the only Russian who lived in the house all the time the Germans were here, and every time I went out of my room I wondered if on my return I'd find the sack of potatoes under my mattress. Fortunately, they never imagined I'd keep potatoes in my bed."

In the village, which is about half a mile away from the estate, the Germans stole as freely as in the museum. They walked into houses, laid their hands on pillows, blankets, mirrors, and walked off. They even stole the 100 boxes of shoe polish and the 200 pairs of shoelaces in the village store and sent them to their families and friends in Germany.

In the sixties Tolstoy became interested in bees. He started an apiary in the woods. So absorbed was he in his new venture that he went away early in the morning and stayed with the bees all day. To the annoyance of his wife, he missed his meals and neglected his writing.

Since then an apiary has been as much a fixture of the Tolstoy estate as the orchard or the birch grove. When the Germans came there were fifty hives on the place. They were already put away in the barn for the winter with enough honey in each hive to enable the bees to feed themselves during the months of enforced idleness. Through swift and skilled action the servants hauled nineteen hives into the woods. The other thirty-one the Germans seized. They took the honey from all of them. In some of the hives they left the bees to starve to death; into others they poured cold water and killed the bees.

Downstairs in the old residence is a room in which Sergey Lvovitch, Tolstoy's eldest son, now seventy-nine, spends his summers. He was in Moscow when the Germans came to Yasnaya Polyana. On October 31 a German officer, who spoke of himself as "physician from headquarters" but wouldn't tell his name, demanded that the room be made available for use by Germans. When one of the directors of the museum protested, saying the room was an invaluable part of the museum and could not be occupied by strangers, the German inquired where Sergey Lvovitch was at that time living.

On learning that Tolstoy's eldest son was in Moscow, he pulled out of his pocket a plier-like master key with which all Germans seemed supplied and started, burglar-like, to open the locked closets and drawers. In them Sergey Lvovitch had kept his linens—sheets, towels, pillowcases, old costumes of the peasants in Yasnaya Polyana. In the presence of Russians the "physician from headquarters" divided this booty with another German physician. The rare old wineglasses in the room he put into his pocket.

German officers and soldiers stole everything that might be of use to them. The saddle on which the novelist went horseback riding every day except when he was ill; fifty-four valuable paintings in the museum; the weighing scales of Tolstoy's wife; curtains from the library and the living room; a rare old wall clock, blankets, pillows—all these were stolen.

Tolstoy was a great admirer of Heine and in his library were the works of the German-Jewish poet in German and in Russian. All these volumes disappeared. Five volumes of a richly illustrated universal geography which had been in the library bookcase were slashed with razors, and the colored illustrations were cut out. One painting in the museum which showed a peasant revolt against a landlord so incensed the German officers that they set it afire on the wall on which it was hanging. The linen canvas burned easily and left a black smudge on the wall.

During their stay in the Tolstoy house the Germans burned furniture, books, pictures, frames, bookcases, including the one Tolstoy had built with his own hands. It was not as though there were no firewood on the place or no woods near by. Lying in long piles on the estate was enough firewood to last two years. The Russians again and again pleaded with the "gentlemen" to spare the furniture, the books, the pictures. It was of no use.

"We don't need wood," Dr. Schwartz, a German military physician, once frankly confessed. "We'll burn everything connected with the name of Tolstoy.'"

They came very close to achieving their end.

In the literary museum the Germans took out of the frames many of the pictures on the walls and replaced them with their own pornographic drawings. The Tolstoy estate literally reeked with pornography during the six weeks of the occupation. After their departure it took no little scrubbing to wipe out their pictorial lewdness. Some of these drawings are preserved as samples of the type of civilization German "gentlemen" officers espoused during their stay in Yasnaya Polyana.

When Tolstoy was a boy his eldest brother, Nikolay, often spoke of a green stick which was buried on the Tolstoy estate. A magic stick it was, and anyone finding it—so ran the story—would bring eternal happiness to himself and others. Tolstoy's brother pointed out in the *stary zakas*—the family forest reservation—the place within the embrace of young trees where the stick was buried. All his life Tolstoy remembered this beautiful story and he picked for his burial place the ground within which lay the magic stick. When he died he was laid to rest there. Visitors from all over the world have journeyed to Yasnaya Polyana to stand with bared head at the coffin-shaped mound of earth which marks Tolstoy's grave. It is the humblest grave I have ever seen, with neither stone nor cross on it, with only ferns at the sides and flowers on the top.

On their arrival in Yasnaya Polyana the Germans converted the literary museum into a dressing station and temporary hospital. Of course there was no need for this. A short distance from the house was the model hospital that the Soviet government had built for the peasantry at Yasnaya Polyana and the surrounding countryside. This hospital would have made a logical dressing station and temporary hospital for the wounded men. But schooled in the idea of hate and nurtured in the passion of spite for everything non-German, and especially Russian, the Germans established a station for wounded soldiers in a wing of the Tolstoy residence.

Many of the wounded died. In Yasnaya Polyana all around the literary museum are open fields. Digging graves in these fields would have been easy; there were no roots of trees or rocks or any other natural impediment. Instead, the Germans chose to bury their dead all around Tolstoy's grave! Again the Russians protested, and again the Germans dismissed them with a contemptuous snap of the fingers! Each one of the graves they marked with a birch cross and on each cross they engraved or painted a swastika.

As soon as the Germans departed, workers on the estate and peasants in the village came out with spades and dug up all the German dead. In the seventy-five graves the Russians found eighty-three German bodies. They hauled the bodies some distance away and buried them in bomb-blown craters.

During their stay in Yasnaya Polyana the Germans kept boasting that not only Tula but Moscow would soon be in their hands. They not only made themselves at home in the old Tolstoy homestead but kept telling the Russians (and so did a certain Demidov, a Russian White Guard who had come with them) that Yasnaya Polyana had only economic and agricultural value. Its historical and literary importance held not the slightest meaning for them. A productive, well-paying farm—presumably the future property of a German landlord—was all they cared to make of it.

But the Russian counterattack which had started in the suburbs of Moscow was rolling westward, and on December 14 the Germans retreated from Yasnaya Polyana. After they were gone the Russian caretakers and scientific workers on the estate exclaimed: "Thank God!" But they rejoiced prematurely. Suddenly a German automobile roared into the yard and out of it sprang three German officers. Carrying cans of gasoline or kerosene, they walked inside the Tolstoy home and locked the door. They ascended the stairs and made their way to Tolstoy's bedroom, his library, and the room they had used as a casino—which had been the novelist's wife's bedroom. In each of these rooms they heaped together hay which they must have fetched from downstairs and firewood made from broken fixtures. Over the heaps they poured oil and set them afire. Deliberately they started fires in three separate rooms so as to have the whole house in flames and make putting it out difficult or impossible. The Russian workers and attendants who had been speculating on the reason for the return of the officers now had their answer. They saw through the closed windows tongues of flame inside the house. They screamed protests which the Germans, as always, ignored. From moment to moment the blaze was gaining in magnitude. The officers reappeared and warned the Russians not to attempt to put out the fire. The house, they said, was mined, and if anyone went inside he would be blown to death. They got into their car and roared away.

Disregarding the warning about mines, the Russians rushed into the house. There were no mines. The Germans had lied in the hope of frightening the Russians from the effort to put out the blaze. Under the direction of Marina Ivanovna Shchegoleva and her brother the Russians started frantically to fight the conflagration. The fire extinguishers of the house were gone—the Germans had stolen or smashed them. The water pump was broken and they couldn't put it to work. So they carried pail after pail of snow and dumped it on the fire. The results were discouraging. Then someone remembered an old abandoned well and hastened to tear off the planks nailed over it. Fortunately there was water inside and pail after pail was poured over the rising blaze. Searching around the garret, they also found two fire extinguishers which the Germans had failed to discover. After four hours of grueling work the flames were under control and the house was saved.

As the novelist's granddaughter told us the story of this last act of German vandalism we wondered why they so desperately sought to reduce to ashes one of the greatest literary sanctuaries of the world. It could not possibly have aided any of their military operations. Nor could the returning Red Army, even had it wished to do so, put it to particular military use. It was not as though the house was a fortress. The Germans couldn't help knowing that the act would only further inflame the already volcanic hate against them in Russia and would shock the whole non-Axis world. Why, then, did they do it?

In part the answer came from an unexpected source. When we went upstairs again, Sofya Andreyevna opened the glass doors of Tolstoy's study and invited us to come out on the balcony.

"From here," explained the granddaughter, "Lev Nicolayevitch could see the two things he loved most—nature and the people——"

She meant the village of Yasnaya Polyana, which rose on a hill immediately across the valley. Below us was a neatly kept lawn studded with a few flower beds. The flowers had been withered by frost. Beyond the lawn was a row of stately firs—the outward edge of the dense grove at the rear of the house in which Tolstoy had loved to stroll.

Far away were fields and sky and the haze of descending night. The scene that unfolded from the balcony, even now in growing twilight, was stirringly impressive. One couldn't help imagining Tolstoy sitting or standing on this balcony contemplating the vast sweep of land and sky ahead and thinking of nature, of people, of the world, or perhaps seeking in his imagination to unwind a scene in a novel or a play or to fashion one of the multitude of immortal characters he created. Somehow this balcony loomed as a particularly hallowed place if only because it was the door from the study to the two things the writer loved most—nature and people.

Yet incredible and absurd as it may seem, the German officers, many of them sons of old Junker families, the elite of the Junker class—had converted this balcony into a latrine!

The very word *German* in the minds of those who knew or who had read or had heard of Germans spells cleanliness. One would imagine that German officers, men of some culture and of no little tradition, would, as a matter of hygiene, abide by ordinary animal decency. But they did not.

Nor can it be charged that they behaved as they did because they were lazy or because the Russian cold was too much for them. Their record in this war tells only too eloquently that whatever else the Germans may or may not be, they are not lazy. As for the cold, the balcony, exposed as it was to wind and blizzard—also it made a superb target for ambushing guerrillas—was one of the coldest places on the premises. One must inevitably conclude that it gave the Germans pleasure to flaunt openly and conspicuously their contempt for Russian greatness, or purely human greatness when it was not of German origin.

When the servants reminded them that Germans were supposed to be "cultured" people, the officers responded by whistling and snapping their fingers as at "dogs of whom they wanted to be rid," as Marya Petrovna, the janitress, tersely expressed herself.

In the deepening twilight we walked across the valley to the hill on which stood the village schoolhouse. This was one of the most famous schools in Russia. Built in 1928, in commemoration of the one hundredth anniversary of Tolstoy's birth, it was equipped with up-to-date laboratories in physics, chemistry, biology, agriculture. It had its own library, the largest of any secondary school in the country, comprising thirty thousand volumes.

It included Russia's classics and Russian translations of the leading writers of the world.

A white two-story brick building with many-windowed classrooms, studios, an auditorium for meetings, concerts, dramatic performances, it was the pride of the peasants in Yasnaya Polyana and the surrounding countryside. Six hundred peasant boys and girls were receiving as all-inclusive an education as was offered by any secondary school in the country. In the yard on the same hill was the timber-built dormitory for pupils from faraway villages. Back of the dormitory and a little closer to the highway was an apartment house for the teachers.

We ascended the broad, winding cement stairway that led from the street to the schoolyard and found ourselves face to face not with a schoolhouse but with a white and gigantic ruin. Not a door, not a window was left. Through an opening which had been a door we walked inside. There were no floors, no desks, no rooms—all was burned. There were only heaps of broken glass, shattered brick, rain-packed dirt, and ashes. Not one of the thirty thousand volumes in the library escaped destruction—not one! All went up in smoke. Not a portrait remained of the great writers, scientists, musicians, many of them—like Mozart and Beethoven—Germans, that had graced the walls of the classrooms, offices, and studios. The twenty-five microscopes which were the pride of the laboratories, like all other equipment, were gone—destroyed or stolen. Grim and spectral was the ruined emptiness of this once-famous schoolhouse. By some miracle the gigantic statue of Tolstoy at the foot of what was once a broad stairway escaped demolition. Out of shadows of descending night it looked like a tall, abandoned, angry Moses of Michelangelo.

Of the dormitory not even the walls are left. Built of wood, it burned to ashes, and only the damaged brick ovens and the stooping brick chimneys remain to indicate the place where it had stood. The Germans had likewise set afire the apartment house for the teachers. Furniture, books, clothes, and other personal possessions which the teachers had not had time to move out were burned. Like the schoolhouse, it is only a hulk of battered and ghostly brick walls.

When the Germans came to Yasnaya Polyana they set up staff headquarters in the schoolhouse. Though there were stacks of wood in the yard, they built their fires of the desks, benches, and books they found inside. . . . When they left they set fire to everything that would burn. They stayed up all night burning the schoolhouse and everything in it. Armed guards were posted at strategic points to prevent villagers at the threat of death from coming up to put out the flames.

At the top of the stairway that wound up from the street to the schoolyard, elevated on pedestals, were the statues of a boy and a girl in physical-culture poses. The arms of the statues were broken off. How the statues escaped demolition remains a mystery.

Across the valley from the schoolhouse, at the edge of a sheltered grove, was the model hospital, which, like the schoolhouse, had been built in 1928 in commemoration of the one hundredth anniversary of Tolstoy's birth. That, too, was set afire; so was the children's home in the village. The Germans abandoned to the torch everything that their hasty retreat permitted. The peasant homes close to the highway were burned. Despite the deep snow and the freezing weather outside, they drove the people outdoors. There were screams, pleas. *"Raus, raus,"* was the only answer the peasant women and children heard in response to their frantic appeals for mercy. Nor would the Germans permit them to save much of their belongings, not even pillows, sheets, and clothes, except what they wore or could hastily wrap around their bodies. . . . The main part of the village escaped destruction by fire only because of its geographic position. It lay back of the highway, and the Germans had not the time to visit on it their incendiary spite and wrath.

In the evening we went to the home of a young peasant woman whose husband the Germans had publicly hanged. When we entered, the house was in darkness, and the woman hastened to light the wick in a small can of kerosene. She was a tall, cross-eyed woman with statuesque shoulders and light brown hair. Peasant fashion, she welcomed us with a profusion of warm words.

In her arms was her little three-year-old girl, Galya, as gay, garrulous, and energetic a child as I had known. Chubby and blue-eyed, she couldn't sit or stand still. She giggled and chattered as though the arrival of guests were the happiest event in her life. Alexander Werth picked her up in his arms. She snuggled close, leaning her face against his, and continued to scream and chatter with mirth. So infectious and diverting was the child's exuberance that in watching her, playing with her, we forgot the tragedy that hovered over the household.

The mother started to tell about the hanging of her husband. On the day it happened she was not at home. She was visiting her sister in some other village. When the news of the catastrophe reached her she started for Yasnaya Polyana. On the way a German in uniform intercepted her. He asked for her permit to go from one village to another. With loud tears she explained why she was hurrying home. He made her stay over and peel potatoes for four hours and then permitted her to proceed on her journey.

When she arrived home—here she halted, floundered, could not continue her story. Her words got twisted, her voice faltered, her throat choked. She smiled, tried to compose herself but could not. She picked up the lighted kerosene can and lifted it to the rear wall, on which hung family photographs. Pointing to one, she said, "This is Kolka," meaning her husband. With his broad, sensitive face, and his cap sitting lightly on his head, he looked more like an intellectual than a peasant. He was thirty years old at the time he was hanged. She held the light for some time, as though beholding something new and appealing in the face of the man she loved. Again she tried to speak—and again the words got lost in her choking voice. But the little girl was as loud in her gaiety as before. Standing in her bed with her hand fastened on the railing, she danced and laughed and chattered as if in a deliberate effort to make us forget the sorrowful words her mother had spoken or left unuttered.

We waited for her mother-in-law, who was in the field helping to thresh the kolkhoz grain and was late in returning home. At last she came. Wiry, of medium height, with an alert manner, she acted younger than her fifty-three years, though her thin, wind-scorched, and furrowed face made her look older. The kolkhoz threshing, she said, was so important that they had to work late into the evening—as long as they could possibly see. That was why she was late in coming home. . . . She, too, was happy to find visitors in her daughter-in-law's home. She lived in the other part of the house, across a dark vestibule.

She sat down beside the smoky light and started to talk. On the day the Germans came several Russian soldiers were in her home. Suddenly they heard a plane over the village; then another; then they heard the cracking of a machine gun.

"You'd better run into the woods, darlings," she said to the soldiers, "and I'll take care of your horses."

They were cavalrymen and had seven horses outside the house.

They left immediately, dashed for the woods, made their way to Tula. Despite machine-gun fire from German planes she managed to save the horses, all of them. She threw a rug over their backs to camouflage them and led them off one by one to a secure place.

Then came the roar of a tank and then another. Looking out of the window, she saw the street filling up with tanks.

"Good heavens," she exclaimed, "we are prisoners of the Germans!"

She heard a knock on the door. Opening it, she saw a German. He spoke Russian and told her six men would be quartered in her house. He asked if any guerrillas lived there, and of course she said no.

The six men came—most of them Finns, "the most terrible people in the world, worse than the Germans." This was not the first time I had heard from Russians that in their treatment of civilians the Finns were as brutal as Germans, sometimes more so. I had heard it from peasants in

Lotoshino, in Pohoreloye Gorodishche. German tutelage had evidently proven only too effective for the Finnish troops who were fighting in Russia.

"I washed their linen," the old woman went on; "I looked after them, and they continually insulted me, beat me, laughed at me."

It was painful to hear of Finns imitating Germans in their depredations in Russian villages, all the more painful for one like myself, who have often traveled in Finland, have cultivated the friendship of Finns, and have always written of the country and the people with respect and warmth. Of course the Finnish people must not be held responsible for the misdeeds of Finnish soldiers in Russian villages.

"They came and went in groups—two and three at a time," the peasant woman continued, and always they were chiefly Finns. They stole all her potatoes, twenty-eight sacks, all the other food that they could find, and if she protested and wept they shouted at her or struck her.

One day as she was walking in the village she saw Germans and Finns nailing a plank about twelve feet long to two telephone poles. She wondered if they were going to put up a screen and show an outdoor motion picture. But she made no inquiries and went home. Then one of the Finns who lived in her house came in with a long rope "as thick as two fingers." Brandishing it before her, he told her he was going to hang her son. Terrified, she started for the door. The Finn barred the way, pushed her back, and rushed out, locking the door behind him. She cried and cried, but what could she do? Then through the window she saw them leading her son and another young man, a stranger in the village, out of the post office. Both were bareheaded and their hands were tied behind their backs. She couldn't imagine they were leading her son and the other youth to the gallows. Her Kolka had never done anything to the Germans. He had been a home-loving man, a loving husband, a dutiful son. . . . Why should anyone, even Finns or Germans, want to hang him or do him the least harm? The Germans and the Finns had not even accused him of wrong-doing. But——

A German automobile in a yard a few houses away was damaged by someone who had thrown a hand grenade at it. With the help of Finns the Germans had rounded up twenty-five Russians from among passers-by and ordered them to the post office. There they picked out four young men, led them aside, and announced that unless the person guilty of damaging the machine delivered himself within twenty-four hours, all four of the hostages would be hanged.

Two of those four, the old woman continued, were her sons, but she didn't know anything of the death sentence until the Finn with the rope had spoken of it. Soon the Germans relented and released two of the four young men they had detained. One of them was her younger son. That very day, without waiting for the expiration of the twenty-four hours of grace, at three-thirty in the afternoon, they carried out the sentence.

She did not witness the hanging; she could not because the Finn had locked her in the room. But people who were there told her that her son was so horror-struck he kept his head down so he would not see the hanging of another man. A German struck him on the face, jerked his head up, and ordered him not to lower it again but to watch the other man die the way he would soon die.

When the execution was over a German came to her and, pointing with his hand to his neck, indicating a hanging, said:

"Your son is *kaput*."

She wept aloud, and when the Finn came he shouted at her not to weep. She couldn't stop, and then he struck her and told her she mustn't weep or he would "poke her eyes out. . . ."

The old woman stopped talking, lowered her face, and shook with grief. The little girl continued to dance and laugh and chatter, but this time her very gaiety only accentuated the horror of the tale we had heard.

Soon the peasant woman composed herself and resumed talking, slowly now and brokenly.

"They hanged my Kolka at three-thirty in the afternoon on November 14, and not until November 17, at eleven in the morning, did they allow my younger son to take down the bodies for burial. And all the time if we looked out of this window—we could see Kolka's body swinging on the gallows."

Again she stopped. I glanced at little Galya still dancing and chattering and laughing—and acting like the happiest child in the world. Yet she, too, from the forward end of her bed could have looked out of the window at the fateful gallows and at the swinging body of her father.

"Ever since then," the grandmother said, "whenever I pass those two telephone poles I lower my eyes. I dare not look up."

Her younger son brought the two bodies into the room in which we were sitting and laid them on the floor beside the open fire. They were frozen solid. The heads were twisted, the arms were wrenched.

"I knelt down and straightened my son's arm. It creaked as though it were not flesh but a piece of wood. . . . Then, as the bodies started to thaw, the faces covered with sweat and I kept wiping it as fast as it came out. . . . We gave them a decent burial."

We walked back to the Tolstoy estate. The air was crisp with frost and brilliant with moonlight. We passed Russian soldiers on foot or on horse-back. After a day's hard drilling they were returning to their barracks or to the homes of peasants with whom they were quartered. Some of them walked arm in arm with girls in uniform. Now and then we heard loud talk, a girl's laughter. Far away, at the other end of the village, girls were singing a sprightly Ukrainian melody. It was good to hear this talk, this laughter—especially this singing. . . . It made one aware of the immense vitality of the Russian peasantry, of their power to rise above the mockery and the horror they had endured in the weeks of the German occupation.

A song means so much in all the Slav countries. "When our song dies," reads a Czech song, "we, too, shall be living no more." This is true of Russians, Ukrainians, Poles, Slovaks, Yugoslavs, as of all other martyred Slav peoples, as of Czechs. A song lifts them above the mud and the misery of the past, above the terror and the agony of the moment.

"The song became our life," writes Nikolay Tikhonov in one of his rhapsodic eulogies of Leningrad under German siege. "Our song was not frightened of our tragedy."

This is as true of Yasnaya Polyana as of Leningrad. The song, more than the word, more than the laughter, more than anything else in Russian peasant life, signalizes the triumph of life and hope over all trials and all visitations, even of such a foe as the German of today.

CHAPTER 36: "THE NEW ORDER"

I FIRST went to Pohoreloye, at the Rzhev front, in August—soon after it was reconquered. Ten weeks later I went there again to wander at leisure in the village and in the district, so as to make a study of the "new order" as the Germans espoused and practiced it during the ten months of their stay there.

I deliberately chose not an urban but a rural section for this purpose, because living conditions, bad as they may be in an occupied village, never can be as bad as in an occupied city. When out of bread, potatoes, cabbage, and other common foods, peasants can hunt for berries which, like mushrooms, grow abundantly in the Russian forests. They can also pick sorrel leaves, nettles, and other edible plants growing wild near their homes. They can do this more readily than can city people, whom the Germans do not permit to leave for the countryside without special permission, which they seldom grant. What is therefore true of a village is at least true of a city. There may be exceptions in some parts of the occupied areas, though I have neither heard nor read of any.

Besides, beginning with the housing and ending with the speech of the people, life in a Russian village is more open to study and observation than in the city. Basic facts are more readily ascertained, easier to verify. So I came to Pohoreloye, which is the largest village in the district and the seat of its central government. There is not a single town here; not a single factory. The people are native peasants who work the collective farms, and the officials are natives and outsiders who minister to their daily needs. There are twenty Soviet precincts in the district and one hundred and fifty-two villages. Before the war almost 43,000 people inhabited these villages.

In their drive on Moscow in the autumn of 1941 the Germans occupied all the villages. So swift was their advance that only a small number of the inhabitants managed to escape to the Russian rear. The rest remained in their homes. In the counterattack in January 1942 the Red Army liberated eighty-eight villages, which had been under German rule about two months. The others, sixty-four in all, including Pohoreloye, remained to endure the German "new order" for ten months. These villages offered a supreme opportunity for a study of the German "new order." Ten months was time enough not only for the outlines but for the body of the new order to crystallize, at least to uncover its distinctive features.

On my arrival in Pohoreloye I found the village deep in mud. In all my years of travel in Russia, including those preceding the Five-Year Plans, I had never waded in deeper mire. The main streets were cobbled but the cobbles were flooded with mud. The place where I ate was only a few blocks from the peasant cottage in which I stayed. Yet though I skirted mud ponds, sprang over mudholes, sought to land my feet on grass, plank, rock, by the time I reached the eating place my rubbers and boots were drenched with mud. There was no dodging or escaping it. It heaved and billowed on the cobbled main street no less turgidly than on the uncobbled side streets.

It was not so before the war. Even in the early spring after the thaw of ice and snow, and in late autumn after the usual heavy showers, the main streets of the village, so the peasants informed me, were comparatively dry. Not the least of the aims of the Five-Year Plans was to pave streets, perfect drainage, lay out sidewalks, so that in the wettest months of the year neither city nor village would present the picture of bleak primitiveness that they once did. Not all the cities and villages had acquitted themselves with credit of this aim. Many of them had not. Yet not one but had at least laid the foundation for its achievement. Pohoreloye had done more—had actually made the leading avenues easily passable even for children in the wettest weather of the year.

But strange and incredible as it may seem, one of the first tasks of the

Germans was to shatter the improvements the village had achieved. They lifted boardwalks, tore up bridges, allowed drainage ditches to fill up with scum and mud. In consequence the autumn rains flooded streets, gardens, courtyards; and the procession of heavy trucks rolling continually to and from the front had churned up the water with dirt and the village swam in mud. Peasants with spades were doing their best to open channels and to drain off the water, but the task seemed beyond the powers of spade and man. Too great was the damage the Germans had perpetrated.

The sight of these rivers of mud in Pohoreloye brought to my mind the Russian village as I knew it in the old days, memorable for its all-embracing, implacable primitiveness. I thought of Tolstoy's *The Power of Darkness*, of Chekhov's *The Peasants*, of Ivan Bunin's beautifully written and harrowing *The Village*, of Pisemsky's record of travels in the countryside with its moaning undertone of sorrow and wrath. Pohoreloye, as it lay before me at the end of October 1942, seemed a living physical image of the saddest and direst primitiveness I had observed or had heard or read of in old Russia and in the Russia prior to the Five-Year Plans.

I was not long in the village before I realized from its physical appearance and from the speech of the people that one of the basic aims of the German "new order" in Russia is to undo the achievements of the Five-Year Plans in education, in mechanical advance, in community improvements, and to roll the country back to the condition it had known long before there ever were Soviets and Five-Year Plans. The return to primitiveness and all that it implies in individual backwardness and helplessness, in social disunity and disintegration, in national dilapidation and impotence, might be the slogan of the German "new order" in Russia.

One would imagine that if only for the purpose of whipping out of the country all the production that it could yield and of making their own life more comfortable the Germans would preserve the institutions and the improvements that the machine age in Russia had achieved.

That is precisely what they did not do in the village or the district of Pohoreloye. They smashed the fruits of the machine age. Evidence of the destruction greeted the eye at every step.

Until 1940 the village had drawn its water from the near-by Derzha River. It is a narrow, swift-flowing stream that gathers in its course much refuse from pastures and also from certain villages. After much planning and struggle Pohoreloye had built an artesian well with an engine, an automatic pump, and a wooden water tank with a capacity of 4,000 pails. Pipes were laid from the tank to strategic parts of the village and the people drew their water from these pipes.

On their arrival in Pohoreloye the Germans destroyed the new water system. They removed the engine, broke the pump, burned the wooden tank. During the ten months of their stay in the village they used the muddy and impure water from the Derzha River. Hygiene as well as good taste should have impelled them to take advantage of the artesian well. But their passion for the obliteration of mechanized installations overpowered all dictates of hygiene and good taste.

The Five-Year Plans had centered their chief energies on industry, with the result that housing, especially in rural areas, had remained but slightly more advanced, except for widespread electrification, than it had been in the prerevolutionary days. But the Plans did emphasize the building of community bathhouses. Pohoreloye had built one with a capacity of 500 bathers a day. An army in the field can make precious use of a large and up-to-date steam bathhouse, and again one would have thought that considerations of hygiene and health would prompt the Germans to preserve such an institution. Instead they destroyed it after they had been in the village a little more than two weeks.

Pohoreloye had a power plant. Every home was wired and electrically lighted. The Germans did not dismantle the power plant. They curtailed its use, allowing only institutions and officers' quarters to be lighted. With the advance of winter they neglected the engine, and it froze solid. On their departure they took the dynamo with them; the engine they battered but had not the time to destroy.

There was a new two-story brick schoolhouse in the village. The lower floor the Germans converted into a horse stable. There were enough barns to accommodate all their horses, but their contempt for Russian education was so prodigious that they put their horses into the classrooms. They did more—something so fantastic that when peasants of the village spoke of it they laughed with amusement as much as indignation. When a horse died they didn't bury it. They did not even drag it outside of the schoolhouse. They rolled it into one of the hallways, covered it with dirt, and left it there to putrefy!

In the afternoon of August 4, 1942, the chairman of the district and other Soviet officials who had acted as guerrillas during the months of occupation returned to Pohoreloye. As they went through the schoolhouse they found in one hallway eight dead horses covered with dirt.

All around Pohoreloye were open fields which would have made suitable cemeteries for the German dead—officers and privates. Within the village were large garden plots without trees or rocks which would have served such a purpose no less suitably. Instead the Germans took down four wooden houses in the center of the village and converted the lot into a burial ground for officers.

When they first swept into the village one of their tanks was stuck in a mudhole on a side street. There was plenty of brush, straw, lumber, brick in the village with which to fill up the mudholes, but the Germans did not bother to use such materials. Instead, they used books. From schools and libraries they hauled books into the streets. Tolstoy, Pushkin, Mark Twain, Dickens—all went into mudholes!

During the ten months of the occupation they made away with all the libraries and all the books they could find. The writings not only of Karl Marx, Friedrich Engels, Lenin, Stalin, and other socialist and Bolshevik leaders and authors, but of poets, novelists, playwrights, Russians and others, went into mudholes or into ovens.

The village of Pohoreloye housed the central library of the district. This library was a source of supply for the twenty other libraries and for the fifty-six schools within its geographic jurisdiction. Not one volume of anything has remained on its shelves or anywhere else. Only in the garret of one building, formerly the newspaper office, was a set of eighteen volumes of Lenin—untouched. The librarian had hidden the books there on the day the Germans came, and evidently they had failed to search the garret. All other books in the entire district, on any subject by whatever author, have disappeared—used up as raw materials or gone up in smoke.

Food and books are the first objects that command the energetic attention and action of Germans when they sweep into a village. The food they confiscate; the books they destroy.

In the ten months of their occupation the Germans opened only one school, in a village named Bezumovo. But the school, as the peasants soon learned, had a military rather than an educational purpose. Only two subjects of study were permitted—mathematics and German. The instructors in mathematics were Russians. The teacher of German was an imported Ukrainian, one of those Ukrainian intellectuals from Galicia who in the name of Ukrainian nationalism have allied themselves with Hitler's "new order" in Europe. This particular Ukrainian, who spoke a faultless Russian, came to the village of Bezumovo not primarily to instruct Russian children in German, but to establish personal relations with the peasantry and to spy on them for information about guerrillas.

The school was in session only two weeks. The Russian counterattack drove the Germans from Bezumovo, and the Ukrainian "teacher" fled with the German Army.

"Nothing they did here," said a Russian peasant, "made any sense. They hurt us, but often they hurt themselves. Why do you suppose they did it?"

Why indeed? In the years and centuries to come, hundreds, thousands of learned books will be written in answer to this one question about Germany and Germans in the era of Hitler's rule and Hitler's wars. Predominant in them is the passion for destruction!

There were twelve flour mills in the Pohoreloye district. Two were old-fashioned windmills; the ten others were operated by water or electric power. Now all are gone, burned to the ground.

Of the 8,600 homes in the district only 1,800 remained after the Germans were driven out. The others they either had burned on their retreat or had taken apart during their stay. The lumber, the furniture, the fixtures they used in the multitude of dugouts they had built all along their far-flung trench lines.

The room in which Paul Winterton, of the London *News Chronicle,* and I stayed in Pohoreloye was furnished with a table, a few chairs, a couch, a bed. I complimented the peasant woman of the house on having salvaged so much of her household goods. She motioned with her hand. She had salvaged nothing. The furniture in my room was hauled from German dugouts after the Germans had gone. That was where the Germans had hauled nearly all the furniture of the village and also the pillows, the blankets, the sheets, the towels. It was only after they were driven out that the people got some of it back, though none of the bedding. Most of the furniture was wrecked in the fighting or set afire by the Germans.

Of the fifty-six school buildings in the district only four remained, and these, after the German retreat, were without roofs, floors, windows, furniture, and equipment. In the sixty-four villages which had been under German occupation for ten months, only one school was left—that is, the walls remained. All the others were taken apart or burned.

The annihilation of the achievement of the Five-Year Plans and the concerted effort to drive Russia back to a condition of primitive isolation and helplessness are only one phase of the "new order" as the Germans practiced it in the rural region of Pohoreloye. The waging of biological, at times zoological, warfare on the civilian population is another equally marked. The exigencies of war did not call for such warfare. There was no special military need for it. It can be regarded only as a deliberate policy of depopulation.

One evening in the office of the Soviet, Winterton and I met a group of local citizens who had returned to the village. Here was Loshshadin, a bright-eyed man bent over with rheumatic pain. Formerly he was a cobbler. Now he is chairman of the village Soviet. Eight persons in his family died during the German occupation. Out of eleven persons only three remained alive. Here is Olga Subbotina, a middle-aged woman with a broad face and an unruffled manner. She is the only survivor of a family of ten. Here is Fyodr Baryshnikov, manager of the government general store. His mother and his fifteen-year-old daughter died during the German occupation. There was not a person I met who had no deaths in his family.

Before the war, the population of Pohoreloye was 3,076. When I visited the village the population had dwindled to 905! Not all of those who are missing are no longer alive. The healthy men of military age are in the Army. Some of the villagers are somewhere in the Russian rear. Still others are in the German rear, driven there by the German Army before its retreat. But most of those who are missing will never come back. They are dead.

As things go in occupied zones the district of Pohoreloye did not lose many lives through executions. Two guerrillas were hanged. An eighteen-year-old girl, a schoolteacher named Lubov Ivanovna Ozerova, ran away from Pohoreloye when the Germans first came. A few days later she returned to gather her personal belongings, mainly her clothes. A Finnish soldier tried to stop her. She protested, argued, proceeded to pack her things. Enraged, the Finn drew out his bayonet and stabbed her to death. A neighbor named Zaitseva wanted to bury the girl in the cemetery. The Finn would not permit it.

"So I had to bury the schoolteacher," said Zaitseva, "in a courtyard."

In the village of Ilynskoye, about nine miles from Pohoreloye, a German officer disguised as a colonel of the Red Army made the rounds of Russian homes one evening and solicited enlistments for a new guerrilla brigade. This brigade, he said, would "fight the Germans with every weapon." So successful was his disguise, so perfect his Russian speech, that no one in the village suspected his duplicity. Eighteen peasants agreed to join his brigade. The "colonel" wrote down their names. The next day the Gestapo rounded up these peasants and shot them.

It was not execution but famine and the diseases that result from undernourishment that exacted the high toll of life from the rural district under discussion. On their arrival the Germans, as usual, confiscated all food and forage. They seized not only grain and vegetables but livestock—cattle, sheep, geese, chickens, rabbits. Yet with all their vigilance and ruthlessness they could not prevent peasants from hiding sacks and boxes of food, usually in a hole in the ground—especially cereals and potatoes. As the months of winter dragged on the peasants managed secretly to feed themselves as best they could from their hidden stores.

When spring came few families were in possession of grains, potatoes, cabbage, dried mushrooms. When they worked for Germans digging trenches, building roads, performing other menial tasks, they were given at most a three-pound loaf of bread every three days and every day a bowl of oat soup with only a few kernels of cereal.

On such a diet hard labor was impossible. Productivity was low. No methods the Germans invented to improve it could make up for the lack of physical energy. Men and women could not wield a spade, an ax, a saw with their wonted and normal effectiveness.

All that the peasants had said in that stormy gathering on the public square of Pohoreloye when I first visited the village was more than confirmed by fresh testimony.

Peasants were encouraged to plant gardens and were assured that the vegetables they grew would be theirs. In the absence of horses, plows, other implements they dug up plots of land with spades, set out cucumbers, potatoes, cabbage, beets, and onions. Germans and Finns made the rounds, watching the Russians work, and again assured them that only they would enjoy the fruits of their labors. Yet when onion tops grew large enough to be picked, when beet leaves grew heavy enough to be plucked for soup, Finns and Germans again made the rounds of gardens and shaking the birch sticks in their hands warned the Russians to touch nothing, not a single onion blade, not a single beet leaf! Only Germans and Finns were to get the benefit of the fresh vegetables. Arguments, tears, reproaches did not help.

Most people had no bread any more, nor potatoes. Now they couldn't eat their own garden vegetables. So they fed, as they say, on "grass," meaning sorrel leaves, young nettles and those not so young, clover heads, clover leaves, other wild plants. They dried the clover heads, ground them into a powder which they used as flour for baking cakes. Often they mixed moss with this flour. They mixed everything that would give them bulk, taste, nourishment.

When berries and mushrooms were in season the peasants were formed into groups of ten and ordered to pick them six days a week for the Germans, one day for themselves. Though always under armed escort, only women and children were allowed to do the picking. Men were not trusted and were not permitted to go to the woods. But now and then a soldier did not hesitate to appropriate for himself the berries and the mushrooms that the Russians were permitted to pick for their own use. There was neither law nor order to prevent any soldier or officer from doing anything with either the food or the household possessions or the lives of the peasantry in the villages.

The only crop the Germans fostered was flax. Frankly they told the Russians they needed flax as raw material for their industry. But other field crops they ignored. They made no effort to sow oats, barley, or buckwheat for the simple reason that they had no seed. They had used up all local seed for food, and none came from their own rear.

In consequence of growing lack of food, starvation was rampant. Russians swelled up and died. They expired in their homes, in barns, in fields. They fell to the ground and never arose.

When the Red Army reconquered the district it brought food for the famished population. In Pohoreloye fifty men and women ate bread with such zest that it killed them. In the months they had been living on "grass" their bodies had become so weaned from normal foods that when they indulged their long-starved appetite for bread, their bellies bloated and they died.

In the prewar days there were no epidemics in the district. During the early fruit season dysentery made its appearance but only among those who insisted on eating green fruit. The Pohoreloye district was swept by typhus which the Germans had brought. The destruction of the artesian well, the bathhouse, and other sanitary improvements, as well as the undernourishment of the population, had provided a fertile field for its propagation. The hospital in the village the Germans had taken apart during their stay there, still further evidence of their passion for the destruction of everything modern in the Russian village. The Russians renovated four wrecked buildings, fitted them out with bunks, and were using them as hospitals, chiefly for patients afflicted with typhus.

At a most conservative estimate, at least one third of the population in the sixty-four villages that had been under German occupation for ten months is no longer alive. All the men and women I met said that if they had had to spend another winter under German rule, they did not know how they could possibly have survived it. They had no food reserves. The Germans had done nothing to encourage the growth or preparation of any. In summer, people ate grass and carrion and died. In winter, they would have only carrion, and even more of them would die.

In the Pohoreloye district the Germans informed the peasantry that the kolkhoz would be dismembered and the land divided on the basis of individual ownership. But they did not mean ownership in the sense in which the word is commonly understood. The peasant could not sell or rent his land. Nor could he work it as he pleased. All tillage would be collective, and as in the days of the *mir* under the Czar the village would be collectively held responsible for taxes and for all other obligations to the German Government. They even resurrected the old Russian word *obshtsnina* (community) to designate their new land system, or rather to mark its identity with the one that had prevailed so overwhelmingly and disastrously in the days of the Czar.

This is not, of course, the same land policy which they have been pursuing in other parts of occupied Russian territory. In many places they are retaining the collectives and the state farms not as Russian enterprises but as the possession of the German state or of individual German landlords.

In Pohoreloye the Germans did not stay long enough to carry out their new land order. But the information they imparted to the native peasantry as to its nature offers fresh testimony of their aim to push Russia back into the primitivism out of which, at an inordinate cost of substance and blood, she had striven to emerge. It cannot be too often repeated that "back to primitivism" with its waste and weaknesses is one of the cardinal principles of the German "new order" in Russia.

From the evidence I have gathered in the village of the Pohoreloye district it is obvious that depopulation is another. Despite war, despite wrath over continuous guerrilla warfare the Germans could have done something, if only to derive increasing benefit for themselves from the Russian lands, to enable the civilian population to feed itself. But their racial haughtiness, their hate of Russians, their nightmarish dread of Russian as of all Slav fecundity override all considerations and dominate all their actions. The more Russians that die from starvation and disease, they reason, the fewer Russians there will be in Europe to dispute the supremacy of the Germans. The more the Russians are flung back into the era of primitivism, with hardly any schools, books, and organizations, or none at all, above all, with hardly any weapons of the machine age to fortify them socially, mentally, and physically for the preservation of their group existence, the more helpless they will be in their resistance to German rule and the more readily they will submit to the will and the mastery of the "superior race."

Primitivism, obliteration of national culture, and depopulation are the pillars of Germany's new order in Russia, as Winterton and I learned of it during our visit to Pohoreloye.

"Have you heard of the village Zolotilovo?" asked a Soviet official in Pohoreloye.

"No," I said.

"Well," he said, "it's an extraordinary village. There's hardly a home left there. People live in dugouts, and we wanted to move them for the winter into villages where they could live in houses. But they refused to go. They'd rather live in dugouts in their own village, they said, than in the best of houses elsewhere. There's an example for you of the tenacity of the Russian man and his love for his native piece of land."

So we went to Zolotilovo.

The roads were so muddy that we dismissed the team of horses with the straw-filled cart which the Soviet had offered us and chose instead to make the journey across lots and on foot. The weather could not have been more auspicious. The rains which for two weeks had daily been drenching the countryside suddenly stopped. The skies cleared, the sun was out full and warm, and the air was permeated with the smells of drying leaves, withered grass, and lingering autumn flowers. Leaving the streets of the village, we descended into a valley and followed the bend of the Derzha River.

The ground was wet and squeaked with moisture, but the grass was thick and firm, and the walking was easy and pleasant. My guide was an official of the Pohoreloye Soviet. We assumed he knew more of the war conditions of the countryside than anyone else. Yet hardly had we started to follow the bend of the river when we heard a woman's voice crying frantically,

"Don't go that way—mines will blow you up—mines—mines!" We halted and looked back. A woman was leaning out of a window in a shattered wooden house on a height overlooking the valley wildly motioning with her hands. "Mines will blow you up!" she cried again as she saw us looking toward her. "Go the other way."

"They've planted them as thick as grass—see?" said the Soviet official with discomfiture. "It's ten weeks since we've driven them out of here, and we've dug out of the ground thousands of mines, but there are lots of them still around."

Indeed there were. As we walked along the abandoned battlefield we stumbled not only on signs put up by sapper troops warning pedestrians of danger, but on actual mines locked in tin cans or lying around with their noses deep in grass as if in defiance of attempts to remove or even to find them. Not only mines but hand grenades, anti-tank grenades, and other explosives were scattered in the weed-festered fields or lay in sequestered rows behind thick bushes. Parties of soldiers were at work clearing the fields, but there was so much to be cleared.

"We find German bodies all the time," said a soldier to whom we stopped to talk. "Every day we find them in their machine-gun nests."

"See how thick these nests are," said the Soviet official, pointing to the rows of little trenches running off the main trench line. "They had centered tremendous fire power here, tremendous," added the official.

"But our artillery," resumed the soldier proudly, "hit them straight in their nests—all along here," and he made a wide swing of his arm. "Just flattened them out. We keep finding new bodies all the time."

"What do you do with the bodies?" I asked.

"Throw them into the trenches and cover them up with dirt."

"Do you take off the clothes or the shoes?"

"Take off nothing except documents—just throw them in, clothes and all—let them rot."

"Let them fertilize our good Russian earth," said the official.

We bade the soldier farewell and walked on. The fields were ribboned with ditches, craters, and shell holes, and were littered with the heaps of ruins that Russian guns had made of German pillboxes and dugouts. In some places the trenches were already filled with dirt, but mostly they were still untouched by human hands. But nature had already shot brush and grass out of the walls and had loosened the clay stiffness so the earth could again come together and close the gash on its face.

The fields, level and stoneless—thousands and thousands of acres of loam mixed with sand and clay—had remained uncultivated. The grasses and weeds, now browned with frost but still thick and tall and upstanding and waving in the breeze, testified to the rich substance of the soil. Formerly, it grew rye, oats, potatoes, cabbage, clover, timothy, and other crops as bountifully as it was now growing the unwanted and ruinous weeds.

"It'll be hard," sighed the Soviet official, "to get rid of these weeds." With a frown of pain he swept his eyes over the vast expanse before him and sighed again. "Nothing is so hard as to clean a field grown to weeds for a whole year."

Getting the land cleared of alien growth is only one of the multitude of tasks that is facing the returning peasantry.

We came to a small wooden bridge high over the stream. Below, on an uplifted mudbank and shiny with whiteness of sun and bleached bone, was a human skeleton. Daily, teams and trucks drove over the bridge, pedestrians by the hundreds and of all ages were crossing it back and forth. Only a short ten weeks earlier armies were in pursuit of each other over the road, the bridge, the fields, all in full view of the bony remains of a man. Yet no one had bothered to bury it or dump it into the stream, as though it were something too evil or too sacred to be touched. In a superstitious age its very presence beside so open a thoroughfare might have become the subject of drama and legend, of terror and miracle.

Head raised high, jaws apart as if flaunting the double row of white teeth, large and even and closely set together and void of the least blemish of age or dissipation, the skeleton lay bent at the waist as though after the bullet had struck the body the man had doubled up with pain and fallen never again to rise.

A German or a Russian? No one could tell. Was it insects and worms, wind and water, or also dog and wolf that had so cleanly picked the young flesh from the sturdy bones? Again no one could tell.

The young and handsome lieutenant who had joined us on the trip bent over the unknown soldier without a tomb and looked closely. Then reflectively, and as if speaking to himself, he said quietly:

"He lived, he labored, he loved—and now——"

I had seen many villages which the Germans had burned. Never had I seen one like Zolotilovo. A few buildings had escaped the flames of the incendiaries, only a few. Of the others not even the shattered brick ovens or the soot-stained chimney stacks remained. All was gone—burned, taken apart, hauled away into the dugouts or used up for surfacing the roads.

I stopped on the very approach to the village before a level, rain-soaked stretch of ground that even in its ruins bore the trace of a former home. Out of the weather-packed rubble rose the bent legs of an iron cot, broken white dishes, a twisted sled, a mass of misshapen tin and brass that had once been a samovar, and a half-burned trough out of which a joyful child might have fed a favorite calf, pig, or lamb. These were all precious possessions in the Russian village, evoking images of home life, of daily toil, of comfort, of family refuge, of human and animal contentment. Now nothing remained. The stately birches in the street were felled and blasted —with tops burned, trunks lacerated, limbs hanging like stubby legs and arms. But the nests of the *grachi* (rooks) remained. Not a standing tree but the top was studded with the coarse and capacious nests of these birds, only too familiar in Russia and not much more highly esteemed by the peasantry than the robber crows!

Zolotilovo had been not only a pretty but a prosperous village—as prosperity is counted in the Russian countryside—so the accountant of the kolkhoz told us. It consisted of fifty-five families, all members of the kolkhoz, and each household had its own cow, usually also one calf, a pig or two, at least four sheep and two lambs, so that people could raise their own wool for winter coats and felt boots. They also kept enough hens to supply them with eggs and now and then with choice fowl for a wedding feast or a holiday repast. Some of the people had grumbled, of course, for it is always in the nature of people to grumble when they cannot get the things they think they ought to have or absolutely must have. But how well off they were—how bountifully well off!

The war struck Zolotilovo with the suddenness of an evil dream. Word reached the village that the Germans were in Pohoreloye, which is only a few miles away. Swiftly people started to kill livestock, to dig holes in the back yard or in the fields and bury the freshly slaughtered sheep and calf. Hurriedly they packed into boxes or wrapped in sacks, or only in straw, leather boots and felt boots, woolen coats and holiday suits, grains and vegetables, bridal gifts and ancestral heirlooms, and buried them in the ground. They worked without letup, young and old worked and worked, because every minute counted and every chunk of meat, every potato, every towel they buried, the enemy couldn't steal.

They'had already heard of the ways of the Germans in war—on the radio, from passing soldiers, from letters of husbands and sons at the front. War had become not only fighting but a business or profession of German general, corporal, orderly, and private. The conquered territory was theirs to plunder at their will. They even invented a way of shipping back to Germany beaten eggs that would be fresh on arrival. They'd break the eggs into an earthen jar in which peasants stored milk, seal the top, and ship it—far, far—to fathers, mothers, sisters—to everybody in Germany. So the people in Zolotilovo were preparing for the worst and were determined to save as much as possible from theft and devastation.

Then the Germans came and stayed ten months!

Now they were gone, but Zolotilovo was a village no more. It had taken years and years of toil and sacrifice, of sweat and blood to build it up and in ten months all was as if blown up in smoke. Before the arrival of the Germans there were fifty-five families in the village. Now only thirty-two remained, and these were no longer without missing members. The population had dwindled from two hundred and seventy to eighty-four! From famine alone at least eighty were known to have died. Some of those who had not yet come back were somewhere in the German rear, but whether alive or dead no one knew. Of those youths who were driven to Germany not one word has been heard since the day of their departure.

Gone now were not only the homes, the barns, the livestock, the implements, the boots, the clothes, the samovars, the mirrors, the ikons, the pictures, the books, the papers, the cats, the dogs, but the dishes and the utensils. Never had they imagined they would be reduced within such a swift few months to such utter "nakedness."

On their return to Zolotilovo they hastened to dig up the places in which they had buried some of their possessions. Not always did they find them. Wind, rain, snow, the tramping of soldiers had obliterated all trace of some of them. Often the Germans had, as it were, sniffed them out and dug them up and emptied them of all stores. It was astonishing how many such hiding places the Germans had dug up.

Often, because of the haste and the carelessness with which the hiding and the packing were accompanied, clothes and foods were left in dampness and when recovered were no longer fit for use. Yet the little they had managed to salvage, especially of winter clothes, was of inestimable value. If only the winter wouldn't be too stern—which was too much to expect in a country noted for stern and sturdy winters.

Sergey Ivanovitch Ivanov—what name could be more Russian? He was fifty-two and the accountant of the kolkhoz. Blue-eyed, dark-haired, with the stubble of a beard so thick and short that from a distance it looked like a winter flap around his chin and cheeks, he had been a soldier in the Czar's army and had spent three years fighting Germans in the first World War. During those three years he had met many Germans in many parts of Russia, Poland, Austria. But those Germans were not anything like the Germans in Hitler's army. Those Germans could be cruel, but they had respect for Russians as human beings. They didn't torment people for the sake of the pleasure it gave them. Nor did they destroy property for the fun of it. They were scrupulously careful to save every nail and every board that fell into their hands.

"But these new Germans," said Sergey Ivanitch, as neighbors called him for short, "the devil only knows where they come from. They go into a house and help themselves to anything they want as though it were all theirs. If they see a cupboard or something else they cannot use, they don't mind breaking it into pieces. If you say anything, they shake their sticks or their fists or strike you on the back or on the head.

"I had a clock in my home, an old clock, an ancestral clock. I'd wind it up all the way and it would stay wound for two weeks. A young German walked into my house, took the clock off the wall, tore the insides out and threw them out of the window—just threw them out. Pigs wouldn't eat the clock. Crows wouldn't touch it. Why did he destroy it? I could understand his wanting to take it home, ship it in a box to his family. Thieves are thieves, and there are no thieves like German thieves. But to throw the inside mechanism of an old family clock into the street and smash it—so it could be of no use to anybody—excuse me, friends—I don't understand it."

He paused, thought an instant, wiped his mouth with a gloved hand, and resumed:

"Consider this: Fritz comes into a house, his boots deep in mud. He could have swept the mud off his boots with a broom as we do before we enter a house. But not Fritz. He must walk in with all the mud on his boots. He struts across the living room, sits down, and turning to the woman of the house, he whistles as to a dog—actually, as to a dog—and when she turns he says, 'Come *matka*—pull the boots off my feet and clean them!' Imagine a soldier asking a woman to pull the boots off his feet! It just gives him pleasure to look on our women as his ready servants or slaves—the devil only knows which he considers them.

"Or—or," Sergey Ivanitch rushed on, "consider this. He comes into the house. It is cold, the fire is low. He wants to build it up. There's nobody to do it for him. He's got to do it himself. He goes outside, fetches a chunk of wood and proceeds to chop it up in the middle of the floor and ruin the floor! Why couldn't he have chopped the wood outside, as we do, as the Germans did in the first World War? Ask him, and he laughs or whistles at you and snaps his fingers as at a dog."

A group of children hooded up in winter clothes gathered about us. Sergey Ivanitch, suddenly becoming aware of them, searched their faces with his eyes and pointing to one little girl with large blue eyes and red cheeks, he resumed almost heatedly:

"Here is little Maria, seven only, going on eight. The Fritzes made her carry two pails of water on a cross pole over her shoulders. See for yourself how small she is. She couldn't carry the water. She bent under the load. She groaned, the poor little thing. She wept. And the Fritzes just stood around and laughed and laughed and kept shouting at her, 'Go on, go on,' as though she were a horse or a cow. . . . Akh, what people!"

The wife of the former chairman of the kolkhoz, a middle-aged woman with soft blue eyes and a calm, melodious voice, invited us into her dugout. It was small and square with a little brick oven and a window into the open through which the sun poured in. Seven people shared the dugout. "We do manage," she said. "We sleep here—here"—pointing to the top of the oven, the bunks, the benches, the floor. "We get along. If only it was not so wet under the floor."

She stepped heavily on the boards, and we heard the splash of water underneath. "It's very wet here. But it's better than living under Germans. It's wonderful not to have them. Do you think they'll come back—can they come back?"

"No, Auntie, we shan't let them," replied the young lieutenant.

"Only men like you, Lieutenant, can stop them."

"We shall stop them; they won't come back," the young officer said with such assurance that the two children who sat at the table listening intently and in silence looked up and smiled at him with affection and fervor.

"I understand," I said, "the Soviet in Pohoreloye has offered people here a chance to spend the winter in homes in other villages where they wouldn't have water under the floor."

"No," replied the woman stanchly, "we don't want to leave Zolotilovo. Nothing is so good as your own home, no matter how humble."

"What did I tell you?" said the Soviet official, and laughed.

"Love of your own village is very strong in you people in Zolotilovo," I said.

"Why shouldn't it be? We go outside and look around, and everything is familiar and friendly, and it is cheering to see it all and to know it belongs here and is yours and nobody else's. No, we don't want to live anywhere else—now that the Germans are gone. If they come back—we won't live long anyway—we shan't want to."

We left the dugout and again came out into the open. The warmth of the weather, the bright sun, the woman's words of fealty and love for the village and all that surrounds it made the wanton ruin of the place all the more incredible. It lay before us on all sides like a cloak of death, all the more grim because so absurdly needless.

We heard a dull and muffled sound and wondered what it was. Observing our curiosity, the woman from the dugout said,

"Someone is grinding flour. Come along and see how we do it now."

We followed her into a shed and saw a little girl, white and powdery with flour, turning a long pole attached to a large wooden roller. She stopped working and the woman said,

"We are inventive people—made it ourselves—out of our own materials with our own hands. We had to after the Germans had burned all the flour mills in our district."

In my years of travel in Russia I had seen many old-fashioned mechanisms, but never had I seen a flour mill like the one I was now examining. In the absence of stones, heavy round blocks of wood were used. One end of each block was studded with shiny bits of sharp-edged iron scrap.

"Where did you get the scrap?" I said.

"Broke up an iron cooking pot—that's where we got it," explained the woman. An iron cooking pot is one of the most valuable assets of a peasant kitchen. Yet it was broken up to provide "teeth" for the wooden rollers!

With the long wooden pole that was attached to the top roller it was an easy contrivance to manipulate and though the flour that poured out of the spout was coarse, it was flour and could be baked into bread, into griddle-cakes, into dumplings.

"We have to be inventive," said the woman, "to live nowadays." Of course, of course one felt like echoing.

We bade the little girl good day and to the accompaniment of a rumbling sound we walked out of the shed. In prewar days the village had ceased to hear such sounds, and no one imagined he would ever hear them again. The hum of the engine had replaced the rumble of the hand mill. The engine was gone, and the hand mill and the flail were back, reminding the populace, indeed all of us, of the impermanence of any new and all human achievement in the face of a power as bent on death and destruction as the Germans of today.

Several small new buildings had already been put up in the village. One in particular attracted my attention. It was set in the center of the village, with the moss sticking out of the wooden timbers that were laid over each other and with the large thatch roof hanging over the walls like an overgrown head over a puny body. It might have been a "doll's house" intended for the amusement of the children. But with the ruin all around it was absurd to imagine the people would bother about a doll's house. It was, in fact, the most precious building in the village—the granary! Its size told the story of the paucity of the crop that had been gathered, more significantly than could the bookkeeper's figures.

The Germans had counted on harvesting the rye which the kolkhoz had planted the preceding autumn. Loudly they told Russians they were glad it was so bountiful and they would gather it for themselves. But it was the one thing which in their hurried flight they had not the time nor the means to destroy. Much of the grain was trampled and shelled by the fighting armies, and when the people returned—that is, those who were alive and were free to come back—their first thought was to gather and husband the precious rye. Nothing like one's own bread after months and months of breadlessness!

They had neither horse nor tractor nor cow nor combine nor dropper any more. Hungry and tired as they were, they fitted out scythes and cradles and went out to gather their bread. They cut every stalk of it. Children raked up fallen ears or picked them up by hand. Nothing went to waste, not a single stalk, not a single ear! They shocked it and dried it and threshed it by flail, precisely as their ancestors had done for ages and ages. Then they built the little granary, and here it was—small and pretty, a symbol of their return to a new beginning and a new life.

"My beloved children!" came a trembling voice from across the street. "My darlings. Oh, my darlings!"

An old woman was hailing us. We had not seen her before. She had emerged suddenly from a dugout and was wading through the mud to our side of the street. She wore high boots, a black kerchief, a flapping sheepskin coat with the bare tan hide on the outside. In her hand she carried a mighty staff as high as her chin. As she was coming toward us, words of endearment tumbled from her lips with a catlike purr. She breathed hard when she stopped, but she was so full of talk and emotion that she couldn't pause for rest. Making the sign of the cross over her body in devout Orthodox fashion, she prayed for us in the Old Slavonic tongue, the official tongue of the Russian Orthodox Church.

"O Lord, O Lord, be good to these dear, wonderful, beloved children of yours. O Lord, O Lord, give them health and strength and happiness. O Lord, O Lord."

She sobbed with rapture, and tears stood in her sunken and beautifully rounded blue eyes.

"I am seventy years old," she raced on, "seventy, my dears. I've been to the city—to Kalinin and to Moscow, and I've had tea in city restaurants, yes, I have, my dears. My old man when he was alive liked to imbibe the bitter drink, but I never cared for it. A swallow of yellow wine I didn't mind, but the bitter drink—I never could stand it. All my life I worked and worked, and in my old age my greatest pleasure was to go to a restaurant in the city and order tea and drink it glass after glass at my leisure. Oh, what a delight it was, my beloved children!"

She paused but only long enough to inhale a deep draught of air. Again she started to make the sign of the cross over her body and to pray for us.

"When—when they came here, the evildoers and the lovers of all sin and wickedness, I thought all was at an end and Akulina Condratyevna Yeogorova would never again sip a glass of tea in a restaurant but would soon be laid into her tomb—that's what I thought, my sweet, dear, precious darlings. I couldn't help thinking it when I saw what *they* were doing with our innocent Russian people, who'd never done them any harm and who were so far, so very far away from their country and their homes and their people. And then—and then—like apostles from heaven—our own soldiers came, men like you, my darling boy"—pointing at the lieutenant—"and they drove away the evildoers who'd mocked and tormented us so long, so long. Oh, what a blessed day it was on this earth and in heaven too—and how I prayed and prayed, oh, how I prayed!"

Her eyes dimmed again, and she wiped them with her rough bare hand and swallowed hard and started again.

"And why are evildoers like them born into the world? Crows eat rye but also insects. Spiders kill flies. Nothing the good Lord has created, however evil, but does some good somewhere, nothing but German evildoers. And when I saw our own men speaking our own tongue telling us we were free to go back to our villages—not to our homes, the evildoers had burned them—but to our villages—and when I ate the bread they gave me, these dear soldiers of ours—delicious Russian bread—I wept and prayed and prayed and wept and praised the Lord for being so good, so gloriously good to an old woman like me. Blessed be the Lord!"

She bowed and bowed and again and again made the sign of the cross over her body.

"And then, my beloved friends, I saw our boys, our deliverers and apostles, lead two of the evildoers somewhere. So I hurried over with my staff and said to the commander, 'Good man, beloved man, let me hit them on the head.' I lifted my staff and got ready to strike, but the commander said, 'No, little grandmother, I cannot let you!' Imagine a Russian commander saying it. So I said, 'Why not? Don't they deserve it? Look at our village how they've ruined and burned it.' And he answered, 'Yes, they deserve it and they'll get what's coming to them, but I cannot let you hit them.' I couldn't help lifting my staff again—the good Lord wanted me to lift it and I felt a great strength come to my arms. Oh, how I could have hit the swine——"

The children who were standing around started to laugh, and so did we, but she rushed on unperturbed:

"And I said again to the commander, 'I'm a woman of seventy, a Russian woman, a God-fearing, God-loving woman, and I have to show the evildoer that even our old women hate them and are ready to fight them when they've got no guns to shoot with.' But he, my darlings, this Russian commander, only said, 'No, Granny, I cannot let you hit him.' So then I said,

'Promise me you won't give them a mouthful of food.' And he said, 'I promise,' but I didn't believe him, and why should we feed any of them?"

"Because we aren't like them," said the young lieutenant.

"Oh, angel, my deliverer—they won't come back, will they?"

"No, they won't, Grandmother."

"If they don't, I can go to Kalinin and to Moscow and drink tea in restaurants, glass after glass and at my leisure, can't I?"

"Of course you can," said the lieutenant.

"And I will, I will. I've got money. I get a pension regularly, every month I get it—ninety rubles, and my brother is a judge in Kalinin—he sent me one hundred rubles. I had a thousand when they came here, but they took it away from me. But I don't care as long as they are gone and don't come back any more. It'll be wonderful to go to restaurants and sit at a table with a white cloth and have tea brought to you and drink it at leisure with sugar or jam and little biscuits. Oh, I will go to Kalinin and to Moscow and I will drink tea in restaurants, lots and lots of tea——"

Again she made the sign of the cross over her body and prayed for us, and there was not only hope and fervor in her prayer but faith and dedication and the will to rise above adversity and evil and redeem the humanity and the dignity of man!

PART EIGHT: RUSSIAN QUESTS

CHAPTER 37: "WILL WE HAVE TO FIGHT RUSSIA?"

In ASIA, in Africa, in South America, wherever I stopped on my way home, Americans and Englishmen were eager for news about Russia. Some of them voiced only too freely uncertainties and anxieties as to Russia's postwar intentions and the possible conflicts that may arise between her and the English-speaking countries. Often enough the anxieties simmered down to the question:

"Will we have to fight Russia?"

Perhaps it is only an accident, yet it is worth recording that Americans were more preoccupied with this question than Englishmen. Rare was the Englishman who asked it, rare was the American who failed to ask it.

These Americans were puzzled and worried. They were glad of Russian successes at the Stalingrad and other fronts. They were happy America was contributing to these successes by shipments of planes, tanks, trucks, and food to the Russian armies. They were hoping for the day when America and England would launch a war against Germany in Western Europe and lift from the Russian armies the enormous military burden they have been shouldering on the European continent. They wanted an amicable, peaceful postwar world in which nations would compose differences not by the sword but by understanding.

But they were concerned about Russia. Reduced to essentials, this concern was over possible Russian ventures into world revolution, into Russian imperialism, or into both.

In the communist parties of the world, including that of America, which, they said, were promoting above all else Russia's foreign policy, they beheld a ready revolutionary weapon which Russia might wield at an opportune moment. In the unpredicted and unpredictable might of the Red armies they perceived the possible resurrection of an aggressive Russian nationalism, which might result in an encroachment on the territorial possessions or political prerogatives of other nations in Europe or Asia that would bring Russia into an armed clash with the English-speaking countries. So they were·wondering about the future, about a possible third world war in which, instead of being allies, America and Russia would be fighting enemies. . . .

Fresh from Russia, I could not help thinking of ever-growing Russian misgivings. How often did I hear Russians say, "They [England and America] promised us a second front in 1942. They did not keep their promise. In North Africa they are fighting at most eight or ten enemy divisions while we are facing 240. They want us to go on battling Germany alone in Europe so we'll bleed ourselves white and they can control the world."

The failure of a second front to materialize in the summer of 1942, when Russia was fighting with her back to the Volga, and in the winter of 1943, when she was driving the German armies from one fortified position after another, stirred no end of mistrust, apprehension, anger. American foreign policy in North Africa and in Spain only enhanced the troubled thoughts and emotions.

Thus Russian suspicion of the postwar intention of her allies is no less pronounced than theirs, and particularly American, of her postwar aims.

These suspicions are mutual and are rooted in old hostilities. For twenty-five years Russia has been haunted by the dread of "capitalist encirclement"; the capitalist countries have been provoked by the presence in their midst of communist parties, which to them mean symbols and agencies of world revolution, and by the challenge of Russia's destruction of private enterprise. The unexpected demonstration of military strength by the Red armies has fanned in these countries the fear of an awakened Russian imperialism, while American foreign policy, especially in North Africa and Spain, has intensified Russian misgivings of "capitalist encirclement" and of a possible *cordon sanitaire* against her.

The plea or the argument that in view of the impending Allied victory Russia's fears of such a move against her are groundless, if only because with Germany in defeat and with the other European nations obliged to face the monumental task of reconstruction none will be in a position to attack her, carries neither meaning nor weight with Russians. They have always appraised the world in terms of long-range possibilities. They know from their own experience and from the experience of Germany under Hitler that in this highly mechanized industrial age a weak nation, if it has the resources or obtains the necessary economic support from the outside, can quickly strengthen itself militarily. During the London Economic Conference in 1933, the few German journalists who were present scoffed at Russian fears of an impending war with Nazi Germany. Loudly they proclaimed that Germany was so poor and weak she could not possibly strengthen herself for war against anybody. I was present at the conference and remember only too well these private discussions. In the light of present-day events, the Russian long-range view of Germany was more than justified.

Maintaining a long-range approach to the postwar world, the Russians will not be persuaded into the assurance that their fear of a cordon sanitaire is baseless and absurd. They point to the Spanish Civil War and to Munich, in which Hitler and Mussolini won monumental victories, as attempts, though unsuccessful ones, of reviving the policy of the cordon sanitaire against them.

It would be to no purpose for us to disregard or fail to recognize the basis of the mutual distrust out of which even in the midst of war comes the American query: "Will we have to fight Russia?" and the Russian query: "Will we have to face another cordon sanitaire and go to war with the capitalist world?"

These queries demonstrate among other things the tendency of Russia and America to put the worst possible construction on each other's acts and utterances. Regrettable as this is, it can hardly be otherwise in the light of past hostilities and newly awakened fears in both countries. No doubt an Allied invasion on the European continent mighty enough to compel Germany to withdraw sixty to eighty divisions from Russia would allay anxieties and premonitions which the absence of such a front cannot fail to inflate.

On the face of it nothing appears more fantastic than the postulate of a forthcoming war between Russia and America. Because of geography alone this would be the most unfightable of all wars. Supposing Russian armies seized Alaska? Then what? Supposing American armies seized Kamchatka, Vladivostok, or all of eastern Siberia as far as Lake Baikal? Then what?

The only hope Russia would have of fighting a *victorious* war against America—the faintest of hope—would be to enlist as fighting allies at least all or most of Latin America. No Russian would be foolhardy enough to imagine that the republics to the south of us, predominantly Roman-Catholic and given to a system of large landholdings, would join Russia or any other nation in a war against the United States.

The only hope America could harbor of defeating Russia—the faintest of hope—would be to enlist as allies not only Latin America but England and other European nations. But what would move America to such an act? An uprising of the American Communist Party against the Washington Government? It would be drowned in blood the day it broke out. Would America initiate, join, or lead a holy crusade against Russia, together with England, South America, and other countries, including possibly China? It seems incredible that the Americans, who are worried about a possible war with Russia, would care to sanction or undertake such a crusade—no matter how repugnant to them are Russia's social and economic theories and practices. It seems equally incredible that the British public could be drawn into the support of such a war.

Of course unforeseen and unpredictable situations might arise which would precipitate a violent diplomatic clash between Russia and America. Nothing is impossible in a world so ridden with conflict and trouble and absurdity. But again the question arises—how is America to win a war against Russia, or Russia against America, even if either of them initiates or leads a crusade against the other, together with all the allies they can muster?

It is this writer's judgment that no power or powers in the world can ever *really* conquer either America or Russia. If, therefore, we have no assurance or hope of conquering Russia, and if Russia can have no assurance of conquering America, to what avail would be a war of the one against the other?

Yet nothing would be more foolhardy than to give a decisive and unequivocal answer to the above questions or to prophesy unqualifiedly regarding Russian or Anglo-American postwar policies. Too many have been the prophets about Russia whose prophecies lie rotting in dust. No one really knows what the future holds, because no one really knows how, when, where, and under what circumstances the war will end. One can only attempt an appraisal of Russia's position after the war in terms of her immediate needs and fears and in terms of her development in the past twenty-five years, particularly since 1928. That year looms more and more as the most momentous in Russian history because it witnessed the launching of the First Plan, and the coming of the machine age to the most remote parts of the inhabited regions.

CHAPTER 38: AFTER THE WAR—WHAT?

THE SOVIETS CAME into power in 1917 when the first World War was still being fought. Within a year the Central Powers collapsed not through Bolshevik or any other revolutionary propaganda but through the superior fighting powers of the Allies. Defeat and hunger stirred uprisings in Germany, Austria, Hungary. Communist regimes sprang into being. To Lenin's government in Moscow, engaged at the moment in battling for its life against internal Russian opposition and against foreign intervention, it seemed that revolution was about to sweep the world. On November 7, 1919, on the occasion of the second anniversary of the Soviet regime, Lenin said: "Victory of Soviets all over the world is assured; it is only a question of time."

But in 1921 the Soviet revolutions in Europe were over. Not one of the communist regimes outside of Russia survived. All were drowned in blood. Within a year Fascism reared its head in Italy, and Russia remained the only non-capitalist state in the world. But its power remained unshaken. It survived a brutal civil war, a foreign invasion, a disastrous famine. It lost millions of lives; industry was shattered, agriculture was devastated; the people were divided; they were hungry and ragged. But the Soviet regime rose to undisputed supremacy.

The question arose as to how Russia was to live. How would it build industry, advance agriculture, learn to read and write, strengthen national defense, promote art, science, education? How would it function as a people, a nation, a government?

The answer was Nep (New Economic Policy), which legalized certain forms of private enterprise in city and village. Thus with one stroke military communism in Russia ceased to exist. What was more, Lenin, in spite of all his hopes and prophecies of world revolution, was obliged to reconcile himself to the fact that it had failed to come and that Russia must seek her salvation, at least for the moment, within herself.

The full implications of Nep became more manifest after Lenin's death in 1924. The Bolsheviks split openly into two hostile factions. One, headed by Trotsky, insisted that even while going on with a program of internal development, Russia must base her eventual salvation on world revolution or perish; the other, headed by Stalin, maintained that Russia could survive and grow and develop only if instead of basing her fate on world revolution she strove to build within herself a socialist society. Here, then, was the battle of which I have already written of Trotsky's "permanent revolution" against Stalin's "socialism in one country."

Stalin won. The exile, imprisonment, and eventual execution of various opposition groups signalized among other things the Russian break with the Trotskyite idea.

What Russia is today and what she does in the war are largely the outcome of Stalin's policy of "socialism in one country." In 1938, according to Stalin, the output of Russian industry was 908.8 per cent of what it had been in 1913, a monumental achievement which dramatizes the full mean-

ing of the formula "socialism in one country." The Plans, the mighty
industrial centers, the collective farms, the new railways, airways, waterways, the many new cities; the Red Army, the Soviet Navy; the schools,
the colleges—all of Russia's economic, military, and spiritual power of the
year 1943, including Russia's rediscovery of her past and of her own
Russian soul, derives from a policy which has divorced Russia from
reliance on world revolution and to an increasing extent also from dependence on the outside world.

The war, let it be re-emphasized, has subjected this Russia to the severest
test any nation could possibly face—the test of national survival in a
conflict with the mightiest and most highly mechanized army in Europe.
She has survived the test. Thereby alone she has justified and, in Russian
eyes, almost sanctified herself. Survival, after all, is as much the first law
in the life of a nation as it is in the life of an individual. Whatever the outside world may say or think of Russia, of Stalin's "socialism in one country,"
of the mammoth cost of its fulfillment, the stupendous fact is that in the
most crucial moment in Russian history it has saved the nation and the
people from the cruelest fate they could ever know.

Not that Stalin or any of his followers in Russia have altered their
views on the eventual collapse of capitalism and on the imminence of revolutions that will destroy it. Most manifestly they have not. They continue to
regard capitalism as the source of all major evils in the world, including unemployment, racial persecution, international wars. Their original indictment of capitalism as a system of economics and as a social order they
have neither abandoned nor mitigated. They will have none of it in Russia.
On this subject there has been no hint of repentance or change. On the contrary, there has been only reaffirmation and rededication. Indeed, socialism,
at least in the sense of collectivized ownership of property, is now a powerful and major ingredient in Russian nationalism.

The question is whether in the postwar world Russia will abandon
"socialism in one country" for world revolution.

Of course if the Russian Communist party were to dissolve the Comintern and thereby dissociate itself from the communist parties in other
countries, esteem for Russia among non-communists in America and England would rise high. Russia would lose the monopoly of the communist
idea. But let us not hasten to assume that the rebelliousness which now
empties itself into the communist parties would cease to exist. Most
emphatically it would find an outlet elsewhere, in a new organization
or in the Trotskyite Fourth International. America, despite a system of
free land such as no other nation in Europe has known, despite the opportunities for advancement which the common man in Europe has likewise never experienced, America throughout her history has had re-

bellious groups of a more or less violent nature. Before the communists appeared on the scene, there was the I.W.W. There were, to a feeble extent there still are, the socialist and anarchist groups. There were the Molly Maguires. Stemming from a different social source, there was the Know-Nothing party. There is the Ku Klux Klan. There are the Christian Fronters. It is more than likely that when no longer obliged to base themselves pre-eminently on the support of Russia's foreign policy, the communist parties or those that take their place, whatever their name and whoever their leaders,. may become far more aggressive in their radicalism than they are now. Of course if this is properly understood by Americans it will be only a challenge to demonstrate that the existing American system has the flexibility, the strength, the survival value constructively to dissolve and eliminate the source of the rebelliousness or to reduce it, as has nearly always been the case, to more or less loud but unruinous magnitude. But that is something else again.

Yet communist parties or no communist parties, in Russian history, especially since 1928, in Russian orientation toward "socialism in one country," in the rediscovery and glorification of her past, in the manifestation and enunciation of Russian nationalism, without the curb of the nationality of other people in Russia, in economic and social stabilization—above all, in Russia's need for peace—a desperate need—there is in this writer's judgment a firm basis for full and fruitful co-operation with America and England. There is little likelihood that this co-operation, so pregnant with advantage for Russia, far more than to America, would be jeopardized by Russian ventures into world revolution.

Because of the devastation she has endured, the heavy casualties she has suffered, the stupendous sacrifices the people have faced, there is nothing Russia wants and requires more than peace and a chance to bind up and heal national and personal wounds. Further fighting, whether for nationalist or revolutionary purposes, must entail further devastation, further casualties, further sacrifices, further national and individual blood-letting. Most emphatically the Russian people, though ready for battle unto the end to throttle Germany's ambition to conquer and subjugate them, have no thought of military conflict for any purpose other than national defense. They yearn for peace and for the opportunity to live again in some degree of comfort. "If only the war would soon end" is the wish and the prayer of the Russian masses.

Of course the French Revolution brought forth a Napoleon. But under existing Soviet conditions no general however brilliant and ambitious can hoist himself to supreme power and embark on a campaign of military or revolutionary conquest. Ordinary soldiers and officers of lower rank who distinguish themselves in the fighting receive more frequent and more eloquent eulogy in the press, on the radio, and at meetings than do the most brilliant generals. Not until a general has won a significant victory is his name even mentioned in the press. Unlike England and America, Russia neither dramatizes nor personalizes the achievements of military commanders. Indeed, in no other fighting nation are the deeds and reputations of generals so little publicized.

During the battle of Stalingrad, when the Russians were fighting with their backs to the Volga, foreign correspondents searched far and wide for information as to the man or men who were leading the Russian troops. Their inquiries remained unanswered. They could not even verify whether or not the German and the British radio stations were justified in speaking of Timoshenko as the general in command of the Russian armies. Only later, when victories were achieved, were the names of the military leaders disclosed. But even then foreign correspondents obtained only meager information on the lives and personalities of these leaders.

The most highly dramatized war heroes in Russia are, I must reiterate, never generals or even soldiers but the high-school boys and girls in the guerrilla detachments whom the Germans have hanged. Shura Chekalin, Liza Chaikina, Zoya Kosmodemyanskaya remain the best-known and the most highly revered war heroes of the country. No general can hope to attain the esteem and the adulation that they have evoked in the Russian people.

Besides, political power in Russia is in the hands of civilians, and the system of control is so highly ramified and so vigilant that no military man could hope to wrest it for himself and to use it as did Napoleon for military conquest—not unless something unforeseen happens and the system of control and vigilance collapses, which is unthinkable.

There are other reasons why Russia would not care to antagonize America or England, but especially America, to the point of a break that may result in war. Russia still needs America, especially America, because of American experience in industrial development. America has been Russia's foremost guide and teacher in all her Plans and planning. Without America neither the blast furnaces of Kuznetsk and Magnitogorsk, nor the tractor plants of Stalingrad, Kharkov, Chelyabinsk, nor the automobile factories of Gorki and Moscow would have been possible. By their own admission and in spite of all their immense industrial progress, the Russians still have much to learn from America in technical achievement, in rationalization of labor, in efficiency, in a multitude of other aspects of modern technology.

A short time before the war the Russians had embarked on a gigantic program of commercialized foods. Tomato juice, other vegetable and fruit juices, corn flakes and other cereals, frozen meats, vegetables, fruits were for the first time finding their way into the Russian grocery shops. Popular prejudice against commercialized foods was combated with slogans, posters, electric signs, in the best American manner. The Russian people were beginning to respond with increasing eagerness to these advertising campaigns.

In the large cities the Commissary of Food was opening up cafeterias, quick-lunch counters (and using the American names for them), automat restaurants, automat barrooms, and featuring American food preparation including the "hot dog."

This process of "Americanization," which was only in its beginnings, came practically to a halt on the outbreak of the war. But its continuation is as important in the Russian scheme of things as advanced technological development, and in the pursuit of both, American aid of one kind or another is desirable, or perhaps indispensable. Not a factory director one meets in Russia, not an engineer or an engineering student but dreams of the chance he may someday have of going to America for a visit to American industrial plants.

Nor must one leave out of consideration the age and the personality of Stalin, the most aloof, the least known of the national leaders of the world. He was born in 1879, which makes him now over sixty-three years of age. Whatever his sins and his failings, he would not, one must assume, want to pass out of life when Russia was still engaged in combat, particularly with America and with the bloc of nations that might be her allies. Such combat would subject the Russia and the society with which his name is inevitably linked to further devastation and to the threat of unheard-of ruin.

The Plans and all that they imply in national power, national life, national aspiration, and national survival are the pillars of this Russia and this society, and Stalin has prosecuted them with complete disregard of the sacrifices they entailed in life, comfort, substance. They have saved Russia from conquest and subjugation. They have made Russia the power that she now is in Europe and in Asia. They are also the means of lifting Russia's living standards. In prewar days the amount of available consumption goods was continually rising. While housing remained backward, food staples were no longer a problem—not even eggs, meat, sugar, except perhaps in the more remote communities of the country. "We ate eggs [in prewar days] as freely as potatoes" was an expression I heard over and over in the villages which the Germans had burned and devastated.

Textiles, too, were more abundant. Again and again Russians showed me suits of clothes they had made from English woolens which Russia had begun to import. The cost was prodigious, but writers and engineers who were receiving high rewards for their services did not mind spending money on suits from English cloth. Ilya Erenburg once showed me an armful of letters and diaries that had been gathered from the German dead in the Don and Kuban in which some of the German writers spoke boastfully of the dresses, shawls, ribbons, and fabrics they had looted in Cossack settlements. In prewar days the material standard of living, though low by comparison with that of America, was rising, and had not Russia been obliged to arm as prodigiously as she did, it would have been far more comfortable than it was.

The war has not only halted, it has depressed the standard of living, until civilian Russia is once more reduced to barest essentials, sometimes to less. Further fighting, whether for revolution or imperialism, would depress it still more. Therefore it would be natural to assume, because of his age alone, that Stalin would wish to avoid such a contingency. He would want to resume the task of rebuilding Russia and making an increasing amount of "the fruits of the Revolution" available to the people. A man like Stalin, who has devoted himself exclusively to the task of industrializing the country, unifying its many peoples and nationalities, rousing its purely national consciousness, giving it new strength and new pride in itself, would most manifestly not care at this critical moment in Russia's history to jeopardize the achievements of his life work. He *would* strive with all his powers to preserve and reinforce them.

In an appraisal of Russian postwar policies and actions this purely subjective consideration cannot be disregarded, all the more so because Stalin has always been confident of catching up with and surpassing capitalist countries in production, by means of "socialism in one country" and further Plans. Explicit have been his declarations on this subject. "Only if we outstrip the principal capitalist countries economically," he said in his speech at the 18th Party Congress on March 10, 1939, "can we count on our country's being fully supplied with consumers' goods, on having an abundance of products. . . . What do we require to outstrip capitalist countries economically? First of all, the earnest and indomitable desire to move ahead and the readiness to make sacrifices and invest substantial amounts of capital for the utmost expansion of our socialist industry. Have we these requisites? Of course we have. Further, we require a high technique of production and a high rate of industrial development. Have we these requisites? Of course we have. Lastly, we require time. Yes, comrades, time. We must build new factories. We must train new staffs of workers for industry. But this requires time, no little time. We cannot outstrip the principal capitalist countries economically in two or three years. It will require more than that."

This speech, like the other speeches at the Congress, is concerned preeminently with Russian development, Russian industry, Russian agriculture, Russian education, and everything else that pertains to Russian life and the Russian people. . . .

Such overabsorption in Russia and things Russian has made Russia the most monumental pioneering country the world has ever known. On their creative side the Plans are Russia's method of pioneering her way all over her vast domain into the machine age in the quickest possible time. Despite achievements which have given her a mighty army, a mighty industry and agriculture to support it, the pioneering has no more than attained a good beginning. It has subordinated housing and consumption goods to metal-lurgy and machine-building. It has only begun to penetrate the immense and unoccupied wilderness of Central Asia, Siberia, and the Arctic. Prior to the outbreak of the war, Russia teemed with laboratories and with plans of development. Russia was a land of blueprints—not a town, not a village, however ancient, but had its own far-reaching project of future de-velopment; Russia is still a land with comparatively few highways and railroads, especially in Asia. The emphasis on the family is in part moti-vated by the urge to stabilize, solidify, and sanctify it and in part by the need to populate the newly developed lands and regions and even more those that await development.

"To put masses of people to work on masses of resources" might be one of the slogans of Russian pioneering, and because of the size of the country, the untold and as yet undiscovered resources it holds, Russian national and individual energies can find boundless fulfillment for years and years in purely Russian tasks without the need of plunging into territorial or any other external aggrandizement.

Considering Russian losses in the war, the devastation of occupied re-gions, the depressed standard of living, the desire of the people for peace, Stalin's preoccupation with Russia, his "socialism in one country" and what it has meant and done for Russia and the promise it holds for the future as Russians envisage it; considering also his age and the natural inclination of a man in his position to carry on the mission he has undertaken and lift the standard of living to a point where the people will feel recompensed for the inordinate sacrifices they have made; considering, further, the tasks facing Russia inside Russia—there is more than good and ample reason to assume that Moscow would strive to avoid further conflict, further fighting, further bloodletting. Indeed, it is inconceivable that Moscow would pursue any other policy unless Russia's national security were threatened. To the Russia of today nothing matters so much as national security.

The fact is that for years Russia had been seeking all manner of associa-tions with non-fascist countries for the purpose of staving off war. At the end of 1934 she joined the League of Nations. In May 1935 she signed a treaty of mutual military assistance with France. Simultaneously she concluded a similar treaty with Czechoslovakia. In August 1937 she signed a non-aggression pact with China. Even after Munich Litvinov pleaded, indeed clamored, for a six-power military alliance, so as to present a united military front against Hitler's aggressiveness.

Only when the war broke out and she saw herself menaced by Germany did Russia act to fortify her borders by moving into eastern Poland and the Baltic states and obtaining through war territorial concessions from Finland and Rumania. Denunciations of imperialism did not halt her from occupying borderlands which she had excellent reason to believe Germany was already scheming to use as springboards for an invasion. Had not Germany started the war there is no reason to believe that Russia would have embarked on these military maneuvers.

At the end of the war, whatever else Russia may or may not want, she will insist on the kind of territorial and political security which will give her the protection she feels she must have against possible invasion in the future.

Let it be noted that since the coming of the Soviets Russian thought and action have been motivated by fear of outside attack more than by any other consideration. It is the dread specter that haunted alike and in equal measure Lenin, Trotsky, and Stalin. "Capitalist encirclement and the dangers it entails to socialist society," to use Stalin's expression, has been the challenging refrain of all Soviet oratory, all Soviet diplomacy, and all Soviet internal policy, including the speed and the ruthlessness with which the Plans have been executed. Russia will therefore seek such protection of her frontiers as will rid her of the fear of hostile border states and make her feel territorially secure from attack or in a favorable position to meet it.

We may also be certain that she will oppose any political alignment or coalition in Europe which she will regard as a possible coalition against her. The very word coalition rouses deep emotion in Russia for it brings up the memory of Clemenceau's cordon sanitaire in the early years of Sovietism and the presence on Soviet territory of French, Italian, British, American, and other foreign troops that sought the overthrow of the Soviets.

Will Russia's insistence on this territorial and political security bring her into such severe conflict with America and England that there will follow a rupture of relations and an outbreak of hostilities? In the answer to this question we must search at least in part for the answers to the queries which people in America are asking, "Will we have to fight Russia?" and which people in Russia are asking: "Will we have to face another cordon sanitaire and go to war with the capitalist world?" If the Russian question sounds absurd to Americans, the American question sounds no less fantastic to Russians, who, because of the millions of their dead alone, dread a needless prolongation of the present war and the very mention of a third world war far more deeply than do the people of England and America.

This question becomes all the more complex because of what Germany has done to occupied Europe. She has blanketed it with ruin, degradation, and wrath. The middle classes and the intelligentsia, who in the prewar days were the bulwark of the status quo and therefore the chief fortress against revolution, have been largely stripped of the two values that count most with them—property and nationality. Hitler has proletarianized them. Germans and the German Government have, under one expedient or another, robbed them of precious possessions. Except when they sell their birthright to the "supermen" overlords the road to economic advance is closed to them. Through its master-race practices not only in Slav but in Nordic countries Germany has insulted the person, degraded the dignity of the conquered peoples—of the peasantry and factory workers no less than of the middle class and of the intelligentsia. It has robbed them of personal and national self-expression, for which they have for centuries been fighting.

The longer the war lasts, the more intense the German pressure, the more widespread German expropriations, the more ruthless German degradation of the person and the nationality of the conquered peoples, the more desperate will become their mood, the more resolute their will to revolt. In the midst of their agony and wrath masses of them derive no little consolation from the news and stories that come to them of German defeats, above all of German casualties, and they know that it is largely Russia that is responsible for these casualties and the killing and maiming of the flower of Germany's fighting and plundering manhood. It would be futile to overlook or to underestimate the rapid ascent and the magnitude of Russian prestige in the German-held countries.

If the war lasts long enough and Germany is weakened, these countries, unless held back by an Allied police force, are destined to witness violent uprisings of one kind or another. The emotion of vengeance is so fierce that even an outside police force may fail to halt the uprisings. In France, in Poland, in Hungary, in Rumania, in Holland, in Czechoslovakia, and even in Germany and Italy the amount of social dynamite is continually accumulating. Every day of war adds to the accumulation and the explosiveness of this dynamite.

Who can now foretell the nature of the forthcoming explosions, especially if the war is seriously prolonged and the despair and the wrath of the tormented and disaffected peoples in occupied lands and in Axis countries attain a state of irrepressible revolt? Will the communist parties seek

to capitalize on them? If so, what will England and America do? What will
Russia do? Who can answer these questions now?

There are, of course, documents and pacts which these three powers have
signed. There will be more in the days to come. But like hidden bombs
in a battlefield the many imponderables in the situation—all directly and
exclusively an outcome of Germany's depredations in conquered lands and
the misery and disillusionment the war is bringing to the Axis countries—
may, despite Russia's or anyone else's wishes and efforts, upset the accepted
plans and agreements; unless, of course, honest and united precautions are
devised to prevent such an outcome.

Russia, as well as England and America, has all the reason in the world
to take such precautions. That is why the documents at our disposal offer
for the present, at least, a general guide to mutual postwar aims and intentions. The Atlantic Charter, signed on August 14, 1941, by President Roosevelt and Prime Minister Churchill, at their historic meeting in mid-ocean,
is one of the most important of these documents. Its provisions are as
follows:

First, their countries seek no aggrandizement, territorial or other;

Second, they desire to see no territorial changes that do not accord with
the freely expressed wishes of the peoples concerned;

Third, they respect the right of all peoples to choose the form of government under which they will live; and they wish to see sovereign rights and
self-government restored to those who have been forcibly deprived of them;

Fourth, they will endeavor, with due respect for their existing obligations, to further the enjoyment by all States, great or small, victor or vanquished, of access, on equal terms, to the trade and to the raw materials of
the world which are needed for their economic prosperity;

Fifth, they desire to bring about the fullest collaboration between all
nations in the economic field with the object of securing, for all, improved
labor standards, economic adjustment and social security;

Sixth, after the final destruction of the Nazi tyranny, they hope to see
established a peace which will afford to all nations the means of dwelling
in safety within their own boundaries, and which will afford assurance
that all the men in all the lands may live out their lives in freedom from
fear and want;

Seventh, such a peace should enable all men to traverse the high seas and
oceans without hindrance;

Eighth, they believe that all of the nations of the world, for realistic as
well as spiritual reasons, must come to the abandonment of the use of force.
Since no future peace can be maintained if land, sea or air armaments
continue to be employed by nations which threaten, or may threaten,

aggression outside of their frontiers, they believe, pending the establishment of a wider and permanent system of general security, that the disarmament of such nations is essential. They will likewise aid and encourage all other practicable measures which will lighten for peace-loving peoples the crushing burden of armaments.

At the time the Atlantic Charter was drawn up, England was already at war with Germany. The United States was not. Russia was.

How Russia would respond to the aims embodied in the Charter was the question that immediately arose. Russia's answer came on September 24, 1941, when Ivan Maisky, Soviet Ambassador in London, made the following declaration:

"The Soviet Government proclaims its agreement with the fundamental principles of the declaration of Mr. Roosevelt, President of the United States, and of Mr. Churchill, Prime Minister of Great Britain, principles which are so important in the present international circumstances.

"Considering that the practical application of these principles will necessarily adapt itself to the circumstances, needs, and historic peculiarities of particular countries, the Soviet Government can state that a consistent application of these principles will secure the most energetic support on the part of the Government and peoples of the Soviet Union."

From this declaration it is clear that the Soviet Government accepts and approves the principles of the Atlantic Charter, but hints at possible divergences of viewpoint in their interpretation and application.

On November 6, 1942, the United States was already in the war, an ally of Russia and Great Britain. On that date, in explaining the mutual postwar aims of the Anglo-Soviet-American coalition, Premier Stalin said:

"Abolition of racial exclusiveness; equality of nations and integrity of their territories; liberation of enslaved nations and restoration of their sovereign rights; right of every nation to arrange its affairs as it wishes; economic aid to nations that have suffered and assistance to them in attaining their material welfare; restoration of democratic liberties; destruction of the Hitlerite regime."

Stalin considered further the problem inherent in the different ways of life of England, the United States, and Russia, and its bearing upon a successful collaboration of the three countries.

"It would be ridiculous to deny the difference in ideologies and social systems of the states composing the Anglo-Soviet-American coalition. But does this preclude the possibility and expediency of joint action on the part of the members of this coalition against the common enemy who holds out the threat of enslavement for them? It certainly does not preclude it. More, the existence of this threat imperatively imposes the necessity of joint action upon members of the coalition in order to save mankind from reversion to savagery and medieval brutality."

Then Stalin added this significant sentence: "Is not the program of action of the Anglo-Soviet-American coalition sufficient basis for the organization of a joint struggle against Hitlerite tyranny and for the achievement of victory over it? I think that it is quite sufficient."

This speech emphasized Russia's conception of the war as a joint enterprise of England, America, and Russia for the immediate task of winning the war, and it further emphasized Russia's desire for collaboration with England and America in the postwar world.

In an earlier speech, on November 6, 1941, Stalin said: "Russia's first aim is to free her own territory, and the second aim is to free the enslaved peoples of Europe and allow them to decide their own fate without outside interference in their internal affairs."

These and similar pronouncements in Stalin's other speeches are further reinforced and clarified by Article V of the Anglo-Soviet Pact of May 26, 1942, which reads as follows:

"The High Contracting Parties, having regard to the interests of security of each of them, agree to work together in close and friendly collaboration after re-establishment of peace for the organization of security and economic prosperity in Europe. They will take into account the interests of the United Nations in these objects and they will act in accordance with the two principles of not seeking territorial aggrandizement for themselves and of non-interference in the internal affairs of other States."

Yet it must be emphasized that Russia has abolished private enterprise inside her borders. She has achieved her greatest industrial and military development under a system of collectivized control of property, while England and America uphold private enterprise and under it have achieved their greatest industrial and military development. The mutual suspicions and hostilities that flow from this basic divergence and from others—such as dissimilar interpretation of the terms liberty, democracy, freedom—are certain to create differences in the interpretation of the language in which the Atlantic Charter and Stalin's pronouncements and the Anglo-Soviet Pact are expressed.

The question arises: Can these differences be resolved in conference or must they lead to war? All of Russia's pronouncements spurn the very thought of war; so do all English and all American pronouncements. Noth-

ing would be more disastrous for Russia and for the world than if this war
were to be followed by another with the English-speaking countries alone
or with allies arrayed in combat against Russia and her allies. Such a war
would be fought not only along national but along social or class lines. It
would be simultaneously an international and, in many of the participating
nations, a civil war, than which no war is more barbarous.

I want to repeat the sage words of the neutral diplomat in Russia, whom
I quote in the preface: "Either England, America, and Russia develop a
common language of action in the war and afterwards or God help us all."

With this judgment no one can possibly disagree, least of all the Russians, who, more than any of the allied nations, know the meaning and the
agony of war, and who, in this writer's judgment, are motivated by the
principle and the passion of national security as they understand the phrase,
above everything else. Considering the ideological crystallization which
the war has so obviously emphasized, I cannot imagine the Russian people,
particularly the new youth, ready and willing to fight a war except for national security which to them means first and foremost freedom from the
threat of invasion which has haunted them like an evil dream for over
a quarter of a century. For such a cause millions of them will gladly give
their lives.

But whatever the policies of the various countries, it is well for the outside world, and particularly for America and England, to cease thinking
of Russia as she was in 1917 or in 1932 or in 1937 or even in 1941. Because
so many of her institutions and practices are different and contrary to our
own, Russia is a difficult country to understand; and never so much as now,
when she is attaining crystallized economic and social stability and no less
crystallized a personality. The temptation always is to measure Russia in
terms of our own experience and our own immediate comforts and liberties, as well as the accumulated hates and hostilities of the past quarter of
a century—than which nothing is more futile in any approach to Russia
and the Russian people.

The misinformation about and the hatred of Russia in June 1941 were so
widespread and intense that foreign offices and war ministries had completely misread and miscalculated the powers of Russian industry and agriculture, the spirit of the Russian people, their will and their ability to
fight and die rather than be subjugated. Only a handful of students of Russia had perceived the strength which Russia subsequently demonstrated
and the quality of the sacrifice of which the people proved capable. It was
not prophetic vision but an earnest and sober attempt to appreciate and
understand the realities of the Russian scene which had enabled these stu-

dents to make a proper estimate of the country's military strength. But not only were they disbelieved, they were openly suspected and secretly denounced as "false prophets" and as "subversive elements."

How much understanding is there in an Allied military observer in Moscow who, after the Germans at Stalingrad had already been surrounded, said to foreign correspondents: "Stalingrad is a phony front. . . . The Russians cannot fool me. . . . They are not advancing. There is no offensive. . . ." He was convinced that the Germans were of their own accord retreating to a favored position so that they could swoop on Stalingrad in the spring, seize it with ease, and push on with their further conquest. He could not conceive of the Red Army's inflicting on Germany the most disastrous defeat the Reichswehr has suffered since it plunged Europe into war.

Russia is too important, too powerful, and much too necessary for a decent reconstruction of the world in Europe and in Asia for anyone to permit hatred of Russian politics, contempt for Russian bureaucracy, irritation over Russian personal and ideological unpleasantness to obscure his understanding of the Russia and the Russians of today.

CHAPTER 39: **AFTER TWENTY YEARS**

LATE AT NIGHT the plane landed at an American airfield in Africa. Eight months earlier I had come to the same airfield. Then I was on my way to Russia. Now I was on my way back home to America.

The first time I was here I marveled at the magnificent development which American enterprise had achieved within a surprisingly brief space of time in the wilderness of the Dark Continent. Here were miles and miles of asphalt roads; here was a modern American community with faultless plumbing, with dining rooms, a motion-picture theater, a barbershop, a commissary with an abundance of consumption goods. Here were a tennis court, a football field—everything to maintain a comfortable American standard of living, including an up-to-date hospital with drugs and equipment for any emergency.

I had not completed the multitude of inoculations prescribed for me, and my New York physician had given me a neatly wrapped package of serums so I could have them injected somewhere along the way. The steward of the clipper had put them in the icebox. But when I reached this African airport I had been off the clipper for a day, and the young American doctor would not hazard using my serums. Instead he used his own. "Oh yes," he said, "I can spare them—even if we are so far away from home."

Above all, here was a crowd of young Americans with the cheer, the buoyancy, the energy, the spirit of enterprise one associated with the gaiety of American college life and with the strenuousness of American pioneering. I left this community with the thought that, whatever the faults of American civilization, its skill in translating pioneering projects into living realities was hardly rivaled anywhere in the world. Here was a weapon of civilization, a method of making nature yield its riches for the benefit of man, which was fraught with uncertain yet stupendous potentialities not only for Africa but for all undeveloped lands.

Now I was back. The community was almost unrecognizable—so immense had been its growth in the months of my absence. It had become an army post. There were more miles of asphalt road, more buildings, more dormitories and dining rooms, all as commodious, cheerful, livable, as the ones I had known on my first trip. Here were crowds of young Americans all in uniform, from many states, of many European strains, and their cheer and good fellowship were no less marked than when I first knew the place. Comradely were the relationships between officers and men—as comradely, I thought, as the most solicitous American fathers and mothers would want them to be.

Here, at every turn, was further evidence of American enterprise and American skill to whip life and comfort out of nature's wilderness. To me, fresh from Moscow, this was all the more impressive because the Plans, which have given the Red Army the mighty sinews of war with which to fight the Germans, would have been impossible without the American enterprise and the American industrial achievement of which this army post was so illustrious an example.

Two miles away from the post was the Atlantic Ocean with one of the most superb bathing beaches in the world. Regularly a truck left the post for the beach to take passengers there and to bring others home. Native Negroes, chiefly women and girls, had a little market place on the beach. They sold bananas, oranges, coconuts, pineapples, and other tropical fruits. The sight of these fruits after many months in Russia, where they were nowhere to be had nor seen, was an exhilaration. I bought an armful, more than I could eat. Merely looking at the clusters of bananas, the oranges, the pineapples, and the coconuts made one realize how immeasurably rich and happy life could be, and was, in this faraway part of the world. Here one could lie out in the sun, swim, listen to the rhythmic roar of the ocean, eat choice tropical fruit, and be in the company of genial people who were glad to talk to you or have you talk to them—and forget the war and the trials and desperations of mankind. By comparison with the land from which I had come it all seemed fabulously, enchantingly unreal!

Vividly I recalled the Russia from which I had arrived. There was deep snow in Moscow when I boarded the plane. There was even deeper snow, biting frost, and icy wind in the open steppe when the plane landed in Kuibyshev. The Englishmen who were at the airport and who offered to drive me to the hotel in the city where I was to spend the night were wrapped in furs and felts and looked unlike any Englishmen I had ever known or seen anywhere in the world. Hardy was the Russian climate, hardy were the people who lived in it.

I remembered my last days in Moscow and the Russians whom I had gone to see before my departure to bid them farewell. It was almost Christmas, yet apartment houses were still without heat. Schools, hospitals, libraries, and theaters were kept warm, but not apartment houses nor offices. The city administration was faced with the choice of using fuel and electric energy for industry, especially for the manufacture of munitions, or for heating homes and offices. Ammunition came first. Even the censors in the Foreign Office were sitting in their overcoats.

Among the people I visited was an elderly school director and his wife, also a schoolteacher. They lived in a two-room apartment, which, considering Russian housing, made adequate living quarters. When I entered the apartment the man and his wife were sitting in their overcoats. They were apologetic for the low temperature in their home and expressed the hope that they would soon have heat—the hope but not the assurance.

They insisted I stay for supper. I knew they lived on their modest rations, but there is no refusing Russian hospitality. The wife cooked soup, fried griddlecakes, and as a special delicacy opened two cans of Kamchatka crab. The cans, I was certain, had come from the reserves which many Muscovites keep for possible emergencies, and I sought to persuade my hostess not to open them. But my words were drowned out in an avalanche of expostulations, reproofs, and cheering assurances that reserves did not matter any more anyway, for the Germans would never again threaten Moscow. . . .

So we ate hot soup and bread and griddlecakes and Kamchatka crab. We drank tea, not with sugar but with small pieces of candy, and we talked at length of America, of the world, and of the subject which rarely fails to come up in Russian conversation—namely, the sacrifices Russians were making in the war. The schoolmaster's family had suffered no casualties, but he and his wife spoke of other families, neighbors and friends, who mourned the passing of more than one man from their midst. Particularly did the schoolmaster stress the large number of orphans in his school. "We are winning victories," he said gloomily, "but we are paying a terrible price, more terrible than you imagine." I had heard no sadder and more challenging words in all the months I spent in Russia.

Nor was he the only one who spoke in such terms. Happy as Russians were over their victories (in the winter of 1943), they were only too acutely aware of their sacrifices. The fighting was on Russian soil; and it was Russian towns, Russian villages, Russian cities that were being devastated. Not for a single moment did they forget the stupendous price they were paying. Too many, too constant, too poignant were the personal reminders, especially when they came home from work. They could not escape the photographs on the walls of a father, a son, a husband, a brother who will never return.

Few Russians wear mourning. There is no law against it. When there is so much bereavement in the country, there is no use showing it in public. That also is why there are seldom any public funerals. I did not see one in Moscow all the time I was there. But if people did wear mourning in the capital or in Leningrad the streets and public squares would be black.

"Rzhev," I heard an officer say, "is a slaughterhouse. We are killing plenty of Germans, but they are killing lots of our men—the flower of our youth."

It is not only at Rzhev that the flower of Russia's youth is perishing, and Russians know it only too well.

There is nothing more horrifying than to listen to the tales of people from Leningrad as to what had happened in that city in the winter of 1941–42. Surrounded and beleaguered by Germans, people had so little bread and other foods that many died from famine. They died in the shops, in the streets, in their homes. They died while walking to their place of work, while sitting down to a bowl of hot soup, while lying down to rest.

"You ride in a streetcar," related a college student, "and beside you sits a man or woman, and suddenly you hear something heavy fall to the floor. You turn and look and—there lies a corpse. . . ."

Desperate with hunger, people often did not report the dead in their homes, so that they could use the ration cards and obtain an additional slice or two of bread. Perhaps half a million people in Leningrad died from the famine. But the city held. The people swallowed their woe. Leningrad did not fall. Neither the Finns nor the Germans held a triumphal march along the Nevsky. Russians died from famine but not on German-built gallows.

Here on the beach of the American Army post in faraway Africa were warmth and cheer and good fellowship and an abundance of everything, including the choicest fruits in the world. But in Russia there were cold and undernourishment, toil and sweat, battle and death.

It was just twenty years since I had first gone to Russia to write a series of stories on the Russian village and the Russian peasantry under the Soviets. How different was the village then from what it was in the summer of 1942! How different was all Russia, city and village, peasant and worker, intellectual and official, Bolshevik and non-Bolshevik, the men and the women, the children and the youths! What a bewildering and galloping transformation had come over the very land of Russia—in the once-gay and fat Ukraine, in the hot and once-pastoral Central Asia, in the rough and turbulent Urals, in the immense and haunting Arctic . . . everywhere.

The vocabulary, the emotions, the aspirations, the very noises in the air, not only of guns but of machines, were new and epochal. Lenin had dreamed of the day when there would be 100,000 tractors on the land. There were over half a million of them in 1940, nearly all manufactured on Russian soil by Russian men and women! Poorly fed, poorly clad, poorly housed (the civilians, not the army), Russia in 1942 and 1943 was fighting the most momentous and the most murderous war in her history, largely with her own weapons, from bayonet and rifle to armored tank and four-engined bomber—fighting, bleeding, dying—but never losing heart and with no thought of defeat! And in the Volga villages which I had visited school children were reading not only Pushkin and Tolstoy but Dickens and Mark Twain!

The transformation had wrung from the people the highest of sacrifices. They had been cold and hungry, weary and bewildered, masses of them had been. But they worked. They had no choice. If they did not work, they did not eat. If they grumbled, they were denounced. If they sabotaged or were suspected of sabotage, they were jailed, exiled, shot. Concentration, or, as the Russians say, labor, camps, were crowded. Jails were crowded. Late in the night men and women were wakened out of sleep and hauled away—to some faraway and dreary exile or to their death. The days of the First Plan were among the most violent and the most agonizing the Russian people had known since the coming of the Soviets. In more than one sense it was a continuation and a reincarnation of the battle and the psychology of the civil war.

Those were days of tumult and woe, but also of toil and more toil and still more toil. The slogan "socialism in one country" became the new gospel—not merely an article of faith but a plan of the most stupendous pioneering Russia or any other country, including America, had ever witnessed. It engulfed every person in the land. It rolled over every region, every hamlet, every hut. It struck hard at any obstacle, whether river, forest, swamp, or sea; man or woman; intellectual, peasant, or worker; communist or counterrevolutionary.

Despite inexperience and privation, "socialism in one country" was forging a new Russia, not a Russia of black ties and white ties, of silk shirts and silk gowns, of ballrooms and drawing rooms, or even of good plumbing; not a Russia of free speech and free assembly except for those who made themselves an unswerving and implacable part of the behest and the will of the Kremlin—but a Russia of engines and blast furnaces, of iron and steel, of gigantic factories and mechanized farms; a Russia that might not have enough glass to replace broken windowpanes and enough leather for new shoes, but which would possess the fire and the dynamite, the guns and the shells, the tanks and the planes, also the will and the skill to save herself from the tortured extinction which, with the support of Junkers and landlords, workers and peasants, Hitler's Germany would attempt to inflict on her.

With the rise of the factories, with the sweep of the tractors and the combines over the rich and fertile steppes, came a sobering of social thought, a stabilization of social ideas. Morality and the family struck new and deep roots, as deep as Russia has ever known. Gone were the literary firebrands who had heaped scorn and anathema on the men and women who beheld in art something more than the fighting slogan and the fighting passion of the moment. Some of these firebrands met their fate before a firing squad.

Pushkin became the hero, almost the deity of the people. One by one the great dead of the past were lifted to living glory. Even the iconoclast Pisarev, who loathed Pushkin, became a voice of challenge and importance. Compared to America and England the people were poorly dressed. They lived in crowded homes. But the country resounded with the lilting tunes of folk melodies, the best that her many-nationed people had created in the past ages. *Predki* (ancestors) and *stariki* (old people) became words and personages of adulation and emulation. Young people gloried in the love letters they wrote and did not mind reading them to gatherings of friends, whether soldiers in a dugout at the front or girls in a dormitory of the rubber factory in Moscow.

Private enterprise was gone—hardly a shred of it remained—and youth, especially, was ready to spill its blood, oceans of it if necessary, to prevent its return. Russia was rediscovered, Mother Russia, old Russia, the Russia of yesterday and of the faraway and hazy past, of legendary heroes and legendary triumphs. There were glory and goodness in this old and once-despised Russia. It had sinned and glutted, but it had worked and fought. It had also suffered and dreamed. It had bequeathed to future generations an array of scientists and statesmen, field marshals and conquerors, poets and novelists, literary critics and playwrights, and, yes, churchmen who were patriots, which the new Russia was now only too happy to embrace, to make its own, to revere and to love with all the fervor of its tumultuous heart.

All this happened under the slogan and the dispensation not of "world

revolution," but of "socialism in one country." It is this Russia that gave
birth to the sixteen-year-old Shura Chekalin, the twenty-two-year-old Liza
Chaikina, the eighteen-year-old Zoya Kosmodemyanskaya, whose heroic
martyrdom has made them as renowned as ancient folk heroes and who are
now Russia's new saints!

How different is the Russia of 1943 from the Russia of 1917, of which I
had only read, and the Russia of 1923, which I had seen! Uncertainty and
chaos have given way to sober reflection and to deep-rooted stability. Class
hate and class warfare, so flaming in former years, have yielded to love of
country, love of the past, love of the people—and to immeasurable faith
in the future. Talk and meetings, and more talk and more meetings, have
given way to the most strenuous and the most gigantic pioneering that the
world has known—pioneering for coal and iron, for steel and copper, for
oil and potash, for wheat and cotton, for cows and sheep, for new cities
and new schools, for a new life and a new destiny.

This Russia still has much to learn and to make her own. Superb as is the
democratic eloquence of her constitution—which is assiduously studied in
schools, factories, army camps, collective farms—much of it, very much,
still awaits fulfillment in everyday life. In political self-expression the
pioneering is only in its infancy. Though the former method of elections
is gone, though now not the show of hands but the secret ballot elects into
office Party secretaries, Soviet officials, trade-union functionaries, civil liberties, in the American and English sense of the word, are still a constitutional promise. In his speech of March 10, 1939, Stalin declared: "Since our
Party is in power, they [Party members] also constitute the commanding
staff of the leading organs of the state."

In other words, the Party is supreme and will tolerate no political opposition and no outside interference in any of its plans and programs. Thus,
though Russia has won the battle of social and racial equality, the elevation
of the individual to a condition of political self-expression, with the right
to criticize the dictatorship or to speak his mind freely and adversely of its
leaders, is still a matter of the future.

Nor in this writer's judgment is there much likelihood of there being
serious relaxation of this condition until Russia has at least lifted production to a degree which permits a comfortable material standard of living
and which in time of war insures ample industrial output for the fighting
armies. All of Russia's industrial pioneering under the Soviets was achieved
under the planning, the discipline, and the severity of a dictatorship. Only
a dictatorship could have imposed on the people the sacrifices and the privations which the fulfillment of the Plans required. There are, of course,
other motives and reasons for the dictatorship, but without it Russia never
could have packed into the brief space of thirteen years, 1928–41, the

amount of industrial, pioneering it achieved. With the devastation the country will be facing at the end of the war, the dictatorship, in the effort to rehabilitate the country and push on with new plans of industrial pioneering, will not relinquish any of its powers and prerogatives. It knows what it can do, what it has done, under its own system of control, and it will take no risks with any other methods, particularly if Russia should be facing fresh hostilities in the outside world.

Yet stupendous have been the internal achievements of Russian pioneering, especially, I must repeat, since 1928. The methods, I wish to emphasize, are other than those which the American pioneers espoused and practiced. The times are different, the historical setting, the immediate national needs, the international relationships, the technological compulsions, the ideological aspirations are also different. But the one consciously directed ultimate aim —large-scale development of internal resources—is not dissimilar to that which America's "rugged individualism" sought and strove to fulfill. Most manifestly Russian collectivized pioneering under a dictatorship would have been unthinkable without the experience and the achievements of America's "rugged individualism."

The greatest and most momentous triumph of this Russian pioneering is written large and red in blood and valor on the steppes and in the forests of Mother Russia.

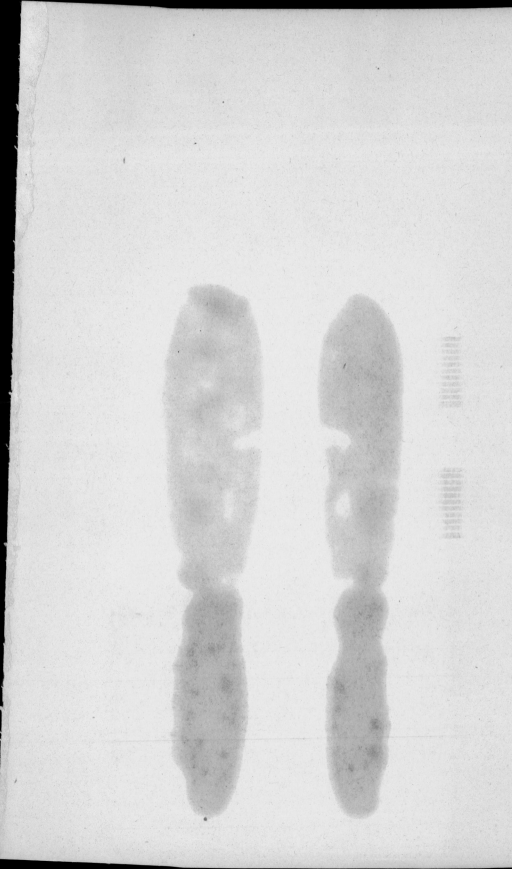

DATE DUE	

GAYLORD   PRINTED IN U.S.A.